Natural and Man-made Catastrophes – Theories, Economics, and Policy Designs

Natural and Man-made Catastrophes – Theories, Economics, and Policy Designs

S. Niggol Seo
Muaebak Institute of Global Warming Studies
Seoul
South Korea

WILEY Blackwell

This edition first published 2019
© 2019 John Wiley & Sons

All rights reserved. No part of this publication may be reproduced, stored in a retrieval system, or transmitted, in any form or by any means, electronic, mechanical, photocopying, recording or otherwise, except as permitted by law. Advice on how to obtain permission to reuse material from this title is available at http://www.wiley.com/go/permissions.

The right of S. Niggol Seo to be identified as the author of this work has been asserted in accordance with law.

Registered Offices
John Wiley & Sons, Inc., 111 River Street, Hoboken, NJ 07030, USA
John Wiley & Sons Ltd., The Atrium, Southern Gate, Chichester, West Sussex, PO19 8SQ, UK

Editorial Office
9600 Garsington Road, Oxford, OX4 2DQ, UK

For details of our global editorial offices, customer services, and more information about Wiley products visit us at www.wiley.com.

Wiley also publishes its books in a variety of electronic formats and by print-on-demand. Some content that appears in standard print versions of this book may not be available in other formats.

Limit of Liability/Disclaimer of Warranty
While the publisher and authors have used their best efforts in preparing this work, they make no representations or warranties with respect to the accuracy or completeness of the contents of this work and specifically disclaim all warranties, including without limitation any implied warranties of merchantability or fitness for a particular purpose. No warranty may be created or extended by sales representatives, written sales materials or promotional statements for this work. The fact that an organization, website, or product is referred to in this work as a citation and/or potential source of further information does not mean that the publisher and authors endorse the information or services the organization, website, or product may provide or recommendations it may make. This work is sold with the understanding that the publisher is not engaged in rendering professional services. The advice and strategies contained herein may not be suitable for your situation. You should consult with a specialist where appropriate. Further, readers should be aware that websites listed in this work may have changed or disappeared between when this work was written and when it is read. Neither the publisher nor authors shall be liable for any loss of profit or any other commercial damages, including but not limited to special, incidental, consequential, or other damages.

Library of Congress Cataloging-in-Publication Data has been applied for

ISBN: 9781119416791

Cover Design: Wiley
Cover Image: © Zloyel/iStockphoto; © Vladimir Vladimirov/iStockphoto;
© prudkov/iStockphoto; © Galyna Andrushko/Shutterstock;
© Fotos593/Shutterstock

Set in 10/12pt WarnockPro by SPi Global, Chennai, India
Printed in Singapore by C.O.S. Printers Pte Ltd

10 9 8 7 6 5 4 3 2 1

Contents

List of Figures *ix*
List of Tables *xi*
About the Author *xiii*
Preface and Acknowledgments *xv*

1 The Economics of Humanity-Ending Catastrophes, Natural and Man-made: Introduction *1*
1.1 Fables of Catastrophes in Three Worlds *1*
1.2 Feared Catastrophic Events *3*
1.3 Global or Universal Catastrophes *7*
1.4 A Multidisciplinary Review of Catastrophe Studies *11*
1.5 Economics of Catastrophic Events *16*
1.6 Empirical Studies of Behaviors Under Catastrophes *18*
1.7 Designing Policies on Catastrophic Events *21*
1.8 Economics of Catastrophes Versus Economics of Sustainability *25*
1.9 Road Ahead *26*
      References *26*

2 Mathematical Foundations of Catastrophe and Chaos Theories and Their Applications *37*
2.1 Introduction *37*
2.2 Catastrophe Theory *39*
2.2.1 Catastrophe Models and Tipping Points *40*
2.2.2 Regulating Mechanisms *42*
2.3 Chaos Theory *43*
2.3.1 Butterfly Effect *44*
2.3.2 The Lorenz Attractor *45*
2.4 Fractal Theory *46*
2.4.1 Fractals *46*
2.4.2 The Mandelbrot Set *49*
2.4.3 Fractals, Catastrophe, and Power Law *50*
2.5 Finding Order in Chaos *55*
2.6 Catastrophe Theory Applications *60*

2.7	Conclusion *61*
	References *62*

3	**Philosophies, Ancient and Contemporary, of Catastrophes, Doomsdays, and Civilizational Collapses** *67*
3.1	Introduction *67*
3.2	Environmental Catastrophes: *Silent Spring* *69*
3.3	Ecological Catastrophes: The Ultimate Value Is Wilderness *73*
3.4	Climate Doomsday Modelers *76*
3.5	Collapsiology: The Archaeology of Civilizational Collapses *79*
3.6	Pascal's Wager: A Statistics of Infinity of Value *82*
3.7	Randomness in the Indian School of Thoughts *85*
3.8	The Road to the Economics of Catastrophes *88*
	References *89*

4	**Economics of Catastrophic Events: Theory** *95*
4.1	Introduction *95*
4.2	Defining Catastrophic Events: Thresholds *98*
4.3	Defining Catastrophic Events: Tail Distributions *100*
4.4	Insurance and Catastrophic Coverage *104*
4.5	Options for a Catastrophic Event *110*
4.6	Catastrophe Bonds *114*
4.7	Pareto Optimality in Policy Interventions *119*
4.8	Events of Variance Infinity or Undefined Moments *125*
4.9	Economics of Infinity: A Dismal Science *129*
4.10	Alternative Formulations of a Fat-tail Catastrophe *132*
4.11	Conclusion *135*
	References *137*

5	**Economics of Catastrophic Events: Empirical Data and Analyses of Behavioral Responses** *145*
5.1	Introduction *145*
5.2	Modeling the Genesis of a Hurricane *147*
5.3	Indices of the Destructive Potential of a Hurricane *149*
5.4	Factors of Destruction: Wind Speeds, Central Pressure, and Storm Surge *151*
5.5	Predicting Future Hurricanes *153*
5.6	Measuring the Size and Destructiveness of an Earthquake *156*
5.7	What Causes Human Fatalities? *159*
5.8	Evidence of Adaptation to Tropical Cyclones *162*
5.9	Modeling Behavioral Adaptation Strategies *166*
5.10	Contributions of Empirical Studies to Catastrophe Literature *171*
	References *172*

6	**Catastrophe Policies: An Evaluation of Historical Developments and Outstanding Issues** *177*
6.1	Introduction *177*

6.2	Protecting the Earth from Asteroids *178*	
6.3	Earthquake Policies and Programs *181*	
6.4	Hurricane, Cyclone, and Typhoon Policies and Programs *182*	
6.5	Nuclear, Biological, and Chemical Weapons *187*	
6.6	Criteria Pollutants: The Clean Air Act *191*	
6.7	Toxic Chemicals and Hazardous Substances: Toxic Substances Control Act *198*	
6.8	Ozone Depletion: The Montreal Protocol *201*	
6.9	Global Warming: The Kyoto Protocol and Paris Agreement *203*	
6.10	Strangelets: High-Risk Physics Experiments *207*	
6.11	Artificial Intelligence *209*	
6.12	Conclusion *210*	
	References *210*	

7 Insights for Practitioners: Making Rational Decisions on a Global or Even Universal Catastrophe *219*

7.1	Introduction *219*
7.2	Lessons from the Multidisciplinary Literature of Catastrophes *221*
7.3	Fears of Low-Minds and High-Minds: Opinion Surveys *228*
7.4	Planet-wide Catastrophes or Universal Catastrophes *230*
7.5	Making Rational Decisions on Planet-wide or Universal Catastrophes *234*
7.6	Conclusion *241*
	References *241*

Index *249*

List of Figures

Figure 1.1 Deadliest earthquakes during the past 2000 years. *4*
Figure 1.2 Annual number of cyclone fatalities in the North Atlantic Ocean since 1900. *19*
Figure 2.1 Geometry of a fold catastrophe. *40*
Figure 2.2 The Lorenz attractor. *46*
Figure 2.3 The first four iterations of the Koch snowflake. *48*
Figure 2.4 The Mandelbrot set. *50*
Figure 2.5 Exponential growth under a power law utility function. *54*
Figure 2.6 Population bifurcation. *56*
Figure 4.1 Number of victims from natural catastrophes since 1970. *96*
Figure 4.2 Pareto–Levy–Mandelbrot distribution. *103*
Figure 4.3 Annual insured catastrophe losses, globally. *106*
Figure 4.4 The government cost of federal crop insurance. *110*
Figure 4.5 Spreads for CAT bonds versus high-yield corporate bonds. *117*
Figure 4.6 Outstanding CAT bonds by peril (as of December 2016). *118*
Figure 4.7 A trajectory of carbon tax with uncertainty. *124*
Figure 4.8 A family of Cauchy distributions with different scale parameters. *127*
Figure 5.1 Hurricane frequency in the North Atlantic: 1880–2013. *149*
Figure 5.2 Changes in power dissipation index (PDI) and sea surface temperature (SST) from 1949 to 2009 in the North Atlantic Ocean. *151*
Figure 5.3 The fatality–intensity relationship of tropical cyclones in South Asia. *160*
Figure 5.4 The surge–fatality relationship of tropical cyclones in the North Indian Ocean. *164*
Figure 6.1 History of the sulfur dioxide allowance price of clearing bids from spot auction. *198*
Figure 7.1 Catastrophes by the spatial scale of events. *231*

List of Tables

Table 1.1	Deadliest cyclones, globally.	*21*
Table 2.1	Pareto distribution of American wealth.	*51*
Table 2.2	Calculating the Feigenbaum constant for a nonlinear map.	*58*
Table 2.3	Calculating the Feigenbaum constant for a logistic map.	*59*
Table 2.4	Calculating the Feigenbaum constant for the Mandelbrot set.	*60*
Table 3.1	A summary of topics covered.	*69*
Table 4.1	Insured losses from catastrophes by world region in 2016.	*107*
Table 4.2	Growth of the US Federal Crop Insurance Program.	*108*
Table 5.1	Projections of tropical cyclones in the southern hemisphere by 2200.	*154*
Table 5.2	Projections of tropical cyclones in South Asia by 2100.	*155*
Table 5.3	Earthquake statistics, worldwide.	*159*
Table 5.4	Estimates of intensity, income, and surge effects.	*163*
Table 5.5	An NB model for the cyclone shelter program effectiveness (number of cyclone fatalities).	*166*
Table 5.6	Probit choice model of adopting a tropical cyclone adaptation strategy in southern hemisphere ocean basins.	*168*
Table 5.7	Probit adoption model of adaptation strategies to cyclone-induced surges and cyclone intensity in South Asia.	*170*
Table 6.1	Historical budgets for US NEO observations and planetary defense.	*180*
Table 6.2	Tropical cyclone RSMCs and TCWCs for ocean regions and basins.	*184*
Table 6.3	NFIP statistics on payments, borrowing, and cumulative debts.	*186*
Table 6.4	Treaties on nuclear, biological, and chemical weapons.	*190*
Table 6.5	NAAQS for criteria pollutants, as of 2017.	*193*
Table 7.1	Top fears of average Americans.	*229*
Table 7.2	Nobel laureates' ranking of the biggest challenges (2017).	*230*
Table 7.3	Global-scale or universal-scale catastrophes.	*233*
Table 7.4	Elements of a rational decision on global-scale catastrophes.	*237*

About the Author

Professor S. Niggol Seo is a natural resource economist who specializes in the study of global warming. Born in a rural village in South Korea in 1972, he received a PhD degree in Environmental and Natural Resource Economics from Yale University in May 2006 with a dissertation on microbehavioral models of global warming. Since 2003, he has worked on various World Bank projects on climate change in Africa, Latin America, and Asia. He has held Professor positions in the UK, Spain, and Australia from 2006 to 2015.

Professor Seo has published five books and over fifty international journal articles on global warming economics. He has been on the editorial boards of three international journals: *Climatic Change* (Stanford University), *Food Policy* (University of Bologna), and *Applied Economic Perspectives and Policy* (Oxford University Press). Among the academic honors he has received is an Outstanding Applied Economic Perspectives and Policy Article Award from the Agricultural and Applied Economics Association in Pittsburgh in June 2011.

Preface and Acknowledgments

This book entitled *Natural and Man-made Catastrophes – Theories, Economics, and Policy Designs* lays the foundation for the economic analyses of and policymaking on truly big catastrophes that may end humanity or even the universe but, at the same time, may occur randomly to utterly shock the world.

Such global-scale or universal catastrophes analyzed in the book include blackhole-generating strangelets, artificial intelligence that surpasses the human brain capacity, asteroids that may collide with Earth, killer robots, nuclear wars, global warming that could end all civilizations on the planet, ozone layer depletion, toxic chemicals, criteria pollutants, extreme tropical cyclones, and deadly earthquakes.

To build the economics of humanity-ending catastrophes, the author takes a multidisciplinary approach. The book provides a critical review of the scientific theories of catastrophe, chaos, and fractals in Chapter 2; of the philosophical, environmental, and archaeological traditions of societal collapses and doomsdays in Chapter 3; of economic models and markets of catastrophic events in Chapter 4; of empirical global catastrophe data and empirical modeling experiences in Chapter 5; of past policy interventions and future policy areas on catastrophes in Chapter 6; and of surveys of opinions from varied social groups on fears and challenges, as well as practical insights in Chapter 7. The book showcases many instances where a concept or theory developed in one discipline is appropriated by other disciplines in a revised form.

Of the aforementioned range of catastrophic events, the most catastrophic events during the past century to humanity have been tropical cyclones and earthquakes as far as the number of human fatalities is concerned. A single event of these catastrophes has killed as many as about half a million people. Besides these two catastrophes, humanity has gained substantial experience of other catastrophes caused by toxic chemicals, ozone layer depletion, air pollutants, and global warming. In building the economics foundation of humanity-scale catastrophes, this book takes full advantage of the evolving literature on the empirical economic analyses of these recurring disaster events.

The first chapter starts with "The Economics of Humanity-Ending Catastrophes," although the book is multidisciplinary in character. Here, the economics broadly suggests that a decision-making agent in market places, whether an individual, a community, a nation, or an international entity, makes its decisions on catastrophic events optimally, that is, by maximizing the net benefit from alternative solutions. At the heart of the economics, hence, lie the behavioral alterations of an economic agent faced with catastrophe situations, which are called by multiple names in the book, including adaptation behaviors, regulating mechanisms, policy interventions, and virus–antibody relationships.

In the final chapter, the book provides a set of practical guidelines for making rational decisions on a random catastrophe that may terminate humanity. After presenting multiple opinion surveys on people's greatest fears and challenges, the author provides a classification of catastrophic events based on the scale of damages. A rational decision making is then sketched which highlights the roles of science, psychology, religion, economics, an adaptive system, and an ultimate stop-control.

In the preparation of the book, many individuals kindly provided advice, encouragement, and critical comments. The author must start by thanking the late Benoit Mandelbrot, Martin Weitzman (Harvard), and William Nordhaus (Yale) for their inspiring works on the economic aspects of catastrophe events. For the empirical models and data discussed in the book, I would like to thank Laura Bakkensen (University of Arizona), Kerry Emanuel (Massachusetts Institute of Technology), and Robert Mendelsohn (Yale) for their work on hurricanes. I would like to acknowledge comments from Michael Frame (Yale) on fractal theory, Eli Tziperman (Harvard) on chaos theory, Guy D. Middleton (Newcastle University) on the archaeology of societal collapses, and Khemarat Talerngsri (Chulalongkorn University) on disaster events in Thailand.

Finally, I would like to express my appreciation toward John Wiley & Sons' publishing team and especially Andrew Harrison who advised on the proposal of the book. I am also thankful to many anonymous referees who kindly read through the proposal and provided valuable comments.

S. Niggol Seo
Muaebak Institute of Global Warming Studies
Seoul, South Korea

1

The Economics of Humanity-Ending Catastrophes, Natural and Man-made: Introduction

1.1 Fables of Catastrophes in Three Worlds

Since the beginning of human civilizations, humanity has feared catastrophes and has endeavored to prevent them, or cope with them if not stoppable. It is not an exaggeration to say that fears and horrors of catastrophes are deeply inscribed in the consciousness of human beings. As such, an enduring literature of catastrophes, natural and man-made, is easily found in a rich form in virtually all fields of mental endeavors including science, economics, philosophy, religion, policy, novels, poetry, music, and paintings.

The author has grown up listening to many fables and myths of catastrophes, some of which will be told presently, and is convinced that the readers of this book have heard similar, perhaps the same, stories growing up. Many stories of catastrophes may have been culturally passed on from generation to generation, some of which are a local event while others are larger-scale events.

Of the three fables, let me start with a fearful tale of a catastrophe that has been transmitted in the Mesopotamian flood tradition and the biblical flood tradition (Chen 2013). The great deluge myth goes that there was a great flood catastrophe a long time ago, which was caused by the fury of a heavenly being. All humans, animals, and plants were swept away to death by the deluge.

An old man, however, was informed of the catastrophic flooding days ahead, owing to the services he had rendered during his lifetime, and was instructed to build an ark. He built and entered the ark with his household members, essential goods, and animals. His family would be the only ones to survive the catastrophe, being afloat for 150 days in the deluge.

This myth of flood catastrophe has been passed down millennia as an early-warning fable for an imminent catastrophe on Earth, called popularly a judgment day. In that fateful day, only a handful of people will be permitted to escape the doomed fate. This fable or myth has left enduring imprints on many cultures and civilizations, including academics (Weitzman 1998).

When it comes to the tales of catastrophes, not all of them are loaded with fear and invoke imminence of a judgment day. Some tales are rather humorous and even make fun of the doomsday foretellers.

In the Chinese literature Lieh-Tzu, there was a man in the nation of Gi who was worried greatly that there was no place to escape if the sky fell. His panic was so much that he could neither eat nor sleep. On hearing his anxiety, a person who pitied his situation told him, "Since the sky is full of energy, how could it fall?" The man from the Gi nation replied, "If the sky is full of energy, shouldn't the Sun, Moon, and Stars drop because they are too heavy?" The concerned neighbor told him again, "Since the Sun, Moon, and Stars are burning with light, in addition to being full of energy, they will remain unbroken even if they should fall to the ground." The man from the Gi nation responded, "Shouldn't the Earth be collapsed then?" (Wong 2001).

In the East Asian culture, there is a popular word "Gi-Woo" which comes from the "Gi" nation and "Woo" which means worry and anxiety. The word is used in a situation in which someone is worried about something too much without a sound basis. The fable of Gi-Woo is a humorous depiction of a human tendency to worry too much beyond what is reasonably needed.

In the third type of fable of catastrophes, tellers of the fable take a different approach from the two aforementioned fables – that is, a rational and intelligent approach on the catastrophic risk. Recorded in the Jataka tales, the Buddha's birth stories, there was a rabbit who always worried about the end of the world. One day, a coconut fell from a palm-tree and hit the rabbit who, startled, started to run, screaming the world is breaking up. This intriguing tale goes as follows (Cowell et al. 1895):

> Once upon a time, a rabbit was asleep under a palm-tree. All at once he woke up, and thought: "What if the world should break up! What then would become of me?"
> At that moment, some monkeys dropped a cocoanut. It fell down on the ground just back of the rabbit. Hearing the noise, the rabbit said to himself: "The earth is all breaking up!" And he jumped up and ran just as fast as he could, without even looking back to see what made the noise.
> Another rabbit saw him running, and called after him, "What are you running so fast for?" "Don't ask me!" he cried. But the other rabbit ran after him, begging to know what was the matter. Then the first rabbit said: "Don't you know? The earth is all breaking up!" And on he ran, and the second rabbit ran with him.
> The next rabbit they met ran with them when he heard that the earth was all breaking up. One rabbit after another joined them, until there were hundreds of rabbits running as fast as they could go.
> They passed a deer, calling out to him that the earth was all breaking up. The deer then ran with them.
> The deer called to a fox to come along because the earth was all breaking up. On and on they ran, and an elephant joined them.

This tale of a frightened rabbit does not end here: there is a remarkable turnaround in the tale, which the author has saved, along with the rest of the story, for the final chapter of this book. It is quite sufficient to point out that we all – that is, the author and the readers who picked up this book on humanity-scale and universal catastrophes – are frightened rabbits. We are much scared about the possibility of the world's break-up owing to numerous uncontrollable mishaps, including nuclear wars, a gigantic asteroid collision, strangelets, singularity, killer robots, and global warming (Dar et al. 1999; Hawking et al. 2014).

1.2 Feared Catastrophic Events

The list of catastrophic events that are feared by people and societies is hardly short (Posner 2004). Some of these events have received extensive attention from researchers and policy-makers in the past, while others are emerging threats, therefore not-well-understood phenomena (for example, refer to the survey of American fears by Chapman University 2017). Some events have inflicted great harm on humanity over and over again historically, while other events are only a threat with a remote possibility. Some catastrophes are caused primarily by the force of nature, while others are primarily manmade.

Historically, catastrophic events are locally interpreted (Sanghi et al. 2010). A catastrophic event is one that wreaks havoc on a local community. The local community can be as small as a rural village, a town, or a city. A local catastrophe is most often a natural disaster, such as earthquakes, droughts, floods, heat waves, cold waves, tornadoes, and hurricanes.

Examples of a local catastrophe include an earthquake that strikes a city. Among the strongest earthquakes recorded are the 1960 Valdivia earthquake that hit the city of Valdivia in southern Chile, the 1906 San Francisco earthquake, the Great Kobe earthquake in 1995 in Japan, the 1950 Assam–Tibet earthquake, the 2004 Indian Ocean earthquake, and the 2011 earthquake off the Pacific coast of Tohoku in Japan.

The numbers of fatalities that resulted from the deadliest earthquakes in history make it obvious to the reader why these events are catastrophic events. The Shaanxi earthquake in China in 1556 killed 830 000 people; the Indian Ocean earthquake in 2004 resulted in the deaths of 280 000 people in South Asia; the 2010 Haiti earthquake was reported to have killed about 220 000 people; the Great Kanto earthquake in 1923 in Japan killed about 105 000; and the Kobe earthquake in Japan in 1995 killed 6434 people (Utsu 2013; EM-DAT 2017).

The deadliest earthquakes recorded in history are shown in Figure 1.1. Labels are attached to the vertical bars with more than 100 000 deaths. It is noticeable that the high-fatality earthquakes occurred most often at the centers of civilizations: Mongolian earthquakes at the time of the Mongol empire, Roman earthquakes during the time of the Roman empire. Also, high-fatality earthquakes occurred in high population centers: the Indian Ocean earthquake, Kashmir, and Chinese cities such as Shaanxi and Tangshan.

As is clear in Figure 1.1, the high casualty events have not let up in recent decades despite progresses in technological and information capabilities. The 2011 Tohokhu earthquake in Japan claimed about 16 000 lives; the 2010 Haiti earthquake was reported to have killed about 220 000 people (according to the Haitian government); the 2008 Sichuan earthquake claimed about 88 000 lives; the 2005 Kashmir earthquake 100 000 lives; and the 2005 Indian Ocean earthquake 280 000 lives. As such, earthquakes remain one of the most catastrophic events that people are concerned about today.

An earthquake occurs as a result of the movements and collisions of the lithosphere's tectonic plates (Kiger and Russell 1996). The Earth's lithosphere, i.e. a rigid layer of rock on the uppermost cover of the planet, comprises eight major tectonic plates and many more smaller plates. By connected plates, an earthquake in Japan can induce

4 | *1 The Economics of Humanity-Ending Catastrophes, Natural and Man-made: Introduction*

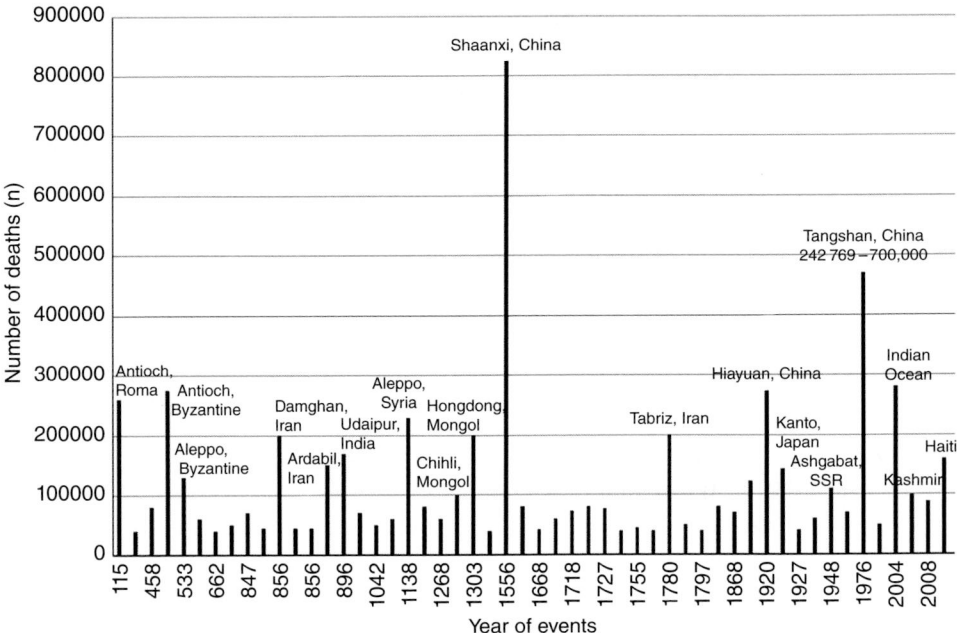

Figure 1.1 Deadliest earthquakes during the past 2000 years. *Source*: Utsu (2013), EM-DAT (2017).

another earthquake in New Zealand. Therefore, an earthquake catastrophe can occur at a regional or subglobal scale.

A hurricane is another catastrophic natural event that is feared and has received much policy attention (Emanuel 2008). It is another example of a local catastrophe. A hurricane, or a tropical cyclone as it is called in South Asia and the southern hemisphere and a typhoon in East Asia, is generated in an ocean, moves toward a landmass, and makes landfall on a coastal zone; many also dissipate in the ocean. As soon as it reaches the land, a cyclone weakens and quickly dissipates.

A hurricane's catastrophic potential is often characterized by wind speeds (McAdie et al. 2009). A category 1 tropical cyclone moves at the speed of over 74 mph (119 km h^{-1}) measured as the maximum sustained wind speeds (MSWSs); a category 2 tropical cyclone moves at the speed of over 96 mph; and a category 3 tropical cyclone moves at the speed of over 111 mph. A category 3 tropical cyclone is classified as a severe tropical cyclone, along with category 4 and 5 tropical cyclones.

The destructive potential of a hurricane is approximated by the rate of spinning of the cone of the storm, as well as the size of the cone of winds. Both variables are determined by the minimum central pressure of the hurricane. At sea-level altitude, the pressure stands at 1000 hPa (hectopascals or millibars). The lower the pressure at the center of a tropical cyclone, the faster the rate of spin motion of the cyclone. The lower the minimum central pressure, the more destructive a tropical cyclone becomes.

A catastrophic hurricane event is measured by the number of human deaths as well as the magnitude of economic damages (Seo 2014, 2015a). Economic damages occur most often in the form of destruction of houses and buildings or structural damages to them.

As such, damages are larger in low-income coastal zones with structurally weak houses (Nordhaus 2010; Mendelsohn et al. 2012).

The strongest hurricanes resulted in the number of deaths as large as those from the deadliest earthquakes shown in Figure 1.1. Cyclone Bhola that made landfall along the Bangladesh coast in 1970 incurred 280 000 human fatalities; the 1991 Bangladesh tropical cyclone killed 138 000 people; the 2008 Cyclone Nargis that hit the southwestern coast of Myanmar killed 84 000 people (Seo and Bakkensen 2017).

Cyclone fatalities are relatively much smaller in advanced economies such as the US, Japan, and Australia (see Figure 5.3). Since 1973, there has been no hurricane event in the US that has resulted in the deaths of over 100 people, with the exception of hurricane Katrina which killed more than 1225 people (Blake et al. 2011; Seo 2015a; NOAA 2016; Bakkensen and Mendelsohn 2016).

Another local-scale catastrophic event that is cyclically occurring and is a major concern for countries in the Asian monsoon climate zone is flooding. A monsoon climate is a climate system characterized by an exceptionally high rainfall during the monsoon season and an exceptionally low rainfall during a nonmonsoon season (Meehl and Hu 2006; Goswami et al. 2006; Chung and Ramanathan 2006; Seo 2016d). Overcoming this cycle of heavy rain and drought is an important policy endeavor in the monsoon climate-zone countries such as Thailand and India (Maxwell 2016).

In Thailand, flooding is a regularly occurring natural disaster attributed to the monsoon climate system. A severe flooding event occurs once every few years and often results in a large number of human deaths. The 2017 southern Thailand flooding resulted in over 85 deaths; the 2011 flooding caused 815 deaths; the 2010 floods killed 232 people; the 2013 South Asian floods killed 51 people; and the 2015 South Asian floods killed 15 people in Thailand (EM-DAT 2017).

The total number of deaths caused by floods in 2004 amounted to 7366 globally, 5754 in 2005, 8571 in 2010, 3582 in 2012, and 9819 in 2013. During the 2004–2013 period, the total number of deaths globally caused by floods amounted to 63 207, of which 71% occurred in the Asian continent (IFRC 2014).

Other catastrophic events have a scale of consequences at the national level as well as at the global level. A national-scale catastrophe would affect the population of an entire nation in a direct way. A severe drought event that befalls an entire nation over a sustained period, for example, a year or several years, is one example of such a national catastrophe. All communities across the nation will experience the consequences of the severe drought in a direct way.

The Dust Bowl of the 1930s in the US is one example of a national catastrophic event caused by a severe drought coupled with other factors such as farming practices and storms (Warrick et al. 1975; NDMC 2017). An exceptionally long period of severe and extreme droughts in Ireland during the 1854–1860 period resulted in a nationwide famine and the great Irish migration period to the US (Noone et al. 2017).

Catastrophes caused by earthquakes, hurricanes, flooding, and severe drought are primarily naturally occurring. Another type of catastrophe is primarily caused by humankind's activities – examples include toxic substances and chemicals, criteria pollutants, nuclear accidents, and ozone depletion.

Toxic chemicals and substances are a national health issue, the productions and uses of which can lead to a serious public health crisis as well as a damaged ecosystem (Vogel and Roberts 2011; Carson 1962). Toxic substances are chemical substances and mixtures whose manufacture, processing, distribution in commerce, use, or disposal may present an unreasonable risk of injury to health or the environment (US Congress 1978).

The US Environmental Protection Agency (EPA) created an inventory of existing chemicals, relying on the authority given by Congress through the passage of the Toxic Substances Control Act (TSCA) (Noone et al. 2017). The inventory listed 62 000 chemicals in the first version and has grown to more than 83 000 chemicals to date.

Relying on the authority specified by Section 1.6 on the Regulation of Hazardous Chemical Substances and Mixtures of the TSCA, the EPA attempted to restrict toxic chemicals such as asbestos, polychlorinated biphenyls (PCBs), chlorofluorocarbons (CFCs), dioxin, mercury, radon, and lead-based paint.

However, the US federal agency failed to regulate these toxic chemicals, halted by a series of lawsuits filed by chemical companies as well as a high burden of proof placed on the EPA by Section 1.6 for demonstrating substantial evidence of unreasonable risk (Vogel and Roberts 2011).

Notwithstanding the failures of the federal agency, US state-level regulations on toxic chemicals have increased. Since 2003, state legislatures passed more than 70 chemical safety laws for limiting the use of specific chemicals such as lead in toys, polybrominated diphenylethers (PBDEs) in flame retardants, and bisphenol A (BPA) in baby bottles (NCSL 2017).

Another category of manmade catastrophes could occur through numerous air and water pollutants. Through repeated exposures to smog, acid rain, particulate matter, lead, and other pollutants, an individual may suffer from various chronic diseases for a sustained period, and even face death. Particularly vulnerable to pollutants are those with existing health conditions, the elderly, children, and pregnant women (Tietenberg and Lewis 2014).

According to the World Health Organization (WHO), around seven million people die annually as a result of air pollution exposure, of which three million are due to exposure to outdoor pollution and four million due to exposure to indoor pollution. Of the seven million deaths, about six million deaths occur in South-East Asia and West Pacific regions (WHO 2014, 2016).

The US Clean Air Act (CAA), the signature legislation for regulating air pollutants, which was passed in 1970 and has been revised since then, defines the six most common pollutants as criteria pollutants. These are ground-level ozone, particulate matter, sulfur dioxide, nitrogen oxides, lead, and carbon monoxide (US EPA 1977, 1990). The CAA defines and enforces the ambient air quality standards for the six criteria pollutants, which are explained in depth in Chapter 6.

The sources of emissions vary across the pollutants. Coal-fired, oil-fired, and gas-fired power plants which generate electricity for numerous economic activities are primary sources of air pollutants such as sulfur dioxide, nitrogen oxides, particulate matter, volatile organic compounds, and ammonia (Mendelsohn 1980). A variety of vehicle uses is another primary source of air pollutants such as nitrogen oxides, volatile organic compounds, and particulate matter. Agriculture and forestry as well as manufacturing are also major sources of air pollution (Muller et al. 2011).

A nuclear power plant is another way to produce electricity and energy (MIT 2003). Through human mistakes or an unforeseen series of events, accidents at nuclear power plants have occurred, which led to one of the most catastrophic outcomes in human history. Leaks of nuclear radiation or contacts with radioactive materials led to a large number of immediate deaths or prolonged deaths through cancer.

There have been two catastrophic nuclear accidents categorized as an International Nuclear Events Scale (INES) level 7 event: the Chernobyl disaster and the Fukushima Daiichi accident (NEI 2016). The Chernobyl disaster in Ukrainian SSR in 1986 caused 56 direct deaths and cancer patients estimated as ranging from 4000 to 985 000.

The Fukushima Daiichi nuclear accident in Japan in 2011 was caused by the above-mentioned 2011 Tohoku earthquake and the subsequent tsunami. The earthquake was itself once-in-a century magnitude. The earthquake–tsunami–nuclear disaster event destroyed more than one-million buildings. The government of Japan declared a 20-km evacuation and exclusion zone, from which 470 000 people were evacuated.

Nonetheless, the reality of producing enough energy to support the national economies is that a large number of countries rely heavily on nuclear power plants for energy production. Countries that supply at least a quarter of national energy consumption through nuclear energy are France (76.9%), Slovakia (56.8%), Hungary (53.6%), Ukraine (49.4%), Belgium (47.5%), Sweden (41.5%), Switzerland (37.9%), Slovenia (37.2%), the Czech Republic (35.8%), Finland (34.6%), Bulgaria (31.8%), Armenia (30.7%), and South Korea (30.4%) (NEI 2016).

The permanent members of the United Nations (UN) Security Council and other major countries rely on nuclear energy significantly: the US (19.5%), China (2.4%), Germany (15.8%), Spain (20%), Russia (18%), and the UK (17%).

1.3 Global or Universal Catastrophes

The categories of catastrophic events introduced in Section 1.2 may wreak havoc on the communities that these events befall, but the scale of impacts is limited to a local area or to an entire nation even in a larger-scale shock. It does not mean, however, there would be no indirect effects on neighboring nations or trade partners.

Having said that, concerned scientists have often noticed that the possibility of an even larger-scale catastrophe may be increasing since the middle of the twentieth century. Notably, the ending of World War II through the first use of nuclear bombs in Hiroshima may have signaled at the same time both rapid scientific and technological advances and the possibility of potentially global-scale catastrophic events.

Many observers also noted that truly catastrophic events that can challenge human survival on Earth or even end the survival of the universe itself may be becoming more likely in tandem with the increase in scientific and technological capacities of humanity (Posner 2004; Kurzweil 2005; Hawking et al. 2014).

A catastrophic event that could end life on Earth is a global-scale catastrophe, while a catastrophic event that could end the existence of the universe as we know it now is a universal catastrophe. A global or a universal catastrophe is what humanity is most concerned about when it comes to a probable future catastrophe.

What are global or universal catastrophes? Is a global catastrophe likely at all? As a matter of fact, several such events have been proposed by concerned scientists. Nuclear warfare, a large-size asteroid colliding with the Earth, a high-risk physics or biological experiment for scientific purposes, and artificial intelligence (AI) and killer robots are recognized as causes for a likely global-scale or universal catastrophe.

An asteroid collision with the planet is a probable global catastrophe event (Chapman and Morrison 1994; NRC 2010). It is widely supported that a single asteroid led to the extinction of dinosaurs on Earth 66 million years ago by hitting "the right spot" with oil-rich sedimentary rocks (Kaiho and Oshima 2017).

An asteroid is a small planet that orbits the Sun, most of which is located in the Asteroid Belt between Mars and Jupiter. Asteroids, meteorites (fragments of asteroids), and comets (an icy outer solar system body) refer to different near-Earth objects (NEOs) against which the US' planetary defense activities are directed to prevent a possible collision with the Earth (NASA 2014).

When asteroids, meteorites, or comets are within 30 million miles (50 million kilometers) of the Earth's orbit, they are called NEOs. According to the US National Aeronautics and Space Administration (NASA), a 0.6-mile (1-km)-wide NEO could have a global-scale impact and a 980-ft (300-m)-wide NEO could have a subglobal impact (NASA 2014). The dinosaur-extinction asteroid was 7.5 miles wide (Kaiho and Oshima 2017).

According to NASA, as of 2016, about 50 000 NEOs have been discovered, but it is estimated that three-quarters of the NEOs existent in the solar system are still undiscovered. The discovery of an asteroid is the first and critical step in planetary defense against it, which is done mostly by ground-based telescopes. Deflecting or destroying an asteroid is another stage of the planetary defense mission, the possibility of which increases dramatically when it is discovered early (NRC 2010).

Reflecting the rising concern on possible asteroid collisions, the US government established the Planetary Defense Coordinating Office (PDCO) in 2016 under the leadership of NASA (NASA 2014). Of the total NEOs discovered globally, about 95% of them are discovered by NASA.

Nuclear warfare is cited as another probable global-scale catastrophe (Turco et al. 1983; Mills et al. 2008). A nuclear war between two nuclear powers, e.g. between the US and Russia or between India and Pakistan, has the potential to devastate entire civilizations on Earth.

A series of nuclear explosions will destroy living beings and built structures on the local area of explosions, which itself would not lead to a global-scale catastrophe. However, such nuclear explosions can alter the global atmosphere to cause global-scale freezing, which results in a global catastrophe (Turco et al. 1983). Alternatively, it is projected that nuclear explosions could destroy the ozone layer in the stratosphere, which possibly could result in a global-scale catastrophe (Mills et al. 2008; UNEP 2016).

A handful of countries in the world may have the capability to stage a nuclear war against their foes. As of 2018, nine countries are recognized, at different levels, to have the capabilities to own or build nuclear weapons. Among them are five permanent members of the UN Security Council: the US, Russia, the UK, France, and China. Additionally, four countries are known or believed to have nuclear weapons or have the capacity to make them: India, Pakistan, North Korea, and Israel (UNODA 2017a,b,c,d).

However, many other countries are reported to have the scientific and technological capacities to build nuclear arms, but have complied with the international nuclear treaty (explained below) and withheld their ambitions for developing them (Campbell et al. 2004). The international treaty refers to the Treaty on the Non-Proliferation of Nuclear

Weapons, commonly known as the Non-Proliferation Treaty (NPT), at the UN which aims to contain the competitive buildup of nuclear weapons and prevent a nuclear war.

The NPT entered into force in 1970 and was extended indefinitely in 1995. As of 2018, the NPT has been signed by 191 nations, which is an over 99% participation rate (UNODA 2017a,b,c,d). The NPT has established a safeguards system with responsibility given to the International Atomic Energy Agency (IAEA). The IAEA verifies compliance of member nations with the treaty through nuclear inspections.

However, the threat of a probable nuclear war has not been eliminated. It is notable that many nuclear-weapon regimes have not joined or not complied with the NPT, e.g. India, Pakistan, Israel, and North Korea, while other nations are on their way to developing them, e.g. Iran.

Further, whether the nuclear-weapons regimes including the US and Russia will commit to the NPT's grand bargain for a complete and full disarmament of nuclear weapons has yet to be confirmed, that is, by ratifying the treaty of a complete ban of further nuclear tests (Graham 2004).

Many researchers, but not all, have also cited global warming and climate change as a probable global catastrophe. The observed trend of a globally warming Earth may continue in the centuries to come, and if some of the worst projections of future climate by some scientists were to be materialized, a global-scale climate catastrophe should be unavoidable (IPCC 1990, 2014). However, these worst case scenario projections are treated by the Intergovernmental Panel on Climate Change (IPCC) as statistically insignificant (Le Treut et al. 2007).

The most dismal outlook with regard to the phenomenon of a globally warming planet is that global average temperature would rise by more than 10° C or even up to 20° C by the end of this century (Weitzman 2009). Such levels of global climate change would certainly force the end of human civilizations on Earth, as we know them (Nordhaus 2013).

However, this dismal outlook is in sharp contrast to the best-guess prediction or mean climate sensitivity presented by the IPCC, which has been in the range of 2 to 3° C by about the end of this century (Nordhaus 2008; IPCC 2014; Seo 2017a).

Also, several scientific hypotheses exist on catastrophic climatic warming, of which the author introduces several here. A hockey-stick hypothesis states that global average climate temperatures will run away in the twenty-first century as in the blade of a hockey-stick (Mann et al. 1999; IPCC 2001). The second hypothesis is that an abrupt switch in the global climate system may occur, shocking everyone on Earth, including scientists (NRC 2010). The third hypothesis is that a global catastrophe may occur by way of crossing the threshold or reaching the tipping point of various climate system variables, e.g. a reversal of the global thermohaline circulation in the ocean (Broecker 1997; Lenton et al. 2008).

However, projections of the future climate system by climate scientists are highly uncertain, and are expressed as a wide range of divergent outcomes from a large array of future storylines or scenarios (Nakicenovic et al. 2000; Weitzman 2009). Further, many scientific issues remain unresolved in the climate prediction models called in the literature Atmospheric Oceanic General Circulation Models (AOGCMs) (Le Treut et al. 2007; Taylor et al. 2012).

Notwithstanding the range of uncertainties and scientific gaps that exist even with more than four decades of admirable scientific pursuits, there is a silver lining with regard to the future of global climate shifts. If the Earth were to warm according to the IPCC's middle-of-the range predictions or the most likely projections, people and societies will find ways to adapt to and make the best of changed climate conditions (Mendelsohn 2000; Seo and Mendelsohn 2008; Seo 2010, 2012a, 2015c, 2017a).

The magnitude of damage from global warming and climatic shifts will critically hinge on how the future climate system unfolds and how effectively and sensibly individuals and societies adapt (Mendelsohn et al. 2006; Tol 2009; Seo 2016a,b,c).

Existing technologies as well as those developed in the future will greatly enhance the capacities of individuals and societies (Seo 2017a). Some of these technologies are breakthrough technologies that can replace fossil fuels entirely or remove carbon dioxide in the atmosphere or engineer the Earth's climate system, which include, inter alia, nuclear fusion power generations, solar energy, carbon capture–storage–reuse technology, and solar reflectors (ITER 2015; MIT 2015; Lackner et al. 2012; NRC 2015).

These mega technologies are broadly defined as a backstop technology in the resource economics literature. Although many of these breakthrough technologies can be employed to tackle climate change for the present period, the cost of relying on any of these technologies is more than an order of magnitude higher than the least-cost options available now to achieve the reduction of the same unit of carbon dioxide (Nordhaus 2008).

A catastrophe whose scale of destruction goes beyond the planet has been suggested by scientists (Dar et al. 1999; Jaffe et al. 2000). A salient example is a probable accident in the Large Hadron Collider (LHC), built by the European Organization for Nuclear Research (CERN) for the purposes of testing various predictions or theories of particle physics. It is a 27-km-long (in circumference) tunnel built under the France–Switzerland border at a depth of 175 m (CERN 2017).

The LHC is a particle accelerator built to test theories on the states of the universe during the short moments in the origin of the universe. More specifically, it tests the initial states of the universe right after the Big Bang (Overbye 2013). It was suggested by scientists that the experimental process may create a strangelet or a black matter unintentionally, through which a black hole is created. The entire universe would be drawn to the black hole, if it were to be stable, bringing an end to the universe (Plaga 2009; Ellis et al. 2008).

Scientists overwhelmingly reject the possibility of such a universal catastrophe. A group of researchers called the probability of it absurdly small (Jaffe et al. 2000). An impact analysis group of the CERN experiments reported that there is no possibility at all of the universe-ending catastrophe (Ellis et al. 2008). Many groups of scientists argue that such collisions of particles occur naturally in the universe, leaving no impacts on the universal environment (Dar et al. 1999; Jaffe et al. 2000; Ellis et al. 2008).

Although no actions have been taken to reduce the risk of this universe-ending catastrophe, the forecast of it has not materialized yet. The experiments at CERN led to the award of the Nobel Prize in Physics in 2013 "for the theoretical discovery of a mechanism that contributes to our understanding of the origin of mass of subatomic particles, and which recently was confirmed through the discovery of the predicted fundamental particle, by the ATLAS [A Toroidal LHC ApparatuS] and CMS [Compact Muon Solenoid] experiments at CERN's LHC" (Nobel Prize 2013).

The list of catastrophes presented up to this point paints quite a dismal picture for the survival of humanity and even the universe. Nonetheless, there seems to be a more feared and more likely catastrophe in the minds of many concerned scientists, that is, AI.

AI, i.e. intelligent robots and machines, may become more intelligent and powerful at some point and kill all the living beings, i.e. beings with life, including humans (Hawking et al. 2014).

The all-life-ending catastrophe may be brought on by the lifeless machines and robots. In some areas of human activities and dimensions, robots are already more efficient and intelligent than humans and have replaced human laborers. The day may come quite quickly according to many experts when the brain capacity of robots, measured by such indicators as IQ, surpasses that of humanity. This would be the moment of singularity (Kurzweil 2005).

When the singularity arrives, it would be the greatest marvelous achievement of humanity, but the last one, according to the physicist Hawking (Hawking et al. 2014). The AIs will control humans and may end up killing all humans and even all living beings in the universe, intentionally or unintentionally.

Not all the experts on AI share this perspective. Optimists would argue that robots who are lifeless beings or insentient beings may become friendly neighbors to humanity, all-smiling and supportive as they are at present.

The world's notable entrepreneurs have been pursuing competitively advanced AI machines and robots and their applications to various business fields, examples of which include a self-driving automobile by Tesla motors, an AI healthcare software system by Softbank, and an intelligent personal assistant Siri by Apple.

In many ways, many nations are investing competitively in the development of AI based on the conviction that gaining superiority in AI would make the nation a military superpower in the world. The downside of this competition lies in the fear that the killer robots may become uncontrollable, or even the war robots could start a war without a human order.

In fact, war robots already play a pivotal role in war army combats as well as local police battles. Ethical issues and banning the use of such robots were taken up for discussion at the UN experts meeting on Lethal Autonomous Weapon Systems (LAWSs) (UNODA 2017b).

1.4 A Multidisciplinary Review of Catastrophe Studies

Having presented the first impressions of the range of catastrophic events that this book is concerned with in establishing the economic perspectives, the author, perhaps the reader as well, needs to consider how the book should proceed and what approach should be taken to achieve the goals of the book.

Of the many possible ways that the book can be written, the author has determined to emphasize the generality of the concept of catastrophe across many academic fields of catastrophe studies. This book, consequently, takes a multidisciplinary approach, which should also be appealing to a wide range of academic disciplines and in a wide range of policy circles.

On the other hand, the book is also positioned to make the clearest and the most direct presentation of the economic issues and analyses with regard to catastrophic events. This means that the background of the economic analyses presented in the book will be market places in which an economic agent, whether an individual or a community, weighs the benefit against the cost incurred over a long period of time of a decision for the purposes of achieving an optimal outcome resulting from the decision (von Neumann and Morgenstern 1947; Koopmans 1965).

Studies of and stories about catastrophic events are perhaps as old as the birth of human civilization or humanity's invention of letters. The three tales and fables introduced above were recorded in some of the oldest books that human civilization compiled and transmitted through time until today. Further, catastrophe concepts and studies are quite pervasive across the sciences, mathematics, philosophy, economics, psychology, policy sciences, and even literary works, which will be made clearer in this book.

Scientific descriptions and mathematical formulations of catastrophe and chaos emerged during the latter half of the twentieth century. Taking advantage of his predecessor's works on structural stability, catastrophe theory was presented during the 1960s and 1970s by French mathematician René Thom who formulated it in the context of structural stability of a biological system (Poincaré 1880–1890; Thom 1975). Catastrophe was defined as a sudden dramatic shift of a biological system in response to a miniscule change in a certain state variable (Zeeman 1977).

Thom's works became known as the catastrophe theory because he presented a list of seven elementary catastrophes that would become widely appropriated by applied scientists and economists of catastrophes. Seven generic structures of catastrophe, each of which is expressed as a form of a potential function, were fold catastrophe, cusp catastrophe, swallowtail catastrophe, butterfly catastrophe, hyperbolic umbilic catastrophe, elliptic umbilic catastrophe, and parabolic umbilic catastrophe (Thom 1975).

In another literature, the chaos theory surfaced by a stroke of serendipity and was developed to depict the systems that are in chaos or disorder, in which chaos was defined as the absence of an order in the system, or a disorderly system, or an unpredictable system (Lorenz 1963; Strogatz 1994).

As it has turned out over the course of its development, the literature of the chaos theory has become as much about the scientific endeavors to find an order in a chaotic, disorderly system as it was about the absence of order, disorder, or unpredictability of a certain system (Tziperman 2017).

Edward Lorenz is generally credited with the pioneering experimental works that led to the establishment of the field of chaos theory. As a meteorologist at the Massachusetts Institute of Technology, he was working to develop a system of equations that can predict the weather of, say, Cambridge, Massachusetts a week ahead of time (Lorenz 1963). Through his experiments with the computer simulation of the weather system, he came across the finding that a miniscule change in an initial point or any point in the system leads to a widely strange outcome in the predicted weather, a phenomenon that he later called "butterfly effects" (Lorenz 1969).

Continuing to work on his weather system, Lorenz presented a simplified system, that is, a system of three ordinary differential equations, the set of outcomes of which has been known to represent the chaos theory. The Lorenz attractor, i.e. the solutions to the Lorenz system, is deceptively simple mathematically; however, it so richly expresses a disorderly system or an unpredictable system (Gleick 1987; Strogatz 1994). The Lorenz attractor is the system with the absence of order in that it shows neither a steady state nor a periodic behavior, i.e. two known types of order in a system (Tziperman 2017).

Another important contribution to the theory of catastrophe or chaos came from the theory of a fractal developed separately by Benoit Mandelbrot (Mandelbrot 1963, 1967,

1983, 1997). From the studies of crop prices, coastal lines, financial prices, and others, Mandelbrot defined a fractal to be a figure that has a self-similar figure infinitely as its component or at a larger scale and in which this self-similarity is repeated in ever-larger scales of the figure (Frame et al. 2017).

In the fractal image, you can zoom in on the figure over and over again and find the same figure at a smaller scale forever. It is interpreted that a fractal is an image of an infinitely complex system and a fractal is often described as a "picture of chaos" (Fractal Foundation 2009). For instance, it would be impossible in a fractal world to measure the length of the British coastline correctly (Mandelbrot 1967).

The self-similarity, also referred to as self-affinity, is the central concept of the fractal theory, which manifests as scale invariance in the statistical literature that defines a power law tail distribution, also called the Pareto–Levy–Mandelbrot distribution, as well as a fat-tail distribution (Pareto 1896; Mandelbrot 1963, 1997; Gabaix 2009). The power law distribution arises in many economic and noneconomic processes and has been relied upon in the study of a highly volatile system such as financial market crashes or a highly uncertain catastrophe event such as the end of human civilization caused by global warming (Mandelbrot 1997; Taleb 2005; Weitzman 2009).

Fractal theorists argue that a fractal is very common or "everywhere" in nature; that is, one can encounter a fractal easily in such things as trees, rivers, cauliflowers, coastlines, mountains, clouds, seashells, and hurricanes. The fractal theorists strived to formulate a fractal image as a set of simple equations, the best known of which are the Mandelbrot set and the Julia set (Mandelbrot 1983; Douady 1986).

At this point, one may wonder: Is chaos the world as it is or is there an order that is simply elusive to untrained observers? As noted above, scientists had the same curiosity very early in the literature and the search for an order in a chaotic system, say, a disorderly order, has increased over the course of the literature with as much prominence as chaos itself.

The Feigenbaum constant is broadly thought to be a ground-breaking discovery in the chaos theory in that it unveils a hidden order in a chaotic or disorderly system (Feigenbaum 1978). Feigenbaum was examining a population bifurcation diagram, that is, a diagram of successive bifurcations of a biological population in which bifurcation points hinge on the rate of population growth. Feigenbaum made a major discovery in the field of the chaos theory that the population bifurcations in the diagram occur in an orderly way at a constant rate of 4.669.

To put it more precisely, he discovered the exact scale at which the population diagram is self-similar, which is the scale in the fractal image. In other words, if we make the population bifurcation diagram 4.669 times smaller at the point of a bifurcation point, then it will look exactly the same as the next point of bifurcation (Tziperman 2017).

Long before these catastrophe sciences and models ever existed, there had been already voluminous works on conceptualizations of a catastrophe. In the philosophical and theological traditions, inquiries on catastrophic events had been framed with reference to the end of the world or the beginning of the world as we know it presently. Numerous theories or even haphazard forecasts of an ultimate doomsday had been proposed in association with human activities.

In Chapter 3, the author provides a wide-ranging review of selected theories and works in the ancient and contemporary philosophical, broadly defined, traditions. The chapter starts with environmental and ecological classics by Rachel Carson and Aldo Leopold.

This is followed by the review of climate doomsday modeling works and the archaeology of civilizational collapses, so-called collapsiology.

An environmental classic by Rachel Carson entitled *Silent Spring* is filled with the sentiments of doom and death caused by humanity's environmental and ecological degradations through unregulated chemical uses (Carson 1962). In her book, Carson laments that "Everywhere was a shadow of death," "the haunting fear that something may corrupt the environment to the point where man joins the dinosaurs as an obsolete form of life," and "It was a spring without voices." *Silent Spring*, one of the most influential environmental books in history, does not, however, rely on a formal theory or conceptualization of a catastrophe.

Quite a different perspective on human civilizations and their existence was put forth by Aldo Leopold, which is ecocentric (Leopold 1949). In his much-acclaimed and influential book *Sand County Almanac*, Leopold proposes a new ethical perspective in which the ultimate value lies in the wilderness or wildness of things.

Leopold writes that "the ultimate value … is wildness. But all conservation of wildness is self-defeating, for to cherish we must see and fondle, and when enough have seen and fondled, there is no wilderness left to cherish." And in another chapter, he declares that "In wildness is the salvation of the world" (Leopold 1949).

Leopold wields a double-edged sword: on the one side, he sees little value in mankind's works and establishments; on the other side, he sees no danger in destructions of mankind's works and establishments by the ineluctable forces of nature. In his unique perspectives, it seems as if a true catastrophe lies only in humanity's excessive interventions in the holistic existence of natural worlds.

Recently, renewed enthusiasm in catastrophes has emerged among archaeologists and scientists. A group of researchers have re-examined past collapses of once-glorious civilizations, including the Maya civilization in Mesoamerica, the Mycenaean civilization in ancient Greece, the Moche civilization in northern Peru, and the Western Roman Empire (Diamond 2005; Gill et al. 2007; Kenneth et al. 2012; Drake 2012).

The common feature in this emerging literature is that the past civilizations' collapses are attributed to abrupt climatic shifts at the times of those collapses (*New Scientist* 2012). These archaeologists and scientists rely on newly available archaeological data thanks to climate change and global warming research, such as ice-core temperature data, cave stalagmites, carbon isotopes, and sea-surface temperatures (Le Treut et al. 2007).

However, the archaeological literature of civilizational collapses by and large refutes the climate doomsday assertions by the aforementioned researchers based on unmodified associations between societal collapses and changes in climate conditions. In the collapsiology or the archaeology of collapses, the fall of a civilization is explained as "a highly complex operation" which is certain to be "distorted by oversimplification" (Wheeler 1966).

Collapsiologists offer intelligent discussions on past societal, civilizational collapses that take into account complexities in social, economic, and cultural systems (Middleton 2017). One of the definitions widely adopted by them of a societal collapse is a rapid political change and a reduction in social complexity. Society's collapses are identified through various empirical measures, including fragmentation of a state into smaller entities, desertion of urban centers, breakdown in regional economic systems, and abandoning prevalent ideologies (Schwartz and Nichols 2006).

An insightful conclusion from the archaeologists is that a civilization does not collapse. The Mayan civilization, for example, underwent many periods of states' collapses through outsiders, internal conflicts, Spanish armies, and Christianity, but the Mayan civilization itself has survived. Millions of Mayan descendants live in Central America today. What collapsed are the Mayan states, not the civilization (Middleton 2017).

Leaving aside a collapse archaeology, Chapter 3 takes up ancient philosophical and theological traditions on catastrophes for discussion. A conceptual formulation of a catastrophe is abundantly found in the ancient philosophical traditions that delved into the question of the end and beginning of the universe. In the chapter, the author sketches two such traditions that have had enduring influences.

An Indian school formulated that things occur randomly, that is, things and beings arise without a cause. In the same way, things and beings perish without a cause, at random. The philosophical arguments of this school are expounded in detail in the Buddha's sutra entitled "The all-embracing net of views" (Tipitaka 2010). This school was called a "fortuitous originationist" as the sages of this school proclaimed that self and the world originate fortuitously, by chance.

In the world of fortuitous originationists, as beings come into being randomly, i.e. without a cause, they decease randomly, i.e. without a particular cause. Stated another way, according to this school, life begins catastrophically and it ends catastrophically, with neither a cause nor a warning.

The aforementioned sutra, told by the Buddha, elaborates other philosophical views, which were prevalent in ancient India, that are pertinent to the literature of philosophical traditions on catastrophic events. These include the views of eternalists as well as the views of annihilationists (Tipitaka 2010).

Directing our attention to the western philosophy, a prominent tradition that explicates conceptualizations of a catastrophe is Blaise Pascal's probability theory or probabilistic thinking on eternity (Pascal 1670). In formulating his probability theory, Pascal refers to an extreme case in a probability distribution: an extremely low probability event which, nonetheless, is rewarded with infinity of welfare or wealth, if it were to turn out to be true.

In Pascal's so-called wager argument, this extreme event is a belief in God. In Pascal's wager, it is argued that the existence of God is an extremely low probability event, a highly unlikely event. But, the belief in the existence of God is worthwhile, *at all costs*, because it safeguards or salvages one from the catastrophic event in which God does exist and shall send nonbelievers to an eternal hell. According to the wager, it would be a winning gamble even if one were to lose all things in this life and perhaps in the next lives eternally.

Pascal's world is one in which the concern of a truly catastrophic event dominates all worldly decisions. Applied to a policy decision on a probable catastrophe, the wager theory would imply that all available resources are directed to prevent a catastrophic event whose damage is in infinity. This infinity of a catastrophe's destruction would include "a catastrophic end of human civilizations as we know it or the catastrophic end of all life on the Planet," an expression popularized in the global warming policy literature (Weitzman 2009).

In theory as well as reality, there are a whole range of options one can take to avoid such a catastrophic end of human civilizations, and most often there are alternatives that demand much less sacrifice to achieve whatever goals an individual or society hopes to achieve (Nordhaus 2011). From another perspective, even the truly catastrophic event such as the one Pascal put forward can be prevented by wisely employing the alternatives available and is not expensive (Seo 2017a).

1.5 Economics of Catastrophic Events

Economists have long been interested in the characteristics of catastrophes, differences among catastrophic events, and effective ways to hedge against or prevent such events. What distinguishes the economics of catastrophes from other fields of studies is its exposition and utilization of individuals' incentives in the face of catastrophic challenges which are translated into market systems and securities bought and sold among individuals. In Chapters 4 and 5, the author explains the theoretical aspects of catastrophe economics in the former and empirical models of catastrophe economics in the latter.

A catastrophe is often defined by a threshold, also called a critical value, which is also sometimes called a tipping point with an implication of a catastrophe theory. A threshold approach is applied to various pollution control policies (US EPA 1990). For example, the US CAA regulates air pollutants using the National Ambient Air Quality Standards (NAAQS), in which the ozone's NAAQS is set at the threshold of 0.070 ppm in 8-hour average concentration and the primary threshold for particulate matter (PM10) is 150 µg m^{-3} in 24-hour average (US EPA 2017).

Another way to define a catastrophe is through the characteristic of a tail in a statistical distribution. Take climate change predictions for the year 2100 by numerous climate centers. The predictions would report a range of outcomes whose distribution can be statistically derived nonparametrically. A long-tail event is an unlikely event, but when it is also fat it can dominate any rational decision-making (Schuster 1984; Weitzman 2009).

A long-and-fat-tail distribution is relevant to catastrophe economics in that it can arise from a high degree of uncertainty about the future. The author explains a classification of statistical distributions by tail properties. An event, or a random variable, can be classified into one of three types according to the tail behavior: a thin tail distribution, a medium tail distribution, and a fat tail distribution (Schuster 1984).

The author then examines the broader economics literature of risk and catastrophes which have a large uncertainty or occur with a very low probability. A large volume of economics literature is available on behaviors of investors who manage a high-risk asset or a high-risk portfolio (Markowitz 1952). A rich economics literature is also available on market securities that are contingent on occurrences of catastrophic events such as fires, hurricanes, earthquakes, and droughts.

An insurance is the earliest financial instrument that was devised to help individuals hedge against a catastrophic event which is largely uncontrollable (Shiller 2004). By pooling the same risk across a large number of individuals, an insurance company can pay for the catastrophic damage in return for insurance premiums paid by the insured.

Of the weather-related catastrophic losses, insured losses accounted for about 30% in 2016 of total economic losses of US$175 billion. In North America and Europe, insured losses accounted for about 50%, while they accounted for only 10% in Asia (Swiss Re Capital Markets 2017). The author explains the catastrophic coverage in the US crop insurance and subsidy program in detail in Chapter 4 (Sumner and Zulauf 2012).

An options contract and a futures contract are other financial instruments that are widely used in respective markets for managing specific risks or catastrophes. An option contract specifies the rights of the purchaser to sell or buy at a written price an underlying asset at a future date, which may or may not be exercised by her/him (Black and Scholes 1973; Fabozzi et al. 2009). The investor will consider the possibility of a catastrophic event in deciding whether or not to exercise the option contract.

An investor who owns a financial or commodity asset, e.g. a residential property, can purchase a relevant option contract for the asset in order to minimize the risk from a precipitous price fall, as was experienced in the subprime mortgage crisis in the late 2000s (Shiller 2009). A large number of different types of options are traded in the commodity and financial markets, including crude oil options, natural gas options, corn options, gold options, and S&P 500 options (CME Group 2017).

Catastrophe bonds (more commonly called CAT bonds) are a new financial derivative that was devised specifically to deal with very rare or unprecedented catastrophe events for which traditional insurance products are inadequately developed. The CAT bond is based on the concept of reinsurance, i.e. the insurance of an insurance product (Edesess 2014). An insurance company purchases a reinsurance from a reinsurance company, which is contingent on an occurrence of an exceptionally catastrophic event and an exceptionally large amount of insurance claims.

The concept of a CAT bond was developed during the 1990s after Hurricane Andrew that hit the US in 1992, from which multiple insurance companies went bankrupt. Since then, outstanding CAT bonds have increased steeply, from US$20 billion in 2007 to US$70 billion in 2015 (Swiss Re Institute 2017). Besides the reinsurance companies, issuers of CAT bonds have expanded to an insurance company, a government entity, pension funds, a corporation, and a nongovernmental organization. Mexico issued CAT bonds in 2006 for earthquakes, the first national sovereign to do so. In 2014, the World Bank issued the first CAT bonds for natural disaster risks in 16 Caribbean countries (World Bank 2014).

With the background of the range of market securities that are designed to deal with the catastrophe risks, how should a government intervene and design a policy for a specific catastrophic event of public concern? In the economic policy literature, a rational decision-making framework has long been the foundation of policy interventions (von Neumann and Mrogentern 1947). In this approach, the government should intervene in a way that the benefit of a policy remedy over the cost is maximized over a long-term timeframe (Ramsey 1928; Koopmans 1965).

In the rational expectation approach on policy interventions, the tail distribution of a catastrophe event is captured by the variance of a random event (Nordhaus 2011). Let's take for consideration the degree of global warming by the year 2100. The higher the uncertainty, the larger the variance of the distribution. The lower the uncertainty, the smaller the variance of the distribution. The probability density function of the random event contains the full information on whether the tail is long and fat or short and thin (Nordhaus 2008).

A policy decision is made on such variables as the penalty imposed on the emissions of carbon dioxide, a primary Earth-heating molecule. The levels of the carbon dioxide penalty are determined from the range of outcomes based on the estimation of variance in the random variable. The carbon penalty, in the form of carbon tax or carbon price, is determined in the matter to achieve an optimal social welfare, with the policy measure permeating through the economy (Nordhaus 1994).

Is this rational social welfare optimization approach adequate for addressing catastrophic events? Since the late 2000s, it has been one of the most debated questions, especially in the context of global warming policy. Arguing that the approach is a misleading policy principle, critics suggested a generalized precautionary principle as an alternative (Weitzman 2009).

The critics' arguments are based on the concept of a fat-tail distribution and formalized as the dismal theorem. The crux of the arguments is that an individual is willing to sacrifice an infinity of resources in order to prevent a truly catastrophic event such as future global warming because the tail distribution of a global warming prediction is fat (Wagoner and Weitzman 2015).

The author elaborates a general structure of the integrated assessment model (IAM) which is an empirical policy model for the social welfare optimization policy approach. The first and most well defined of the IAMs, the Dynamic Integrated model of Climate and Economy (DICE), is explained with major policy results. These results are compared with the results from the generalized precautionary principle.

The author then critically examines the dismal theorem's model structure, critical assumptions, model parameters, and missing components, which shows that a modification of each of these aspects in the model leads to a different conclusion contradictory to the precautionary principle (Nordhaus 2011; Pindyck 2011; Yohe and Tol 2012; Seo 2017a).

The highlight of the critique is a great number of adaptation strategies and technological changes that are available for a long-term policy issue such as global warming (Seo 2015b, 2017a). Incorporating these aspects into the dismal model structure, the author offers an enhanced richer climate policy model, whose implications in the context of catastrophe management cannot be understated.

1.6 Empirical Studies of Behaviors Under Catastrophes

Catastrophic events are in most cases unexpected. That is, they occur with an element of surprise. They are referred to as a once-in-a-century event, or a once-in-a-millennium event, and so on. As such, it is often very difficult to avoid catastrophic consequences once the event is set in motion. For example, once an earthquake hits a high-rise building in a city, it is very difficult for the dwellers to escape from the collapsing building. For another, if a large asteroid were about to hit a city, it would be nearly impossible for the residents to escape in time from horrendous disasters.

Notwithstanding, an examination of the historical damages and fatalities caused by deadliest natural events such as earthquakes and tropical cyclones reveals that they have fallen substantially over time. Figure 1.2 shows the number of deaths annually caused by tropical cyclones in the North Atlantic Ocean from 1900 to 2016 (Blake et al. 2011; NOAA 2016). As revealed, the number of fatalities has fallen markedly during the twentieth century. With the exception of 2005 when Hurricane Katrina hit, the number of fatalities rarely exceeded 10 annually.

How has the reduction in the number of fatalities from natural catastrophes occurred? Although it varies across different natural or man-made events in the degree of difficulty,

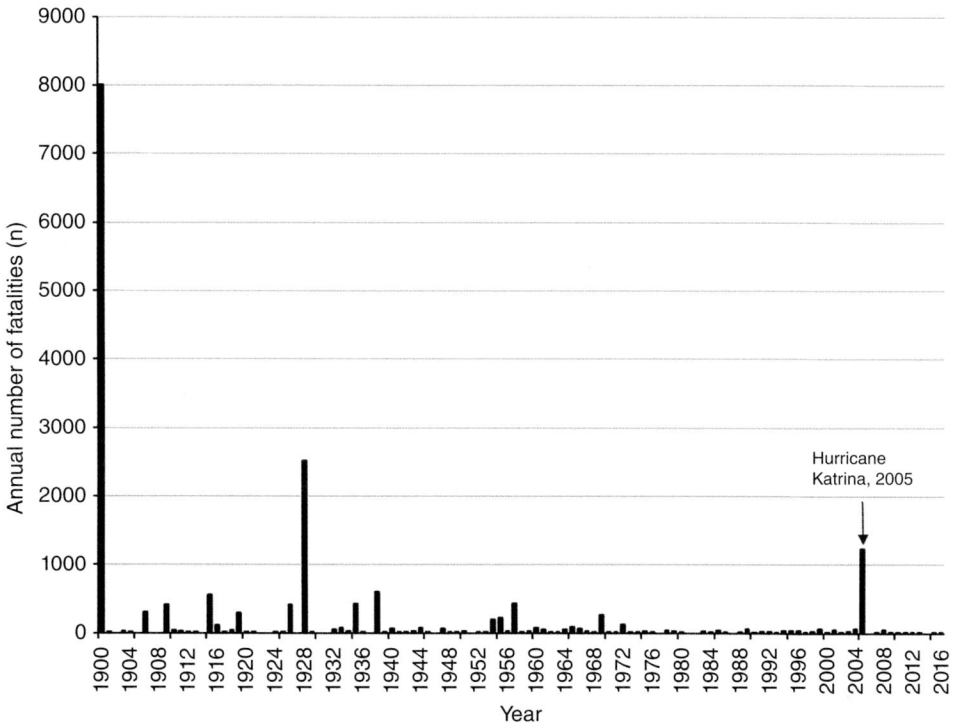

Figure 1.2 Annual number of cyclone fatalities in the North Atlantic Ocean since 1900. *Source*: Blake et al. (2011), NOAA (2016).

it is generally difficult to take reactive actions to a catastrophic event, that is, *post* the event. However, it does not mean that individuals and societies cannot be prepared for a catastrophic accident, that is, *ante* the event.

Among the preparatory actions and strategies that can be taken, some actions are taken well before the event in time, while others are taken just before the event. Adaptation can take place long before the event, just before the event, at the time of the event, and even without any direct association with the event (Seo 2017a).

Chapter 5 is devoted to the review of empirical economic studies on catastrophic events, as a continuation of the economics of catastrophes. The chapter utilizes the historical data of the two deadliest natural events in the past century in terms of the number of human fatalities: tropical cyclones and earthquakes (EM-DAT 2017; Swiss Re Capital Markets 2017).

The empirical data, empirical economic models, and questions raised and addressed through these models in Chapter 5 are without doubt fundamental inputs or aspects of the economics of catastrophes presented in Chapter 4. Empirical results in Chapter 5 augment or lessen various arguments and theories offered in the previous chapter.

With reference to the tropical cyclone literature and that of earthquakes, the author illustrates the fascinating complexity in the sciences of these natural events: How is a hurricane generated? What are the indicators of destruction and deaths? Is it possible to predict hurricanes in the year 2100? What are the best indicators of earthquake size, magnitude, and destructiveness? (Richter 1958; Emanuel 2008; McAdie et al. 2009; Knutson et al. 2010).

The complex science questions can be put in the economics context: What are the primary causes of human deaths through hurricane events? The traditional measures of hurricane intensity are expressed by way of hurricane wind speeds. The Saffir–Simpson scale of hurricane intensity distinguishes five categories of hurricanes based on MSWSs (NOAA 2009). Other measures that are based on maximum wind speeds are accumulated cyclone energy (ACE) and the Power Dissipation Index (PDI) (Emanuel 2005).

However, it has been reported that central minimum pressure of a cyclone better explains the destructiveness of hurricanes than maximum wind speeds (Mendelsohn et al. 2012). A further study in cyclone-prone zones of South Asia shows that it is neither wind speeds nor central pressure that is a primary killer of people there, but the surge of seas during cyclone events (Seo and Bakkensen 2017).

Economists have asked how sensitive the total economic damage is to an increase in hurricane intensity (Pielke et al. 2008). In the US, a very large elasticity of hurricane damages was reported with an increase in the maximum wind speeds or a decrease in minimum central pressure (Nordhaus 2010; Mendelsohn et al. 2012; Seo 2014).

Simultaneously, hurricane studies have found a large income elasticity of hurricane economic damages as well as of hurricane fatalities. That is, the higher the income of the region of a hurricane landfall, the lower the number of fatalities and the smaller the magnitude of economic damages. In the southern hemisphere ocean, a one-unit increase in income decreases the number of fatalities from a cyclone by 4.85% (Seo 2015a). In South Asia, a one-unit increase in income decreases the number of human fatalities by 3% (Seo and Bakkensen 2017; Seo 2017b).

The large income effect is ascribed to behavioral factors, exogenous technological changes, and induced technological changes. Behavioral factors include better awareness of hurricane threats, better knowledge of effective evacuation strategies, and moving out of hurricane-prone zones. Exogenous technological changes are economy-wide technological advances that help effective responses, which include resilient houses and buildings, ownership of automobiles, and communication technologies.

Recent studies have highlighted the important role that induced technologies have played in reducing the number of fatalities to cyclones. Induced cyclone technologies include satellite monitoring of cyclones, early warning systems, and cyclone trajectory projection technologies. The higher the income, the higher the adoption of these technologies. The higher the adoption of induced technologies, the lower the number of fatalities (Seo 2015a; Seo and Bakkensen 2017).

One of the salient findings of the behavioral economics of cyclones was its analysis of the effect of a long-term policy intervention in Bangladesh. A recent study of tropical cyclones in South Asian countries that included Bangladesh, India, Myanmar, and Sri Lanka showed that the cyclone shelter program run by the government of Bangladesh since the 1970s has become highly effective in reducing the number of fatalities in the event of severe tropical cyclones. It showed that the number of fatalities fell by 75% due to the cyclone shelter program, given the same degree of severity of a cyclone (Seo 2017b).

The cyclone shelter program was initiated by the government of Bangladesh in order to reduce the extremely large number of cyclone fatalities in the country (Khan and Rahman 2007). Cyclone Bhola in 1970 killed 280 000 people, while the 1991

Table 1.1 Deadliest cyclones, globally.

Rank	Year	Cyclone ocean basin	Country most affected	Name	Number of fatalities
1	1970	Indian Ocean	Bangladesh	Cyclone Bhola	Up to 500 000
2	1975	Western North Pacific	Philippines	Typhoon Nina	229 000
3	1991	Indian Ocean	Bangladesh	1991 Bangladesh Cyclone	138 866
4	2005	Indian Ocean	Myanmar	Cyclone Nargis	138 366
5	1975	Indian Ocean	India	Andhra Pradesh Cyclone	Up to 14 204
6	1963	Indian Ocean	Pakistan	Severe Cyclonic Storm Three	11 520
7	1998	North Atlantic	Atlantic	Hurricane Mitch	11 374
8	1999	Indian Ocean	India	Odhisha Cyclone	At least 10 000
9	2013	Western North Pacific	Philippines	Typhoon Haiyan	6329
10	2005	North Atlantic	US	Hurricane Katrina	1833
11	1959	East Pacific	Mexico	1959 M Hurricane	1800
12	1982	East Pacific	Mexico	Hurricane Paul	1696
13	1976	East Pacific	Mexico	Hurricane Liza	1263

Source: Fatality data are from various sources including the Joint Typhoon Warning Center (2017), Indian Meteorological Department (2015), National Disaster Risk Reduction and Management Council, the Philippines (NDRRMC 2017), and Bakkensen and Mendelsohn (2016).

Bangladesh tropical cyclone killed 138 000 people (Seo and Bakkensen 2017). The program subsequently received international support from the World Bank to restore existing shelter and build new shelters (World Bank 2007; Paul 2009).

It should be emphasized that cyclones, also called hurricanes in North America and typhoons in East Asia, are the most catastrophic natural events, measured by the number of human deaths, that humanity has experienced since the beginning of the twentieth century, with earthquakes being another deadliest catastrophe. In the twentieth century, many earthquakes killed more than 100 000 people (see Figure 1.1) and four tropical cyclones resulted in more than 100 000 human fatalities (see Table 1.1).

Therefore, empirical studies of these deadliest events reveal ample insights for the analyses of other catastrophes that neither are well described nor have quality empirical data, but are of much public concern, e.g. nuclear explosions, destruction of the ozone layer, asteroid collisions, a catastrophic failure in physics experiments, and AI. These are elaborated and then integrated into the economic models of catastrophic events presented in Chapters 4 and 5 and policy analyses in Chapter 6.

1.7 Designing Policies on Catastrophic Events

As made clear in Figure 1.1 on earthquake fatalities, Table 1.1 on cyclone fatalities, and many other empirical catastrophe data, catastrophic shocks, both natural and man-made, have long been one of the fundamental, perhaps ineluctable, aspects of human existence. As such, mankind's efforts to deal with and reduce the risk and fatal consequences from catastrophe events have long been as much an essential aspect of human societies. A large array of policies undertaken historically to deal with catastrophic events, failed or successful, is also available for those who are concerned about future catastrophes.

In Chapter 6, the author provides a wide-ranging review of policy principles and measures that were adopted and implemented to address catastrophic risks, which include asteroids, earthquakes, cyclones, nuclear wars, criteria pollutants, toxic and hazardous chemicals, destruction of the ozone layer, global warming, high-risk physics experiments, and AI.

The US government's policy interventions for protecting the Earth from NEOs such as asteroids, meteorites, and comets are very recent and centered on the programs run by the PDCO under NASA (NRC 2010; NASA 2014). The PDCO was established in 2016 and coordinates all the efforts related to protection against NEOs, the government budget for which has increased sharply in the past few years.

Policy concerns on asteroid collisions with the Earth are rather novel, even though multiple impacts of large asteroids are recorded in history. Up until now, the issue is discussed and addressed at a national level, despite a global-scale threat posed by a large asteroid, with more than 95% of all NEOs discovered by NASA.

The protection problem from asteroids is unique because asteroid protection has a best-shot technology production function (Hirshleifer 1983; Nordhaus 2006). This means that not all countries need to mobilize resources to protect the Earth from asteroids.

By contrast, another global environmental challenge has been tackled by extensive global cooperation: the ozone layer depletion. The global policy efforts to address the problem of the ozone layer depletion are encased in the Montreal Protocol on Substances that Deplete the Ozone Layer, a global environmental treaty signed in 1987.

The ozone layer in the stratosphere plays a vital role in protecting the Earth from the ultraviolet A (UVA) radiation from the Sun, the depletion of which is found to cause skin cancer and is related to other health effects such as cataracts. Scientists reported during the 1970s that a widely applied coolant for refrigerators and air-conditioners, i.e. CFCs, was destroying the ozone layer in the stratosphere (Molina and Rowland 1974). In Montreal in 1987, countries agreed to phase out CFCs, and this was updated multiple times, most recently in Kigali, Rwanda in 2016 (UNEP 2016).

The success of the Montreal Protocol as a global treaty is widely recognized. Remarkably, the participation rate is 100% of all UN members. Under the Montreal Protocol, the nations have phased out nearly all Ozone Depleting Substances (ODSs), and the total column ozone is projected to recover to the benchmark 1980 levels by the middle of the twenty-first century over most of the globe with the full compliance of the Protocol (WMO 2014). However, it remains unclear whether the recent Kigali Amendment would be ratified by as many nations and be as successful as the previous Amendments, owing to its non-UVA emphasis (White House 2013).

The international policy roundtable on nuclear weapons and wars is where global-scale catastrophic consequences are explicitly elaborated (Turco et al. 1983; Sandler 1997). In addition, it is an international policy area where an international treaty is clearly established. The NPT entered into force in 1970 and was extended indefinitely in 1995. As of 2018, the NPT is signed by 191 nations, but with notable nonmembers such as India, Pakistan, and Israel (UNODA 2017a).

Unlike other global-scale challenges, there are only a handful of countries, nine countries more precisely, that are known to have the capacity to build nuclear weapons or own them. However, there are many other countries that have the technologies but are restrained from developing nuclear weapons by reliance on a grand bargain in which nuclear weapon states provide a nuclear umbrella and commitment to a full and complete disarmament of nuclear arms. The stability of this grand bargain in the future remains to be seen (Campbell et al. 2004).

Chapter 6 describes the details of the NPT, including the safeguards system, and the implementations of the treaty by the IAEA. Other than the NPT, other treaties and agreements on other weapons of mass destruction at the UN level are discussed, such as the Biological Weapons Convention (BWC) and the Chemical Weapons Convention (CWC) (UNODA 2017b,c).

Local-scale and national-scale catastrophic events are addressed by local/national laws and regulations, calling for no international cooperation through a global treaty or agreement. Earthquakes, hurricanes, flooding, severe droughts, toxic substances, and air pollutants are such events whose policies and experiences are detailed in Chapter 6.

The US earthquake policy responses are anchored by the National Earthquake Hazards Reduction Program (NEHRP) authorized by the US Congress in 1977, which was amended in 2004 (US Congress 2004). The objective of the NEHRP is to "reduce the risks to life and property from future earthquakes in the United States."

The NEHRP is run by the National Institute of Standards and Technology (NIST) as the lead agency for the program, in collaboration with other federal agencies such as the Federal Emergency Management Agency (FEMA), the National Science Foundation (NSF), and the United States Geological Survey (USGS) (FEMA 2016).

Each of these agencies plays a distinct role. The NIST is charged primarily with the responsibilities of developing earthquake-resistant design and construction practices; NSF with supporting basic research programs; USGS with monitoring and analysis of earthquakes; and FEMA with implementing effective earthquake risk reduction tools.

Governmental responses to hurricane disasters in the US are anchored at the National Flood Insurance Program (NFIP) that was established by the National Flood Insurance Act of 1968. As declared in the name of the program, the NFIP is a federally subsidized insurance program for residents in hurricane-prone zones (Knowles and Kunreuther 2014).

Like other federal subsidies, the cost of the NFIP has increased drastically since its introduction in the 1970s for several reasons, one of which was a remarkable population shift to hurricane-afflicted zones and coastal counties. The astounding shift of population meant that more people and property are placed in vulnerable zones to hurricanes, which eventually resulted in a drastic increase in the NFIP's borrowing authority. As of 2012, the NFIP was in debt by US$17.7 billion (King 2013).

For the purposes of addressing the financial burden of the NFIP, the US Congress passed in July 2012 the Biggert Waters Flood Insurance Reform Act (BW12) as a comprehensive reform of the NFIP. The BW12 introduced a major reform of the NFIP by phasing out insurance subsidies and discounts and forcing a system of insurance premiums that reflect the flood risk, but these transformational provisions were rolled back subsequently (FEMA 2012).

The laws and regulations that regulate toxic chemicals and hazardous substances in the US include the TSCA of 1978, the Federal Insecticide, Fungicide, and Rodenticide Act (FIFRA) of 1948, and the Comprehensive Environmental Response, Compensation, and Liability Act (CERCLA) of 1980 (US Congress 1978, 1980, 2012).

The US EPA, entrusted with the authority under the TSCA, made numerous attempts to restrict or ban toxic chemicals such as PCBs, CFCs, dioxin, asbestos, hexavalent chromium, mercury, radon, and lead-based paint.

But, the EPA's interventions were largely unsuccessful, faced with legal challenges from chemical companies. In the case of asbestos, the EPA issued a regulation that would have banned almost all uses of asbestos, based on an extensive risk assessment over 10 years as a carcinogen. Asbestos producers immediately challenged the regulation in the court which ruled against the regulation and the EPA, citing a lack of "substantial evidence" for both unreasonable risk and the least burdensome approach to remove the risk (Vogel and Roberts 2011).

Other laws and regulations that aim to limit or remove harmful chemicals include the CAA and the Clean Water Act. Atmospheric concentrations of six criteria pollutants are regulated by these laws: ground-level ozone, particulate matter, sulfur dioxide, nitrogen oxides, lead, and carbon monoxide. For these so-called criteria air pollutants, the US EPA has established the NAAQS, a set of thresholds for the concentrations of each of these pollutants (US EPA 1977, 1990, 2014).

With a view to achieving the pollution thresholds/targets at the least cost to the society, market-based policy instruments to control emissions of these gases were subsequently devised, which include the sulfur dioxide allowance trading program in the US (Stavins 1998; Burtraw and Szambelan 2009). The market price of the sulfur dioxide allowance, a measure of cost of the pollutant to society, fell to zero by the early 2010s, after peaking at US$1200 per ton in 2005 (Schmalensee and Stavins 2013).

In a global environmental-policy setting, the Kyoto Protocol and the Paris Agreements purport to limit the emissions of planet-warming gases such as carbon dioxide, methane, nitrous oxides, and fluorinated gases such as hydrofluorocarbons (HFCs), hydrochlorofluorocarbons (HCFCs), and sulfur hexafluoride (SF6) (UNFCCC 1997, 2015).

International negotiators have pushed for meeting various targets/thresholds: the 1990 level of carbon dioxide emissions in the Kyoto Protocol, the 2° C threshold for global temperature increase, and the carbon budget for attaining the temperature threshold. These targets were adopted by the negotiators to avoid a dangerous anthropogenic interference with the climate system, which was the declared goal of the United Nations Framework Convention on Climate Change (UNFCCC) in Rio de Janeiro in 1992 (UNFCCC 1992).

The negotiations of almost three decades on global warming tell the important lesson in addressing a truly global problem: it is extremely difficult to agree on an international treaty with a near-universal participation. This is due, inter alia, to disparate impacts of global warming across countries, unequal policy effects, accounting for historical emissions, and rich countries' climate aids (Seo 2012b, 2015c, 2017a). A recent announcement by the Trump administration to pull out from the Paris Agreements signals that the Paris Agreements may not be a forceful international treaty to achieve its proclaimed goal, although it was hailed as a turning point in climate policy by negotiators (White House 2017).

The most catastrophic outcome may occur through AI or high-risk physics experiments. The former would be life-ending, while the latter would be universe-ending, if the worst-case scenario of each event should materialize. Given the extreme nature of the worst-case scenarios, it is not very surprising that there is not much to discuss when it comes to policy experiences of these events.

Nonetheless, the concern of experts regarding these events has increased quite sharply since the 2010s and some have even suggested a specific policy measure to address certain aspects of the problem (Marchant et al. 2011). The UN experts meetings on LAWS are being organized (UNODA 2017e).

How the global community should deal with singularity in AI or strangelets that can create a stable black-hole is the question at the heart of this book. These are extremely unlikely events, but their consequences, if realized, would be truly catastrophic. The book provides an answer to this important question through a multidisciplinary review of the literatures and examinations of mathematical and probabilistic models. In Chapter 7, the final chapter, a policy framework for dealing with these catastrophes is summarized.

1.8 Economics of Catastrophes Versus Economics of Sustainability

Learning from the wide-ranging review of literature on catastrophic events including life-ending and universe-ending catastrophes, this book elucidates the economics of truly catastrophic events. It offers an ensemble of ideas and quantitative models that can be utilized to support rational decisions on global-scale random catastrophes.

In the economics literature, this field, i.e. the economics of catastrophes, has received little attention in the past (see, for example, Stavins 2012; Tietenberg and Lewis 2014; Mankiw 2014). In the simplest terms, the economics fields of environmental and natural resources where environmental and natural catastrophes are researched are concerned with an optimal allocation of environmental or natural resources in which markets play a central role. In certain situations of externalities, mutual bargaining among concerned parties may arise to settle the externality cost (Pigou 1920; Coase 1960). Governmental policy interventions are called for when such bargaining incurs too high transaction costs or the goods that are in dispute are publicly consumed goods (Samuelson 1954; Nordhaus 1994). A policy instrument is to be chosen from the gamut of policy options in a way to minimize the cost of policy actions given the targeted benefits of regulations (Montgomery 1972; Hartwick and Olewiler 1997). The benefits, and sometimes costs, of a regulation must be measured through the empirical methods that account for a large pool of market behaviors that can be observed (Mendelsohn and Olmstead 2009; Freeman et al. 2014).

In the broader literature of environmental, ecological, and natural resources, there is a group of researchers who are primarily concerned about whether an optimal decision-making can lead to an unsustainable economy or whether sustainability itself should be given priority over optimal uses of resources (World Commission on Environment and Development 1987). This field is recognized as sustainability science (Solow 1993; Hartwick and Olewiler 1997).

The literature of sustainability economics or sustainability in general emphasizes the collapse of a concerned system due to unsustainable practices in which the concerned system may be a society, or an economic system, an ecological system, or a particular ecological population (Ehrlich 1968; Meadows et al. 1972; Costanza 1991; Daly 1996).

What is the difference between the economics of sustainability and the economics of catastrophes that this book describes? The two fields are mostly nonoverlapping. The economics of catastrophes is distinct in that it addresses the decision problems with regard to a very low probability event with a truly shocking consequence in terms of both human deaths and destructions. Put differently, it is concerned about catastrophes that seem to strike randomly, utterly shocking the society.

By contrast, the economics of sustainability is concerned with unsustainable practices or systems. These practices and systems are common and pervasive in the society or industries. Further, the concept of "unsustainability" is loosely defined and utilized in the literature of sustainability. In particular, the concept of unsustainability does not cover truly catastrophic outcomes such as life-ending catastrophes, universe-ending catastrophes, fat-tails, chaos, singularity, and fractals.

More concretely, the threat of an asteroid collision does not result from unsustainable practices or systems, and neither does the threat of AI or black-hole-generating strangelets created from high-risk physics experiments. These catastrophic challenges and others described throughout this book cannot be addressed adequately through the concepts of sustainability.

1.9 Road Ahead

This concludes the introduction to the book. Intriguing theories and models of catastrophe, chaos, fractals, and order await you on the road ahead in the upcoming chapter.

References

Bakkensen, L.A. and Mendelsohn, R. (2016). Risk and adaptation: evidence from global hurricane damages and fatalities. *Journal of the Association of Environmental and Resource Economists* 3: 555–587.

Black, F. and Scholes, M. (1973). The pricing of options and corporate liabilities. *Journal of Political Economy* 81 (3): 637–654.

Blake, E.S., Landsea, C.W., and Gibney, E.J. (2011). *The Deadliest, Costliest, and Most Intense United States Tropical Cyclones from 1851 to 2010 (and Other Frequently Requested Hurricane Facts). NOAA Technical Memorandum NWS NHC-6*. Silver Spring: NOAA.

Broecker, W.S. (1997). Thermohaline circulation, the Achilles' heel of our climate system: will man-made CO_2 upset the current balance? *Science* 278: 1582–1588.

Burtraw D, Szambelan SJ (2009) U.S. emissions trading markets for SO_2 and NOx. *Resources for the Future Discussion Paper 09–40*. Resources for the Future, Washington, DC.

Campbell, K.M., Einhorn, R.J., and Reiss, M.B. (ed.) (2004). *The Nuclear Tipping Point: Why States Reconsider their Nuclear Choices*. Washington, DC: Brookings Institution Press.

Carson, R. (1962). *Silent Spring*. Boston: Houghton Mifflin.
CERN (European Organization for Nuclear Research) (2017) The Accelerator Complex. CERN, Geneva. https://home.cern/about/accelerators.
Chapman, C.R. and Morrison, D. (1994). Impacts on the earth by asteroids and comets: assessing the hazard. *Nature* 367: 33–40.
Chapman University (2017). *Survey of American Fears: Wave 4*. Orange, CA: Chapman University.
Chen, Y.S. (2013). *The Primeval Flood Catastrophe: Origins and Early Development in Mesopotamian Traditions*. Oxford: Oxford University Press.
Chicago Mercantile Exchange (CME) Group (2017). *CME Group aAll Products – Codes and Slate*. Chicago: CME Group http://www.cmegroup.com/trading/products/#pagenumber=1&sortasc=false&sortfield=oi.
Chung, C.E. and Ramanathan, V. (2006). Weakening of North Indian SST gradients and the monsoon rainfall in India and the Sahel. *Journal of Climate* 19: 2036–2045.
Coase, R. (1960). The problem of social costs. *Journal of Law and Economics* 3: 1–44.
Costanza, R. (1991). *Ecological Economics: The Science and Management of Sustainability*. New York: Columbia University Press.
Cowell, E.B., Chalmers, R., Rouse, W.H.D. et al. (trans.)(1895). *The Jataka; or, Stories of the Buddha's Former Births*. Cambridge, UK: Cambridge University Press.
Daly, H.E. (1996). *Beyond Growth: The Economics of Sustainable Development*. Boston: Beacon Press.
Dar, A., Rujula, A.D., and Heinz, U. (1999). Will relativistic heavy-ion colliders destroy our planet? *Physics Letters B* 470: 142–148.
Diamond, J. (2005). *Collapse: How Societies Choose to Fail or Succeed*. New York: Viking Press.
Douady, A. (1986). Julia Sets and the Mandelbrot Set. In: *The Beauty of Fractals: Images of Complex Dynamical Systems* (ed. H.-O. Petgen and D.H. Richter). Berlin: Springer.
Drake, B.L. (2012). The influence of climatic change on the Late Bronze Age Collapse and the Greek Dark Ages. *Journal of Archaeological Science* 39: 1862–1870.
Edesess, M. (2014). *Catastrophe Bonds: An Important New Financial Instrument*. Roubaix: EDHEC-Risk Institute, EDHEC Business School.
Ehrlich, P.R. (1968). *The Population Bomb*. New York: Sierra Club/Ballantine Books.
Ellis, J., Giudice, G., Mangano, M. et al. (2008). Review of the safety of LHC collisions. *Journal of Physics G: Nuclear and Particle Physics* 35 (11).
Emanuel, K. (2005). Increasing destructiveness of tropical cyclones over the past 30 years. *Nature* 436: 686–688.
Emanuel, K. (2008). The hurricane–climate connection. *Bulletin of the American Meteorological Society* 89: ES10–ES20.
EM-DAT (2017). *International Disaster Database*. Brussels: Centre for Research on the Epidemiology of Disasters (CRED), Université Catholique de Louvain.
Fabozzi, F.J., Modigliani, F.G., and Jones, F.J. (2009). *Foundations of Financial Markets and Institutions*, 4e. New York: Prentice Hall.
Federal Emergency Management Agency (FEMA) (2012). *Biggert-Waters Flood Insurance Reform Act of 2012 (BW12) Timeline*. Washington, DC: FEMA.
Federal Emergency Management Agency (FEMA) (2016). *The FEMA National Earthquake Hazards Reduction Program: Accomplishments in Fiscal Year 2014*. Washington, DC: FEMA.

Feigenbaum, M.J. (1978). Quantitative universality for a class of non-linear transformations. *Journal of Statistical Physics* 19: 25–52.

Fractal Foundation (2009). *Educators' Guide*. Albuquerque, NM: Fractal Foundation.

Frame, M., Mandelbrot, B., and Neger, N. (2017). *Fractal Geometry*. New Haven: Yale University http://users.math.yale.edu/public_html/people/frame/fractals.

Freeman, A.M. III, Herriges, J.A., and Cling, C.L. (2014). *The Measurements of Environmental and Resource Values: Theory and Practice*. New York: RFF Press.

Gabaix, X. (2009). Power laws in economics and finance. *Annual Review of Economics* 1: 255–293.

Gill, R.B., Mayewski, P.A., Nyberg, J. et al. (2007). Drought and the Maya collapse. *Ancient Mesoamerica* 18: 283–302.

Gleick, J. (1987). *Chaos: Making a New Science*. London: Penguin Books.

Goswami, B.N., Venugopal, V., Sengupta, D. et al. (2006). Increasing trend of extreme rain events over India in a warming environment. *Science* 314: 1442–1445.

Graham, T. Jr. (2004). *Avoiding the Tipping Point*. Washington, DC: Arms Control Association.

Hartwick, J.M. and Olewiler, N.D. (1997). *The Economics of Natural Resource Use*, 2e. New York: Pearson.

Hawking S, Tegmark M, Russell S, Wilczek F (2014). Transcending complacency on superintelligent machines. *Huffington Post* https://www.huffingtonpost.com/stephen-hawking/artificial-intelligence_b_5174265.html.

Hirshleifer, J. (1983). From weakest-link to best-shot: the voluntary provision of public goods. *Public Choice* 41: 371–386.

India Meteorological Department (IMD) (2015). *Report on Cyclonic Disturbances in North Indian Ocean from 1990 to 2012*. New Delhi: Regional Specialised Meteorological Centre – Tropical Cyclones.

Intergovernmental Panel on Climate Change (IPCC) (1990). *Climate Change: The IPCC Scientific Assessment*. Cambridge, UK: Cambridge University Press.

Intergovernmental Panel on Climate Change (IPCC) (2001). *Climate Change 2001: The Scientific Basis, the Third Assessment Report of the IPCC*. Cambridge, UK: Cambridge University Press.

Intergovernmental Panel on Climate Change (IPCC) (2014). *Climate Change 2014: The Physical Science Basis, the Fifth Assessment Report of the IPCC*. Cambridge, UK: Cambridge University Press.

International Federation of Red Cross and Red Crescent Societies (IFRC) (2014). *World Disasters Report 2014 – Focus on Culture and Risk*. Geneva: IFRC.

International Thermonuclear Experimental Reactor (ITER) (2015) ITER: The World's Largest Tokamak. https://www.iter.org/mach.

Jaffe, R.L., Buszaa, W., Sandweiss, J., and Wilczek, F. (2000). Review of speculative disaster scenarios at RHIC. *Review of Modern Physics* 72: 1125–1140.

Joint Typhoon Warning Center (JTWC) (2017). *Annual Tropical Cyclone Reports*. Guam, Mariana Islands: JTWC.

Kaiho, K. and Oshima, N. (2017). Site of asteroid impact changed the history of life on earth: the low probability of mass extinction. *Scientific Reports* 7: 14855. doi: 10.1038/s41598-017-14199-x.

Kenneth, D.J., Breitenbach, S.F.M., Aquino, V.V. et al. (2012). Development and disintegration of Maya political systems in response to climate change. *Science* 338: 788–791.

Khan, M.R. and Rahman, A. (2007). Partnership approach to disaster management in Bangladesh: a critical policy assessment. *Natural Hazards* 41 (1): 359–378.

Kiger, M. and Russell, J. (1996). *This Dynamic Earth: The Story of Plate Tectonics*. Washington, DC: United States Geological Survey (USGS).

King RO (2013) The National Flood Insurance Program: status and remaining issues for Congress. CRS Report for Congress R42850. Congressional Research Service, Washington, DC.

Knowles, S.G. and Kunreuther, H.C. (2014). Troubled waters: the National Flood Insurance Program in historical perspective. *Journal of Policy History* 26: 325–353.

Knutson, T.R., McBride, J.L., Chan, J. et al. (2010). Tropical cyclones and climate change. *Nature Geoscience* 3: 157–163.

Koopmans, T.C. (1965). On the concept of optimal economic growth. *Academiae Scientiarum Scripta Varia* 28 (1): 1–75.

Kurzweil, R. (2005). *The Singularity Is Near*. New York: Penguin.

Lackner, K.S., Brennana, S., Matter, J.M. et al. (2012). The urgency of the development of CO_2 capture from ambient air. *Proceedings of the National Academy of Sciences of the United States of America* 109 (33): 13156–13162.

Le Treut, H., Somerville, R., Cubasch, U. et al. (2007). Historical overview of climate change. In: *Climate Change 2007: The Physical Science Basis. Contribution of Working Group I to the Fourth Assessment Report of the Intergovernmental Panel on Climate Change* (ed. S. Solomon, D. Qin, M. Manning, et al.). Cambridge, UK: Cambridge University Press.

Lenton, T.M., Held, H., Kriegler, E. et al. (2008). Tipping elements in the earth's climate system. *Proceedings of the National Academy of Sciences of the United States of America* 105: 1786–1793.

Leopold, A. (1949). *A Sand County Almanac*. Oxford: Oxford University Press.

Lorenz, E.N. (1963). Deterministic nonperiodic flow. *Journal of the Atmospheric Sciences* 20: 130–141.

Lorenz, E.N. (1969). Atmospheric predictability as revealed by naturally occurring analogues. *Journal of the Atmospheric Sciences* 26: 636–646.

Mandelbrot, B. (1963). The variation of certain speculative prices. *Journal of Business* 36 (4): 394–419.

Mandelbrot, B. (1967). How long is the coast of Britain? *Science* 156 (3775): 636–638. doi: 10.1126/science.156.3775.636.

Mandelbrot, B.B. (1983). *The Fractal Geometry of Nature*. New York: Macmillan.

Mandelbrot, B. (1997). *Fractals and Scaling in Finance*. New York: Springer.

Mankiw, G.N. (2014). *Principles of Economics*, 7e. Boston: Cengage Learning.

Mann, M.E., Bradley, R.S., and Hughes, M.K. (1999). Northern hemisphere temperatures during the past millennium: inferences, uncertainties, and limitations. *Geophysical Research Letters* 26: 759–762.

Marchant, G., Allenby, B., Arkin, R. et al. (2011). International governance of autonomous military robots. *Columbia Science and Technology Law Review* 12: 272–315.

Markowitz, H. (1952). Portfolio selection. *Journal of Finance* 7: 77–91.

Massachusetts Institute of Technology (MIT) (2003). *The Future of Nuclear Power: An Interdisciplinary MIT Study*. Cambridge, MA: MIT.

Massachusetts Institute of Technology (MIT) (2015). *The Future of Solar Energy: An Interdisciplinary MIT Study*. Cambridge, MA: MIT.

Maxwell D (2016) Thailand: breaking the cycle of flooding and drought. Asian Correspondent (4 October 2016).

McAdie, C.J., Landsea, C.W., Neuman, C.J. et al. (2009). *Tropical Cyclones of the North Atlantic Ocean, 1851–2006*, Historical Climatology Series 6–2. Miami, FL: Prepared by the National Climatic Data Center, Asheville, NC in cooperation with the National Hurricane Center.

Meadows, D.H., Meadows, D.L., Randers, J., and Behrens, W.H. (1972). *The Limits to Growth*. New York: Universe Books.

Meehl, G.A. and Hu, A. (2006). Megadroughts in the Indian monsoon region and Southwest North America and a mechanism for associated multidecadal Pacific Sea surface temperature anomalies. *Journal of Climate* 19: 1605–1623.

Mendelsohn, R. (1980). An economic analysis of air pollution from coal-fired power plants. *Journal of Environmental Economics and Management* 7: 30–43.

Mendelsohn, R. (2000). Efficient adaptation to climate change. *Climatic Change* 45: 583–600.

Mendelsohn, R., Dinar, A., and Williams, L. (2006). The distributional impact of climate change on rich and poor countries. *Environment and Development Economics* 11: 1–20.

Mendelsohn, R., Emanuel, K., Chonabayashi, S., and Bakkenshen, L. (2012). The impact of climate change on global tropical cyclone damage. *Nature Climate Change* 2: 205–209.

Mendelsohn, R. and Olmstead, S. (2009). The economic valuation of environmental amenities and disamenities: methods and applications. *Annual Review of Resources* 34: 325–347.

Middleton, G.D. (2017). *Understanding Collapse: Ancient History and Modern Myths*. Cambridge, UK: Cambridge University Press.

Mills, M.J., Toon, O.B., Turco, R.P. et al. (2008). Massive global ozone loss predicted following regional nuclear conflict. *Proceedings of the National Academy of Sciences of the United States of America* 105: 5307–53012.

Molina, M.J. and Rowland, F.S. (1974). Stratospheric sink for chlorofluoromethanes: chlorine atom-catalysed destruction of ozone. *Nature* 249: 810–812.

Montgomery, W.D. (1972). Markets in licenses and efficient pollution control programs. *Journal of Economic Theory* 5: 395–418.

Muller, N.Z., Mendelsohn, R., and Nordhaus, W. (2011). Environmental accounting for pollution in the United States economy. *American Economic Review* 101: 1649–1675.

Nakicenovic N, Davidson O, Davis G, et al. (2000) Emissions scenarios. A Special Report of IPCC Working Group III. IPCC, Geneva.

National Aeronautics and Space Administration (NASA) (2014). *NASA's Efforts to Identify Near-Earth Objects and Mitigate Hazards*. Washington, DC: IG-14-030, NASA Office of Inspector General.

National Conference of State Legislatures (NCSL) (2017). *NCSL Policy Update: State Statutes on Chemical Safety*. Washington, DC: NCSL.

National Disaster Risk Reduction and Management Council (NDRRMC) (2017). *NDRRMC Disaster Archives*. Quezon City, The Philippines: NDRRMC.

National Drought Mitigation Center (NDMC) (2017). *The Dust Bowl*. Lincoln, NE: NDMC, University of Nebraska-Lincoln http://drought.unl.edu/droughtbasics/dustbowl.aspx.

National Oceanic Atmospheric Administration (NOAA) (2009) Tropical Cyclone Reports. National Hurricane Center, NOAA, Miami.

National Oceanic Atmospheric Administration (NOAA) (2016). *Weather Fatalities 2016*. Silver Spring: National Weather Service, NOAA.

National Research Council (2010). *Defending Planet Earth: Near-Earth-Object Surveys and Hazard Mitigation Strategies*. Washington, DC: National Academies Press.

National Research Council (NRC) (2015). *Climate Intervention: Reflecting Sunlight to Cool Earth. Committee on Geoengineering Climate: Technical Evaluation and Discussion of Impacts*. Washington, DC: National Academies Press.

von Neumann, J. and Morgenstern, O. (1947). *Theory of Games and Economic Behavior*, 2e. Princeton, NJ: Princeton University Press.

New Scientist (2012) Five civilisations that climate change may have doomed. https://www.newscientist.com/gallery/climate-collapse.

Nobel Prize (2013) The Nobel Prize in Physics 2013: François Englert, Peter W. Higgs. https://www.nobelprize.org/nobel_prizes/physics/laureates/2013.

Noone, S., Broderick, C., Duffy, C. et al. (2017). A 250 year drought catalogue for the island of Ireland (1765–2015). *International Journal of Climatology* doi: 10.1002/joc.4999.

Nordhaus, W. (1994). *Managing the Global Commons*. Cambridge, MA: MIT Press.

Nordhaus, W.D. (2006). Paul Samuelson and global public goods. In: *Samuelsonian Economics and the Twenty-First Century* (ed. M. Szenberg, L. Ramrattan and A.A. Gottesman). Oxford Scholarship Online.

Nordhaus, W.D. (2008). *A Question of Balance—Weighing the Options on Global Warming Policies*. New Haven: Yale University Press.

Nordhaus, W. (2010). The economics of hurricanes and implications of global warming. *Climate Change Economics* 1: 1–24.

Nordhaus, W. (2011). The economics of tail events with an application to climate change. *Review of Environmental Economics and Policy* 5: 240–257.

Nordhaus, W. (2013). *Climate Casino: Risk, Uncertainty, and Economics for a Warming World*. New Haven: Yale University Press.

Nuclear Energy Institute (NEI) (2016). *Energy Statistics*. Washington, DC: NEI http://www.nei.org/knowledge-center/nuclear-statistics.

Overbye D (2013) Chasing the Higgs. NYT 4 March.

Pareto, V. (1896). *Cours d'Economie Politique*. Geneva: Droz.

Pascal, B. (1670). *Penseés* (trans. WF Trotter, 1910). London: Dent.

Paul, B.K. (2009). Why relatively fewer people died? The case of Bangladesh's cyclone Sidr. *Natural Hazards* 50: 289–304.

Pielke, R.A., Gratz, J., Landsea, C.W. et al. (2008). Normalized hurricane damages in the United States: 1900–2005. *Natural Hazards Review* 9: 29–42.

Pigou, A.C. (1920). *Economics of Welfare*. London: Macmillan.

Pindyck, R.S. (2011). Fat tails, thin tails, and climate change policy. *Review of Environmental Economics and Policy* 5: 258–274.

Plaga R (2009) On the potential catastrophic risk from metastable quantum-black holes produced at particle colliders. arXiv:0808.1415 [hep-ph].

Poincaré, H. (1880–1890). *Mémoire sur les Courbes Définies par les Équations Différentielles I–VI, Oeuvre I*. Paris: Gauthier-Villars.

Posner, R.A. (2004). *Catastrophe: Risk and Response*. New York: Oxford University Press.
Ramsey, F.P. (1928). A mathematical theory of saving. *Economic Journal* 38 (152): 543–559.
Richter, C.F. (1958). *Elementary Seismology*. San Francisco: W.H. Freeman and Company.
Samuelson, P. (1954). The pure theory of public expenditure. *Review of Economics and Statistics* 36: 387–389.
Sandler, T. (1997). *Global Challenges: An Approach to Environmental, Political, and Economic Problems*. Cambridge: Cambridge University Press.
Sanghi, A., Ramachandran, S., Fuente, A. et al. (2010). *Natural Hazards, Unnatural Disasters: The Economics of Effective Prevention*. Washington, DC: World Bank Group.
Schmalensee, R. and Stavins, R.N. (2013). The SO_2 allowance trading system: the ironic history of a grand policy experiment. *Journal of Economic Perspectives* 27: 103–122.
Schuster, E.F. (1984). Classification of probability laws by tail behavior. *Journal of the American Statistical Association* 79 (388): 936–939.
Schwartz, G.M. and Nichols, J.J. (ed.) (2006). *After Collapse: The Regeneration of Complex Societies*. Tucson, AZ: University of Arizona Press.
Seo, S.N. (2010). A microeconometric analysis of adapting portfolios to climate change: adoption of agricultural systems in Latin America. *Applied Economic Perspectives and Policy* 32: 489–514.
Seo, S.N. (2012a). Decision making under climate risks: an analysis of sub-Saharan farmers' adaptation behaviors. *Weather, Climate and Society* 4: 285–299.
Seo, S.N. (2012b). What eludes international agreements on climate change? The economics of global public goods. *Economic Affairs* 32 (2): 74–80.
Seo, S.N. (2014). Estimating tropical cyclone damages under climate change in the southern hemisphere using reported damages. *Environmental and Resource Economics* 58: 473–490.
Seo, S.N. (2015a). Fatalities of neglect: adapt to more intense hurricanes? *International Journal of Climatology* 35: 3505–3514.
Seo, S.N. (2015b). Adaptation to global warming as an optimal transition process to a greenhouse world. *Economic Affairs* 35: 272–284.
Seo, S.N. (2015c). Helping low-latitude, poor countries with climate change. Regulation. Winter 2015–2016: 6–8.
Seo, S.N. (2016a). Modeling farmer adaptations to climate change in South America: a micro-behavioral economic perspective. *Environmental and Ecological Statistics* 23: 1–21.
Seo, S.N. (2016b). The micro-behavioral framework for estimating total damage of global warming on natural resource enterprises with full adaptations. *Journal of Agricultural, Biological, and Environmental Statistics* 21: 328–347.
Seo, S.N. (2016c). *Microbehavioral Econometric Methods: Theories, Models, and Applications for the Study of Environmental and Natural Resources*. Amsterdam: Academic Press (Elsevier).
Seo, S.N. (2016d). Untold tales of goats in deadly Indian monsoons: adapt or rain-retreat under global warming? *Journal of Extreme Events* 3: doi: 10.1142/S2345737616500019.
Seo, S.N. (2017a). *The Behavioral Economics of Climate Change: Adaptation Behaviors, Global Public Goods, Breakthrough Technologies, and Policy-Making*. London: Academic Press (Elsevier).

Seo, S.N. (2017b). Measuring policy benefits of the cyclone shelter program in the North Indian Ocean: protection from intense winds or high storm surges? *Climate Change Economics* 8 (4): 1–18. doi: 10.1142/S2010007817500117.

Seo, S.N. and Bakkensen, L.A. (2017). Is tropical cyclone surge, not intensity, what kills so many people in South Asia? *Weather, Climate, and Society* 9: 71–81.

Seo, S.N. and Mendelsohn, R. (2008). Measuring impacts and adaptations to climate change: a structural Ricardian model of African livestock management. *Agricultural Economics* 38 (2): 151–165.

Shiller, R.J. (2004). *The New Financial Order: Risk in the 21st Century*. Princeton, NJ: Princeton University Press.

Shiller, R.J. (2009). *The Subprime Solution: How Today's Global Financial Crisis Happened, and What to Do about it*. Princeton, NJ: Princeton University Press.

Solow, R.M. (1993). An almost ideal step toward sustainability. *Resources Policy* 19: 162–172.

Stavins, R. (1998). What can we learn from the grand policy experiment? Lessons from SO_2 allowance trading. *Journal of Economic Perspectives* 12: 69–88.

Stavins, R.N. (2012). *Economics of the Environment: Selected Readings*, 6e. New York: W.W. Norton & Company.

Strogatz, S.H. (1994). *Nonlinear Dynamics and Chaos: With Applications to Physics, Biology, Chemistry, and Engineering*. Boston, MA: Addison-Wesley.

Sumner, D.A. and Zulauf, C. (2012). *Economic & Environmental Effects of Agricultural Insurance Programs*. Washington, DC: The Council on Food, Agricultural & Resource Economics (C-FARE).

Swiss Re Capital Markets (2017). *Insurance Linked Securities Market Update*. Zurich, Switzerland: Swiss Re.

Swiss Re Institute (2017). *Natural Catastrophes and Man-made Disasters in 2016: A Year of Widespread Damages*. Zurich, Switzerland: Swiss Re.

Taleb, N.N. (2005). Mandelbrot makes sense. Fat tails, asymmetric knowledge, and decision making: Nassim Nicholas Taleb's essay in honor of Benoit Mandelbrot's 80th birthday. *Wilmott Magazine* (2005): 51–59.

Taylor, K.E., Stouffer, R.J., and Meehl, G.A. (2012). An overview of CMIP5 and the experiment design. *Bulletin of the American Meteorological Society* 93: 485–498.

Thom, R. (1975). *Structural Stability and Morphogenesis*. New York: Benjamin-Addison-Wesley.

Tietenberg, T. and Lewis, L. (2014). *Environmental & Natural Resource Economics*, 9e. New York: Routledge.

Tipitaka (2010) Brahmajāla Sutta: The All-Embracing Net of Views. Digha Nikaya (trans. B. Bodhi). www.accesstoinsight.org.

Tol, R. (2009). The economic effects of climate change. *Journal of Economic Perspectives* 23: 29–51.

Turco, R., Toon, O.B., Ackerman, T.P. et al. (1983). Nuclear winter: global consequences of multiple nuclear explosions. *Science* 222: 1283–1292.

Tziperman, E. (2017). *Chaos Theory: A Brief Introduction*. Boston, MA: Harvard University.

United Nations Environmental Programme (UNEP) (2016) The Montreal Protocol on Substances that Deplete the Ozone Layer. UNEP, Kigali, Rwanda.

United Nations Framework Convention on Climate Change (UNFCCC) (1992) United Nations Framework Convention on Climate Change. UNFCCC, New York.

United Nations Framework Convention on Climate Change (UNFCCC) (1997) Kyoto Protocol to the United Nations Framework Convention on Climate Change. UNFCCC, New York.

United Nations Framework Convention on Climate Change (UNFCCC) (2015) The Paris Agreement. Conference of the Parties (COP) 21. UNFCCC, New York.

United Nations Office for Disarmament Affairs (UNODA) (2017a) Treaty on the Non-Proliferation of Nuclear Weapons (NPT). https://www.un.org/disarmament/wmd/nuclear/npt/text.

United Nations Office for Disarmament Affairs (UNODA) (2017b) Convention on the Prohibition of the Development, Production and Stockpiling of Bacteriological (Biological) and Toxin Weapons and on their Destruction. http://disarmament.un.org/treaties/t/bwc.

United Nations Office for Disarmament Affairs (UNODA) (2017c) Treaty on the Non-Proliferation of Chemical Weapons. https://www.opcw.org/chemical-weapons-convention/download-the-cwc.

United Nations Office for Disarmament Affairs (UNODA) (2017d) Background on Lethal Autonomous Weapons Systems in the CCW. https://www.unog.ch/80256ee600585943/(httppages)/8fa3c2562a60ff81c1257ce600393df6?opendocument.

United States Congress (1978) Toxic Substances Control Act of 1978. US Congress, Washington, DC.

United States Congress (1980) Comprehensive Environmental Response, Compensation, and Liability Act of 1980. US Congress, Washington, DC.

United States Congress (2004) Earthquake Hazards Reduction Act of 1977 (as Amended in 2004). US Congress, Washington, DC.

United States Congress (2012) Federal Insecticide, Fungicide, and Rodenticide Act of 1947. US Congress, Washington, DC.

United States Environmental Protection Agency (US EPA) (2014) National Emissions Inventory (NEI). 2014. Washington, DC.

United States Environmental Protection Agency (US EPA) (1977) The Clean Air Act Amendments of 1977. US EPA, Washington DC.

United States Environmental Protection Agency (US EPA) (1990) The Clean Air Act Amendments of 1990. US EPA, Washington, DC.

United States Environmental Protection Agency (US EPA) (2017) Criteria Air Pollutants. US EPA, Washington, DC. https://www.epa.gov/criteria-air-pollutants.

Utsu T (2013) Catalog of Damaging Earthquakes in the World (Through 2013). International Institute of Seismology and Earthquake Engineering, Tsukuba, Japan. http://iisee.kenken.go.jp/utsu/index_eng.html.

Vogel, S.A. and Roberts, J.A. (2011). Why the Toxic Substances Control Act needs an overhaul, and how to strengthen oversight of chemicals in the interim. *Health Affairs* 30: 898–905.

Wagoner, G. and Weitzman, M. (2015). *Climate Shock: The Economic Consequences of a Hotter Planet*. Princeton, NJ: Princeton University Press.

Warrick RA, Trainer PB, Baker EJ, Brinkman W (1975) Drought hazard in the United States: a research assessment. Program on Technology, Environment and Man

Monograph #NSF-RA-E-75-004. Institute of Behavioral Science, University of Colorado, Boulder.

Weitzman, M.L. (1998). The Noah's ark problem. *Econometrica* 66: 1279–1298.

Weitzman, M.L. (2009). On modeling and interpreting the economics of catastrophic climate change. *Review of Economics and Statistics* 91: 1–19.

Wheeler, M. (1966). *Civilizations of the Indus Valley and Beyond*. New York: McGraw-Hill.

White House (2013). *The President's Climate Action Plan*. Washington, DC: Executive Office of the President, The White House.

White House (2017). *Statement by President Trump on the Paris Climate Accord*. Washington, DC: White House.

Wong, E. (trans.)(2001). *Lieh-Tzu: A Taoist Guide to Practical Living*. Boston: Shambhala.

World Bank (2007). *Emergency 2007 Cyclone Recovery and Restoration Project*. Washington, DC: World Bank http://projects.worldbank.org/p111272/emergency-2007-cyclone-recovery-restoration-project?lang=en.

World Bank (2014). *World Bank Issues its First Ever Catastrophe Bond Linked to Natural Hazard Risks in Sixteen Caribbean Countries*. Washington, DC: World Bank http://treasury.worldbank.org/cmd/htm/firstcatbondlinkedtonaturalhazards.html.

World Commission on Environment and Development (WCED) (1987). *Our Common Future. WCED*. Oxford: Oxford University Press.

World Health Organization (WHO) (2014). *Burden of Disease from Household Air Pollution for 2012*. Geneva: WHO.

World Health Organization (WHO) (2016). *Ambient Air Pollution: A Global Assessment of Exposure and Burden of Disease*. Geneva: WHO.

World Meteorological Organization (WMO) (2014) Scientific Assessment of Ozone Depletion: 2014. *World Meteorological Organization, Global Ozone Research and Monitoring Project—Report No. 55*. WMO, Geneva.

Yohe, G.W. and Tol, R.S.J. (2012). Precaution and a dismal theorem: implications for climate policy and climate research. In: *Risk Management in Commodity Markets* (ed. G. Helyette). New York: Wiley.

Zeeman, E.C. (1977). *Catastrophe Theory: Selected Papers 1972–1977*. Reading, MA: Addison-Wesley.

2

Mathematical Foundations of Catastrophe and Chaos Theories and Their Applications

2.1 Introduction

Catastrophe is a concept that is certainly beyond the literature of economics, as is the concept of chaos. As a scientific subject, the catastrophe theory and its mathematical models emerged in the 1970s from the study of structural stability of biological systems, whose precursors such as the bifurcation theory in mathematics date even further back to the 1880s (Poincaré 1880–1890; Thom 1972, 1975).

The theory of chaos, on the other hand, originated from the scientific efforts in the 1960s by Edward Lorenz to predict, albeit unsuccessfully, weather and meteorological systems, whose failures were re-interpreted as the first successful establishment of a system of chaos (Lorenz 1963; Gleick 1987).

The catastrophe and chaos theories were formed from the scientists' endeavors to build a set of mathematical equations that can represent a physical or a biological system in a manner pertinent to many fields of research. As such, these theories became quickly one of the intriguing fields of mathematics and applied sciences whose significant products include, among other things, catastrophe models, the Lorenz attractor, fractal theory, Mandelbrot set, and Feigenbaum constants (Zeeman 1977; Mandelbrot 1983; Feigenbaum 1978; Strogatz 1994).

Of the theories and models of catastrophe and chaos, enduring influences can be encountered even today in many research areas, including, inter alia, the fractal theory and power law applications, bifurcation and singularity mathematics, and tipping points (Mandelbrot 1997; Kurzweil 2005; Gabaix 2009). For example, applications of the fractal theory and its power law fat-tail statistics are found in the economic research on financial price crashes as well as uncertain catastrophes (Pareto 1896; Zipf 1949; Mandelbrot 2001; Taleb 2007; Weitzman 2009).

As the three fables of catastrophe introduced in Chapter 1 indicate, it is no exaggeration to say that the concept of a catastrophe existed in one form or another even in the earliest recorded history of humanity. The corpus of Indian, Chinese, and western philosophical traditions reveals that the concept of a catastrophe existed in varied forms at the origins of these traditions and were appropriated in many ways for various philosophical purposes (Pascal 1670; Cowell et al. 1895; Chen 2013).

Because of their conceptual importance in the economics of catastrophic events that will be expounded in this book, the author will in this chapter describe the catastrophe theory and associated catastrophe models, the chaos theory and associated chaos models, and the fractal theory and its applications. In addition to the conceptualizations of these theories, advances in mathematical formulations of these theories will be explained in a way to elucidate and then critique the details of the mathematical catastrophe and chaos models.

This chapter on mathematical theories and models is followed by the next chapter which is devoted to the expositions of the aforementioned selected philosophical, theological, and archaeological traditions concerning catastrophic events and consequences. The next chapter also includes a critical review of selected, but intended to be broad-ranging, classic environmental texts whose influences are still enduring today.

The concepts, models, mathematical formulations, and philosophical constructions of catastrophe and chaos are of interest to the author, and perhaps to the readers of this book as well, primarily because of the recent developments in the public policy domains which have come to increasingly concentrate on catastrophic risks and consequences in policy decisions, whose examples include policy studies of unstoppable global warming, potentially catastrophic consequences of various technological advances such as gene alterations, stem cell technologies, and artificial intelligence (Thomson et al. 1998; Alley et al. 2003; Posner 2004; Kurzweil 2005; Gore 2006; Taleb 2007; Lenton et al. 2008; Weitzman 2009; Nordhaus 2013; Hawking et al. 2014; Bergkamp 2016; NASEM 2017).

The policy conclusions and recommendations from the above-cited studies that place heavier weights on tail catastrophes are often diagonally different from those from the classical public policy studies which have emphasized weighing the costs against the benefits of a public policy intervention and choosing the social welfare maximizing policy option (Koopmans 1965; Arrow et al. 1996; Nordhaus 2008, 2011; Seo 2017).

On top of this trend, noting the remarkable progresses in scientific and technological capacities of humanity during the past century, many also argue that the possibility of a truly catastrophic event at a global scale has been increasing because of human activities (Posner 2004; Kurzweil 2005; Hawking et al. 2014). As mentioned in Chapter 1, such events include the possibility of an unprecedented catastrophe that is associated with advances in capacities to build nuclear and hydrogen bombs (Turco et al. 1983; Mills et al. 2008), advances in genetic sciences to alter human and animal genomes and generate stem cells for medical purposes (USDOE 2008; Nobel Prize 2012), advances in intelligent robots and artificial intelligence (Kurzweil 2005; Hawking et al. 2014), advances in high-risk physics researches such as the Large Hadron Collider experiments (Dar et al. 1999; Jaffe et al. 2000), asteroid collisions (Chapman and Morrison 1994), and capacities to alter global climate systems (NRC 2013, 2015).

The critical importance of the mathematical concepts, theories, and models that will be presented in this chapter will become clearer when the reader has completed reading this book, although it may not be immediately clear at the end of this chapter. To get straight to the point, they lay out fundamental concepts and definitions which are appropriated by the researchers of catastrophes in other fields such as economics, policy studies, and ecological studies.

Stated from a different angle, this means that through in-depth examinations of the mathematical concepts and theories, we will be better positioned to evaluate the usefulness as well as the limitations not only of them in themselves but also of their applications to the aforementioned fields of policy research.

Before we plunge into the details and complexities of the field, it must be acknowledged that the author does not feel it is necessary to provide a review of the entire historical literature of the theories of catastrophes and chaos. Interested readers may refer to Thom (1975), Gilmore (1981), Strogatz (1994), Gleick (1987), Mandelbrot (1997), Rosser (2007), and Frame et al. (2017) for an expansive discussion of the specific literature that one is concerned with.

The objective of this chapter is to present the essential concepts of catastrophe and chaos as well as their mathematical formulations in a scientifically lucid but mathematically rich manner, so that they can serve as a basis for the exposition of the economics of global-scale catastrophic events to be presented in Chapters 4 and 5.

2.2 Catastrophe Theory

The origination of the catastrophe theory in the 1970s is credited by many to French mathematician René Thom. Thom defined the catastrophe in terms of structural stability of a biological system and delineated a mathematical system which is composed of a set of equations in which a catastrophic event is inherent (Thom 1975; Zeeman 1977). He was the scholar who proposed a list of elementary catastrophes, which became wildly popular and is still widely utilized, with colorful names attached to them.

More broadly, the catastrophe theory is interpreted as a special case of the singularity theory in the bifurcation theory of mathematics, which is again part of the mathematical study of nonlinear dynamic systems (Gilmore 1981; Rosser 2007). The great French mathematician Henri Poincaré is widely credited to have invented the bifurcation theory during his inquiry into celestial mechanics.

Poincaré asked whether the orbits of the planets in the solar system would escape without bounds to infinity, be kept within certain bounds, or the solar planets would crash into each other or the Sun. This investigation of the structural stability of the solar system in the event of a small perturbation in the system led to the discovery of the bifurcation theory (Poincaré 1880–1890).

Thom, after his discovery of transversality, furthered his predecessors' works on the classification of different singularities, i.e. catastrophes, and proposed seven elementary catastrophes, with up to six dimensions in control and state variables (Thom 1956, 1972). This later became the catastrophe theory (Thom 1975; Zeeman 1977).

Thom developed a system of equations in which the solutions of the system depend on the parameters in the system. A stable equilibrium is identified when a potential function, to be explained shortly, is at the minimum. A catastrophe is judged to occur when a small change in a certain parameter in the system causes the equilibrium to disappear, which leads to a large and sudden change in the behavior of the system.

This field of research was later named a "catastrophe theory" by Zeeman who coined the term and popularized the field to the public and researchers (Zeeman 1977). In short, a catastrophe is a concept that refers to the situation in which a large and sudden change is induced by a negligible change in the system.

In his paper, Thom provided a list of elementary catastrophes, that is, seven elementary catastrophe models (Thom 1975). He suggested that if the potential function depends on fewer than two state variables and fewer than four control parameters, then there are only seven generic structures of catastrophe. These are fold catastrophe, cusp catastrophe, swallowtail catastrophe, butterfly catastrophe, hyperbolic umbilic catastrophe, elliptic umbilic catastrophe, and parabolic umbilic catastrophe.

2.2.1 Catastrophe Models and Tipping Points

To elaborate on the aforementioned concepts and definitions of catastrophe, let us take for example the fold catastrophe, the simplest catastrophe model. The fold catastrophe has one control parameter (ρ) and one state variable (x). The potential function of the fold catastrophe is defined as follows (Zeeman 1977; Gilmore 1981):

$$F(x; \rho) = x^3 + \rho x. \tag{2.1}$$

The geometry of a fold catastrophe is shown in Figure 2.1. The points are values on the $F - x$ plane, i.e. values of F given x. The relationship between the potential function and the state variable hinges on the value of the control parameter, ρ.

At the negative values of the control parameter ($\rho < 0$), the potential function has two equilibria: one is a stable (minimum) equilibrium and the other is an unstable equilibrium (maximum). As the absolute value of the control parameter becomes smaller while it is still negative, which is denoted by ($0.5 * \rho < 0$) by the author, there are still two equilibria: stable and unstable.

As the value of the control parameter is increased incrementally to get ever closer to zero, the system behaves smoothly. That is to say, there are still two equilibria and the stable equilibrium moves gradually to the left. However, as soon as the value of the control parameter reaches zero ($\rho = 0$), the system fundamentally changes.

A strange phenomenon suddenly occurs at the certain value of the control parameter. That is, the stable and the unstable equilibria of the system suddenly disappear from the system. The value of the control parameter on which this abrupt change in the system is induced is called the bifurcation point. At the positive values of the control parameter ($\rho > 0$), as shown in the figure, there is no longer a stable equilibrium to the potential function and to the system itself.

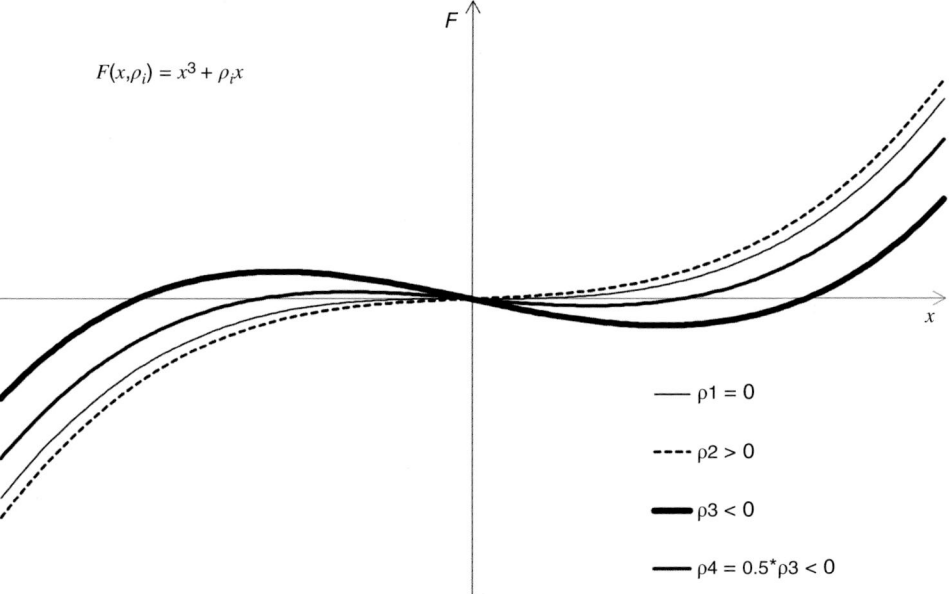

Figure 2.1 Geometry of a fold catastrophe.

If a system, be it biological or physical, exhibits a fold bifurcation, when the value of the control parameter arrives at zero, the stability of the system maintained for the negative values of the control parameter ($p < 0$) is suddenly lost, at which point the biological or physical system makes a sudden transition to a very different world-state in which there is no stable solution.

The bifurcation point is where the catastrophe occurs, abruptly and surprisingly. The bifurcation point is also called by researchers a tipping point or singularity. A bifurcation point as well as a singularity is a key concept for mathematicians, while a tipping point is more frequently referred to by economists and policy scientists (Kurzweil 2005; Lenton et al. 2008).

The other catastrophe models proposed by Thom are cusp catastrophe, swallowtail catastrophe, and butterfly catastrophe, for which there is only one state variable (x). The potential function for the cusp catastrophe has the following form with two control parameters (a, b):

$$F(x; a, b) = x^4 + ax^2 + bx. \tag{2.2}$$

The potential function for the swallowtail catastrophe is as follows with three control parameters (a, b, c):

$$F(x; a, b, c) = x^5 + ax^3 + bx^2 + cx. \tag{2.3}$$

The potential function of the butterfly catastrophe is written as follows with four control parameters (a, b, c, d):

$$F(x; a, b, c, d) = x^6 + ax^4 + bx^3 + cx^2 + dx. \tag{2.4}$$

The elementary catastrophe models that have two state variables are hyperbolic umbilic catastrophe, elliptic umbilic catastrophe, and parabolic umbilic catastrophe. The potential function of the hyperbolic umbilic catastrophe has the following form with three control parameters:

$$F(x, y; a, b, c) = x^3 + y^3 + axy + bx + cy. \tag{2.5}$$

The elliptic umbilic catastrophe is written as the potential function with two state variables and three control parameters:

$$F(x, y; a, b, c) = \frac{x^3}{3} - xy^2 + a(x^2 + y^2) + bx + cy. \tag{2.6}$$

The potential function of the parabolic umbilic catastrophe is written with two state variables and four control parameters:

$$F(x, y; a, b, c, d) = x^2 y + y^4 + ax^2 + by^2 + cx + dy. \tag{2.7}$$

Of the seven elementary catastrophes, nearly all catastrophe theory applications in social sciences and economics, as explained in later sections of this chapter, are either fold catastrophe or cusp catastrophe (Rosser 2007).

For the studies of catastrophe, both physical and economical, many of which will be discussed throughout this book, Thom's conceptualization of catastrophe provides the clearest and the most robust definition, although it had been studied as a bifurcation theory: a catastrophe is an abrupt shift of the system into a "fundamentally" different system that is caused by a miniscule change in the control parameter.

As illustrated in the geometry of the fold catastrophe, a qualitative characteristic of the abrupt change is that a stable equilibrium in the existing system is lost at the tipping point and is never returned in the new world system. At the tipping point, a fundamental change in the existing system occurs suddenly and surprisingly. This feature of Thom's catastrophe theory is also fundamentally important in interpreting the corpus of catastrophe studies in numerous fields.

To elaborate further on the last point, a catastrophe has been explicated in the literature at various spatial scales: a catastrophe at a local context, a national context, a regional context, a global context, or a universal context. Put differently, the spatial size of the loss or damages is an important consideration for the analyses of catastrophic events. The definition by Thom points to a complete transformation of the present system: within Thom's simple world model of catastrophe theory, a catastrophe is universal.

2.2.2 Regulating Mechanisms

Could the new system which emerges at the bifurcation point, i.e. $\rho = 0$, have the possibility to return to the old system in which a stable equilibrium exists? There is nothing that excludes such a possibility in Thom's theory. However, the old system and the new system are relatively defined in Thom's theory. That is, the old system is interpreted as the new system from the standpoints of the new system.

From another point of view, the question of whether the system can be placed in the old system or the new system points to the weakest feature of Thom's catastrophe theory and other catastrophe models.

What do I mean by this? The weakest feature of the catastrophe models is that there is no mechanism at all that regulates the control parameter (ρ) in the system. It is such an important parameter that determines the fate of the system. Nonetheless, it is not attended or taken care of by any person or mechanism in the system. It is a free variable parameter.

This is against the reality in most disaster cases, be it physical, biological, or economical. In reality, an individual or a community would take good care of a parameter of utmost concern in one's life or that of the community. As the control variable approaches the tipping point ever more closely, an individual should adapt to changes in the control parameter or design a policy to prevent the tipping point. Such behaviors of adaptation and policy interventions are not embedded in Thom's theory of catastrophe.

As such is the case, the catastrophe strikes unexpectedly, therefore, abruptly and shockingly. However, in the reality of most catastrophe policy issues, many of which are discussed in this book, a catastrophe is defined, expected, prepared for, and averted, albeit at different levels of such activities. An illuminating example is the whole gamut of adaptation portfolios and strategies for a global-scale climate catastrophe proposed and observed by climate researchers (Mendelsohn 2000; Hanemann 2000; Seo 2016, 2017).

At the scale of biology or geophysics, all mechanisms or processes have at least a regulating or controlling mechanism embedded in them. As a rather crude example, as soon as a lethal virus enters a living organism, an antibody is formed by the immune system of the body to prevent the promulgation of the virus to the extent that it may kill the living organism. Granted there is a host of mechanisms and processes that is way more complex than this simplified version of biology.

This is a critical weakness of the mathematical theory of catastrophe and especially so if a researcher is interested in applying it to the study of socioeconomic systems or contemporary policy issues (for a review of these studies see Rosser 2007). If there is no regulation mechanism of the control parameter at all, all is futile in the world. Put differently, the catastrophe theory as envisioned by Thom and Zeeman is about the nature of a catastrophe which is unexpected, abrupt, and shocking, but not about the causes and interactive mechanisms involved in a catastrophe event. It says little, if not nothing, about the latter.

A more realistic catastrophe model could be formulated by way of adding a regulating mechanism of some sort to the control parameter. That is, we may write the fold catastrophe as follows by adding a regulation function (h) as a function of adaptive mechanisms (ad) and policy interventions (po):

$$F(x; \rho) = x^3 + \rho x, \text{ such that } \rho = h(ad, po). \tag{2.8}$$

We will have an opportunity to revisit this important issue of regulation mechanisms in catastrophe theories and models in the chapters on economics and policy – Chapters 4–6 of this book. There this issue will be rephrased in the context of identifying variables of adaptation and policy interventions in response to a variety of catastrophe events (Seo 2015, 2017).

Let me conclude this section on catastrophe theory with a comment from René Thom on the nature of the catastrophe theory as a scientific theory, which may serve as a warning against an unmoderated application of the theory and models to real-life catastrophes, from his 1977 article entitled "What is catastrophe theory about?" (Thom 1977).

> It would be completely wrong to equate Catastrophe theory with one of the standard scientific theories, like the Newtonian theory of gravitation, or the Darwinian theory of evolution. In such cases, one has to expect that the theory has to get some experimental confirmation, it has to be founded (or at least may be "falsified" in Popper's sense) by experiment. The plain fact is that C. theory escapes this criterion: it cannot be "proved", nor "falsified" by experiment.

2.3 Chaos Theory

In the theory of catastrophe explicated by René Thom and others, a tipping point is where a stable equilibrium is lost and turned into a new world system where there is no equilibrium. Nonexistence of an equilibrium is one feature of chaos. In this way, chaos theory is closely related to the theory of catastrophes, which will be made clear in this section.

The chaos theory is the field of research that examines the systems that are in chaos or disorderly (Gleick 1987; Strogatz 1994). The chaos is defined as absence of order in the system concerned. A chaotic system is described by a set of mathematical equations and presented with a graphical depiction of the chaotic system.

In the early years of the field, researchers engaged in discovering and depicting a chaotic system (Lorenz 1963; Mandelbrot 1963, 1983; Douady 1986). Over time, the field has evolved into scientific efforts for discovering or restoring an order in a previously perceived chaotic system (Feigenbaum 1978; Boccaletti et al. 2000). The author will explain the former in this section and the latter in the following section.

The chaos theory has seen its application to numerous academic disciplines, including, but not limited to, psychology (Masterpasqua and Perna 1997), health sciences (Denton et al. 1990; Lombardi 2000), financial markets (Trippi and Chorafas 1994), and, of course, atmosphere and meteorology (Zeng et al. 1993).

2.3.1 Butterfly Effect

The first true experimenter and founder of the field of chaos theory is a meteorologist, whose name was Edward Lorenz, at the Massachusetts Institute of Technology (Gleick 1987; Emanuel 2011). A tale of serendipity or pure genius has been told of his encounter with the chaos: his quest for an orderly system failed, from which a chaos theory emerged.

Lorenz was interested in constructing a computer model that predicts the weather from a set of conditioning variables, for which he had a computer program written to find a solution simultaneously for a set of 12 mathematical equations (Lorenz 1963). Although he failed to predict the weather events themselves of the Boston area precisely, he was nonetheless able to produce a set of solutions to the system of weather equations. With the 1960s' computing capabilities, it meant that he produced a long sequence of a set of numbers as a printout until a solution set was reached.

When Lorenz wanted one year later to reproduce a particular sequence in the printout, he was forced, for the sake of saving time, to start from the middle of the sequence instead of the beginning of the sequence from last year's printout. He expected that the computer program would produce the same sequence again. Contrary to his expectation, the program resulted in a wildly different sequence of numbers.

He was puzzled, but later figured out that it was due to truncations of numbers. In order to save printout pages, although he stored output numbers up to six decimal points in the computer memory, he recorded them only up to three decimal points in his printout.

He could have given up at that point, but instead began to unravel a novel insight. Even with the differences in numbers by, say, 0.0001 owing to truncations of numbers, the system should have produced a new sequence of numbers very close to the original sequence of numbers. Against all expectations, what occurred in his simulation is that a minute difference in the starting point was leading to a wildly different series of numbers as well as a completely strange destination point.

This phenomenon first reported by Lorenz has been termed the butterfly effect, which is obviously given in the context of weather changes and predictions, which Lorenz attributes to his speech entitled "Does the Flap of a Butterfly's Wings in Brazil Set Off a Tornado in Texas?" at a meeting in Washington in 1972 (Lorenz 1969a,b, 1993).

The essence of the butterfly effect is that a minute disturbance in the atmospheric system caused by a butterfly's wing flapping at some place on Earth causes a major shift in the weather in a remote place on Earth:

> The flapping of a single butterfly's wing today produces a tiny change in the state of the atmosphere. Over a period of time, what the atmosphere actually does diverges from what it would have done. So, in a month's time, a tornado that would have devastated the Indonesian coast doesn't happen. Or, maybe one that wasn't going to happen does (Stewart 2002).

This phenomenon is a common feature to the field of chaos theory. In the literature, it is referred to as sensitive dependence on initial conditions (Strogatz 1994). It is the phenomenon in which a miniscule alteration in the initial conditions gives rise to a drastic change in the behaviors and outcomes of the system.

A deeper trouble foreseen by Lorenz and contained in the chaos theory is that a minor disturbance in a system, be it physical or social, can occur in ways that are practically uncontrollable. A tiny alteration in the initial conditions that determine the current state of a concerned system may occur through, inter alia, experimental noises or discrepancies that are not easily noticeable, background noises that cannot be controlled, tiny inaccuracies in the equipment or laboratories, and an experimenter's habits and preferences. Inasmuch as these noises, discrepancies, and aberrations are uncontrollable at a practical level in a scientific experiment, the states of the system are not predictable with precision.

Lorenz's interpretation points to a major difficulty in scientific endeavors, underneath which is a chaotic system, unpredictable and perhaps even unknowable. Lorenz himself concluded, while laying the foundation for chaos theory, that it is impossible to predict the weather realizations of a chosen region at a particular time.

2.3.2 The Lorenz Attractor

Continuing with his set of equations in the atmospheric convection model for weather predictions, Lorenz came to present a mathematical representation of a chaotic system many years later. By drastically simplifying the atmospheric convection model, he built a three-dimensional system which still has sensitive dependence on initial conditions and extremely erratic dynamics. The original system of equations had 12 equations, from which he drastically reduced the equations of atmospheric convection to three equations. The new system had nothing to do with atmospheric convection and weather predictions any more (Strogatz 1994; Tziperman 2017).

The resultant set of three ordinary differential equations is referred to as the Lorenz system or the Lorenz equation. It has become the foundation of the field of chaos theory as the first system of chaos presented mathematically, which calls for elaborations to be made shortly. The Lorenz equation is written as follows (Lorenz 1963; Strogatz 1994):

$$\dot{x} = \sigma(y - x);$$
$$\dot{y} = \gamma x - y - xz;$$
$$\dot{z} = xy - bz. \tag{2.9}$$

In the Lorenz system, there are three state variables – x, y, z – and three parameters of the system – σ, γ, b. To depict the Lorenz attractor graphically, Lorenz set the parameter values to $\sigma = 10, \gamma = 28, b = 2.5$ (Hirsh et al. 2003).

With these parameters' values, Lorenz drew the solution sets of (x, y, z) on a three-dimensional plane, which is shown in Figure 2.2. With these parameters' values, the solution sets always stayed on a double spiral or two butterfly wings. The double spiral is called the butterfly, the strange attractor, or the Lorenz attractor (Lorenz 1963, 1993). The white marks on the figure are butterflies which trace the solution sets of the Lorenz system.

Recognized as simple but prescient, the Lorenz attractor has come to represent a chaotic system. The reason is that, as shown in the movements of the butterflies in

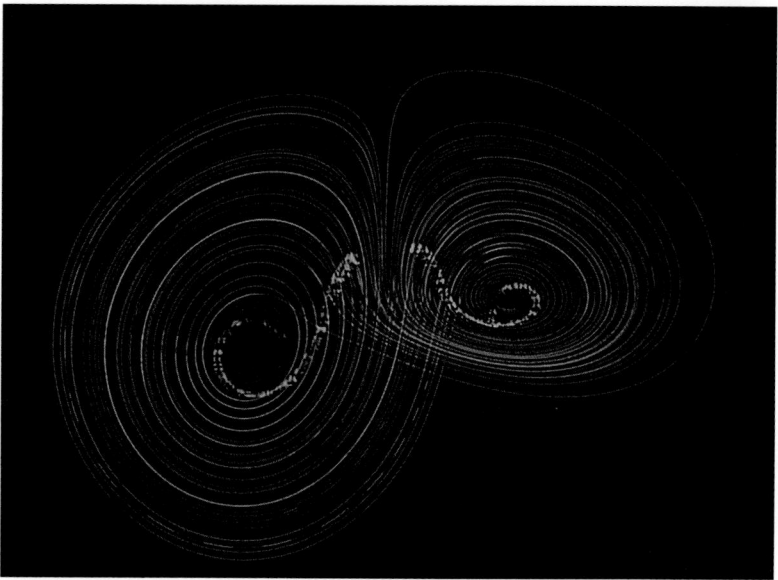

Figure 2.2 The Lorenz attractor, simulated with $\sigma = 10$, $\gamma = 28$, $\beta = 2.5$ from the interactive Lorenz attractor provided online by Cristersson (2017).

Figure 2.2, the Lorenz system goes into a loop in the double spiral and does not rest in a steady or equilibrium state at all, repeating its spiral action randomly, taking on a different loop each time. It is a three-dimensional, nonlinear, nonperiodic, deterministic system.

The Lorenz attractor is a pioneering mathematical system of chaos. That is, it depicts a set of solutions of a chaotic system in which there is no order (Gleick 1987; Tziperman 2017). There is no order in the Lorenz attractor because it fails to exhibit either of the two features of an orderly system. First, the system does not contain a steady-state solution, i.e. a stable equilibrium. Second, it does not contain a periodic behavior or pattern.

2.4 Fractal Theory

2.4.1 Fractals

Another fundamental concept in the literature of the chaos theory is a fractal, the then novel term coined by Benoit Mandelbrot who was at the time employed as a mathematician by the International Business Machines (IBM) corporation. The fractal is interpreted to be an image of chaos, but also occurs, unlike the Lorenz attractor, naturally and therefore is encountered very frequently in our daily lives (Fractal Foundation 2009). The Mandelbrot set, a fractal, quickly became a public emblem for chaos, the reasons for which will become clear to the readers of this book after reading through this section (Gleick 1987; Mandelbrot 2004).

Benoit Mandelbrot was born in Warsaw, Poland in 1924 and, after obtaining his doctoral degree in mathematics through graduate studies in France and the US, worked at

IBM for 35 years. He received the tenure as Sterling Professor of Mathematical Sciences at the age of 75 from Yale University, the oldest person to receive the honor in the university's history (Olson 2004).

One of Mandelbrot's research areas during the early years of his career at IBM was historical US cotton price fluctuations. Initially, he observed that historical cotton prices did not exhibit a normal (Gaussian) distribution, although most studies at that time assumed that distribution, no matter how the historical data were analyzed (Mandelbrot 1963).

He continued to gather all available US cotton price data which were traced back to 1900 and, while analyzing the data, Mandelbrot came to a surprising finding: at whatever scale the data were analyzed, i.e. daily, weekly, monthly, quarterly, or yearly, there was a pattern of the price data that reappeared over and over again.

What Mandelbrot came across in his research on cotton price fluctuations had come to evolve into the theory of a fractal, which is defined as a consistent pattern that appears independent of a scale of analysis. The following quote summarizes nicely Mandelbrot's findings from his 1963 cotton price paper, which will need to be further elaborated in this section and in Chapter 4 of this book:

> The numbers that produced aberrations from the point of view of normal distribution produced symmetry from the point of view of scaling. Each particular price change was random and unpredictable. But the sequence of changes was independent on scale: curves for daily price changes and monthly price changes matched perfectly. Incredibly, analyzed Mandelbrot's way, the degree of variation had remained constant over a tumultuous sixty-year period that saw two World Wars and a depression (Gleick 1987).

At this point, readers might wonder why the fractal has become one of the most prominent theories of chaos. After all, it is a consistent pattern that re-emerges again and again regardless of a scale of analysis. There are multiple routes that lead the fractal theory to a system of chaos. In one route, the fractal's chaos implication lies in the repeatability which goes on endlessly, which has repercussions on scientific measurability.

In his subsequent research, Mandelbrot considered whether it is possible to measure the length of the British coastline precisely, which few had doubted the possibility of doing (Mandelbrot 1967). There are multiple ways scientists commonly rely on to do this. The easiest way one can think of may be that an individual researcher measures it by taking the distance of the British coastline on a map and scaling it up appropriately. However, a map of the British coastline would show many bays, but minor bays that are too small to be drawn on the map would not appear on the map, however refined the map resolution is. So, we can conclude that it is not possible to measure the length of the coastline precisely by reading off the distances on the map.

Alternatively, a scientist can proceed with a brute force method in which s/he takes real distances, piece by piece, directly by walking along the entire coastline and adds them up. But, this method will again fail to produce the correct distance because microscopic bays that exist in between grains of sands cannot be accounted for. In the fractal theory, Mandelbrot argues that there is another coastal line at a smaller scale within the British coastline itself, a fractal. In fact, in the fractal world, there would be an infinite number of fractal images at ever smaller scales (Mandelbrot 1967).

Level 0: 3 line segments of size 1/1.0

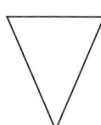

Level 1: 12 line segments of size 1/3.0

Level 2: 48 line segments of size 1/9.0

Level 3: 192 line segments of size 1/27.0

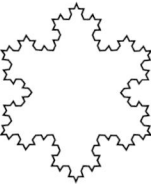

Figure 2.3 The first four iterations of the Koch snowflake, simulated from the interactive simulator provided by Shodor (www.shodor.org).

The fractal image plays a key role in this book because it was developed as a scientific concept but was originated from and has been applied to economic phenomenon in a forceful way. Before we get to the economic interpretations, we need a clearer understanding of a fractal. The concept of a fractal can be visually elucidated by a Koch curve, first constructed by a Swedish mathematician Helge von Koch as early as 1904 (Edgar 2004).

The Koch snowflake or curve, as it was later known, is constructed by the following steps, and the first four iterations are shown in Figure 2.3 (Strogatz 1994):

1) Draw an equilateral triangle.
2) Divide each side into three segments.
3) Draw an equilateral triangle which has the middle segment as the base and which projects outward.
4) Remove the base from the new equilateral triangle.
5) Repeat steps 2–4 for all sides of the original triangle.
6) The Koch snowflake is the limit approached by the repeated applications of the above steps.

The Koch snowflake constructed by the first four iterations of the above procedure are shown in Figure 2.3, denoted as level 0, level 1, level 2, and level 3. Let the line segment in the level 0 figure be S_0, the line segment in the level 1 figure be S_1, and S_2, S_3, … are defined in the same way. Then, the Koch snowflake is $K = S_\infty$.

The Koch snowflake is a fractal, i.e. a self-similar image. A magnification of the Koch snowflake looks exactly the same as the original image. That is to say, the triangle in the top image in Figure 2.3 (level 0) appears in the second image in Figure 2.3 (level 1). The second image appears in the third image in Figure 2.3 (level 2). The third image again appears in the fourth image at the bottom in Figure 2.3 (level 3).

What is the dimension of the Koch snowflake? It is not 1 since the length of the line segment is infinite. Nor is it 2 since there is no area in the figure. It is a fractional dimension. The term "fractal" originates from the word fractional. That is, a fractal is a fractional dimension, e.g. 1.26 dimension which is between 1 dimension and 2 dimension (Strogatz 1994; Tziperman 2017).

The Koch curve has about 1.26 $\left(\approx \frac{\ln 4}{\ln 3}\right)$ dimension (refer to Strogatz 1994 for the calculation). The one dimension is a smooth line or curve. The two dimension is a square. The Mandelbrot coastline or the Koch curve is rougher than a one-dimension curve but smoother than a two-dimension square.

Formally, a fractal is defined as a figure or a pattern that has a self-similar figure or pattern as its component or at a larger scale, and in which this self-similarity or self-affinity is repeated in ever-larger scales of the figure or pattern. This means that you can zoom in the figure and find the same image at whatever scale forever. A fractal is an image of an infinitely complex system, which is why it is sometimes dubbed as a "picture of chaos" (Fractal Foundation 2009).

Unlike the Lorenz attractor, a fractal is not a strange phenomenon or a unique phenomenon to a particular thing or event. A fractal pattern is indeed very common in nature and everyday objects: you can come across it in, say, trees, rivers, coastlines, mountains, clouds, seashells, blood vessel networks, broccoli, and hurricanes. With a little bit of imagination, you can picture an image of self-similarity that is repeated over and over again at different scales of the natural object or phenomenon (Fractal Foundation 2009). What is implicit, not so subtly, in the fractal theory is that everyday objects and events are entangled in an infinitely complex system.

2.4.2 The Mandelbrot Set

Many researchers undertook to create a set of mathematical equations whose solutions result in a fractal, as in the Lorenz system. Among the most successful sets are the Mandelbrot set in the complex plane, along with the Julia sets, introduced in Section 2.4.1 as the public symbol for chaos (Mandelbrot 1983; Douady 1986; Gleick 1987).

In Section 2.4.3 in which the author explains Feigenbaum constants, the Mandelbrot set is an important analytical apparatus of the chaos theory from which Feigenbaum attempted to search for a hidden order (Feigenbaum 1978). So, it is necessary for readers to be acquainted with this mathematical system.

The Mandelbrot set is the set of values of c in the complex plane which are obtained from the following quadratic recurrence equation. For each c, start with $z_0 = 0$ and generate the sequence $z_1, z_2, z_3, \ldots$ by the following basic iteration rule (Frame et al. 2017):

$$z_n \to z_{n+1} = z_n^2 + c. \tag{2.10}$$

The point c, a value in the complex plane, belongs to the Mandelbrot set if the sequence in Eq. (2.10) does not run away to infinity, that is, if z_n remains bounded, however large n becomes. If the sequence does run away to infinity, then the point c does not belong to the Mandelbrot set.

As an exercise, let's take a number in the complex plane, $c = 1$. Then, the sequence diverges as $z_1 = 1$, $z_2 = 2$, $z_3 = 5$, $z_4 = 26$, $\ldots$, etc. Therefore, the point $c = 1$ does not belong to the Mandelbrot set. Let's take instead another number in the complex plane,

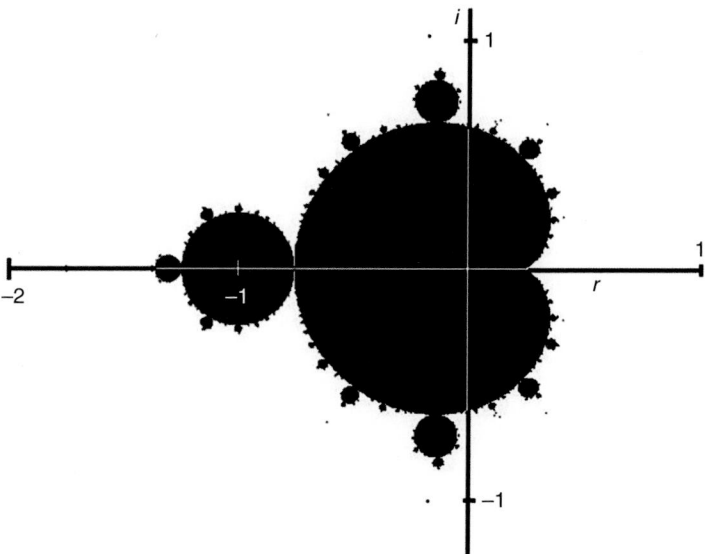

Figure 2.4 The Mandelbrot set, simulated using the simulator provided publicly by Maths algorithms (http://jakebakermaths.org.uk/maths/index.php).

$c = -1$. Then, the iterations of Eq. (2.10) result in the following sequence: $z_1 = -1$, $z_2 = 0, z_3 = -1, z_4 = 0, \ldots$, etc. Therefore, the point $c = -1$ is in the Mandelbrot set.

More formally, the Mandelbrot set, M, can be defined with the limit superior of the sequence z_n (Frame et al. 2017):

$$c \in M \text{ if and only if } \limsup_{n \to \infty} |z_{n+1}| \leq 2. \tag{2.11}$$

The Mandelbrot set is drawn on a complex plane in Figure 2.4 which undoubtedly gives rise to a fractal image. In the complex plane, the horizontal axis is the axis of real numbers (denoted r) and the vertical axis is the axis of imaginary numbers (denoted i). In the figure, black points are the values in the Mandelbrot set while white points are those not in the Mandelbrot set.

2.4.3 Fractals, Catastrophe, and Power Law

One interpretation of the fractal theory is through the power law distribution, or the Pareto distribution after French economist Wilfred Pareto, which is also called the Pareto–Levy–Mandelbrot distribution (Pareto 1896). This interpretation of the fractal reveals a close connection of the fractal geometry to the chaos as well as the catastrophe theory described above (Mandelbrot 1997; Mandelbrot and Hudson 2004; Taleb 2005).

In this subsection and in Chapter 4 again, Mandelbrot's discovery and contributions to the distributional power law are analyzed. In the final years of his life, he emphasized through numerous publications on financial markets that the power law distribution was discovered first during his research on cotton and other financial prices, from which the fractal theory later emerged (Mandelbrot 1963, 1997; Mandelbrot and Hudson 2004). From his study of financial prices, he observed that the distribution of daily price changes reappears in the distribution of monthly price changes, which reappears again in the distribution of yearly price changes (Mandelbrot 1963). That is, even though price changes were highly volatile and their distribution was non-Gaussian, the distribution was independent of scale.

In his own words, Mandelbrot explains the discovery of the power law in financial prices from a theory of self-affinity (Mandelbrot 2001):

> Mandelbrot (1963) derived this power law through a theory, as a necessary consequence of a form of postulated "self-affinity", scale-invariance or scaling. The assumption is that, after suitable renormalization, the same distribution holds for the price changes ΔP over all values of Δt.

More formally, a power law distribution, or Zipf's law, or a Pareto distribution takes on the following form (Zipf 1949; Gabaix 2009):

$$f(x) = k_1 x^\alpha, \tag{2.12}$$

where α is a shape parameter and k_1 is an unremarkable constant. Alternative ways to write the power law are explained in Chapter 4.

Specifically, Zipf's law holds, with k_2 a constant and ξ a power law exponent, that the countercumulative probability distribution has the following form (Zipf 1949):

$$P(x > \tilde{x}) = k_2/\tilde{x}^\xi \text{ with } \xi \cong 1. \tag{2.13}$$

The power law distribution or Zipf's law distribution departs from a Gaussian distribution in a major way in that it exhibits a scaling law or scalability. To clarify this, let's consider the distribution of wealth in America first discovered by Wilfred Pareto as a power law (Pareto 1896). The number of people whose wealth exceeds two million dollars is about one-quarter of the number of people whose wealth exceeds one million dollars. In the power law distribution, this relationship reappears again and again regardless of the level of wealth. For example, the number of people whose wealth exceeds 20 million dollars is about one-quarter of the number of people whose wealth exceeds 10 million dollars (Taleb 2005). The Pareto distribution exhibits the self-similarity, also called self-affinity, that characterizes the fractal geometry.

This scalability or scale invariance is numerically shown in Table 2.1, with $\xi = 2$ and the constant from Zipf's law shown in Eq. (2.13). When the wealth threshold doubles from US$1 to 2 million, the probability that the second threshold (US$2 million) is passed in the wealth distribution is one-fourth of the probability that the first threshold (US$1 million) is passed. As calculated in Table 2.1, the same ratio holds for the increase of the wealth threshold from US$10 to 20 million as well as for the increase of the wealth threshold from US$1 to 2 billion.

Table 2.1 Pareto distribution of American wealth.

	Probability that wealth exceeds $\tilde{x}$	Probability that wealth exceeds $2\tilde{x}$	Ratio: $\left(P(x>2\tilde{x})/P(x>\tilde{x})\right)$
$\tilde{x} =$ US$1 million	$k_2 * 10^{-12}$	$k_2 * 2.5 * 10^{-13}$	0.25
$\tilde{x} =$ US$10 million	$k_2 * 10^{-14}$	$k_2 * 2.5 * 10^{-15}$	0.25
$\tilde{x} =$ US$1 billion	$k_2 * 10^{-18}$	$k_2 * 2.5 * 10^{-19}$	0.25

The scalability or scale invariance does not appear in a Gaussian distribution. In the Gaussian distribution, the above relationship accelerates. That is, the tail probability falls at an accelerating rate. Let's assume that the wealth in America follows a Gaussian distribution. Let's say that the number of people whose wealth exceeds two million dollars is one-quarter of the number of people whose wealth exceeds one million dollars. Then, in the Gaussian wealth distribution, the number of people whose wealth exceeds two billion dollars is far less than one-quarter of the number of people whose wealth exceeds one billion dollars. Stated more formally, the tail probability decays exponentially in the Gaussian distribution.

In the power law tail distribution such as the Pareto distribution and Zipf's law, the variance of the distribution is infinite or, more precisely, undefined. Further, the mean of the power law distribution cannot be defined either since we cannot exclude the possibility of an extremely large event. Since there is no well-defined average of the distribution, there is no standard deviation of the distribution from the average of the distribution (Mandelbrot 2001).

Another way to state the power law distribution is that the Pareto–Levy–Mandelbrot distribution has a fat tail. The tail probability of the power law distribution does not fall quickly, i.e. exponentially, to virtual zero, as it does in the Gaussian distribution which is medium-tailed (Schuster 1984). This means that the probability of realization of an extremely long tail event, a truly catastrophic humanity-ending event which has not been seen before by humanity, does never fall to near-zero. In the fat-tail catastrophe, the unprecedented catastrophe is always there waiting to strike anytime, and when it does it totally terminates the concerned system (Weitzman 2009).

The fractal or the Mandelbrot set is one of the many power law distributions that were identified in many disciplines including physics, biology, and economics. However, the interpreters of the Mandelbrot's fractals as a catastrophe have paid little attention to the critical question of whether the power law distribution is universally pertinent to describing human behaviors or social systems (Mandelbrot and Hudson 2004; Taleb 2007; Gabaix 2009). More narrowly, they have not yet demonstrated whether a power law distribution must exist in a certain human behavioral pattern or a social behavior.

The behavioral implication of the power law distribution in Eq. (2.12) is encapsulated by a constant elasticity parameter, α. From Eq. (2.12), the elasticity is:

$$\varepsilon = \frac{d \ln f}{d \ln x} = \frac{df/f}{dx/x} = \alpha. \tag{2.14}$$

For the power law function, regardless of the level of x, the percentage change in f in response to the percentage change in x remains the same. This behavioral pattern is not pervasive in economic behaviors, perhaps rarely found in human behaviors. To see this, let's take a theoretical utility function, i.e. the Bernoulli utility function, for consideration which has the property of a diminishing marginal utility (Bernoulli 1954):

$$u(x) = \ln x. \tag{2.15}$$

A manipulation of this Bernoulli utility function leads to variable elasticity across the range of the consumption good, shown below, which decreases as the level of the independent variable increases:

$$\varepsilon = \frac{du}{dx}\frac{x}{u} = \frac{1}{u} = \frac{1}{\ln x}. \tag{2.16}$$

In the Bernoulli utility function, the scale invariance of the power law distribution does not show up. The magnitude of elasticity is varied depending on the scale of consumption (x). The higher the consumption level at which the individual is placed, the smaller the consumption elasticity of the utility.

For another example, a demand curve for a consumption good can be estimated and ask whether it exhibits a constant price elasticity. The price elasticity of the demand (D) for the good i can be written as follows:

$$\varepsilon_{P,i} = \frac{\Delta D_i / D_i}{\Delta P_i / P_i} = \bar{c}. \tag{2.17}$$

Recently, US researchers estimated the price elasticities of demands for meat products relying on the 110 290 choices made by 12 255 people during the year-long period from June 2013 to June 2014. Analyzed meat products are ground beef, steak, pork chop, deli ham, chicken breast, and chicken wing (Lusk and Tonsor 2016).

The authors report varying elasticities of the demand for each meat product across the range of the price of the meat as well as across the range of incomes of consumers. For an individual consumer, the higher the price of the meat becomes, the more inelastic the price elasticity becomes. Also, the higher the income of an individual, the smaller the price elasticity of demand is.

A third example of a power law distribution is found in the literature of economic growth models. In an optimal economic growth model, the objective function would be written as $\sum_{t=0}^{\infty} \delta^t u(c_t)$, where δ is pure rate of time preference. If we would pick the utility function that generates a constant interest rate, r, in the economy and a constant economic growth, then the only utility function that is compatible with this growth pattern is a power law utility function which has the characteristic of a constant relative risk aversion (CRRA), γ (Gabaix 2009):

$$u(c) = \frac{c^{1-\gamma} - 1}{1 - \gamma}. \tag{2.18}$$

Macroeconomists have relied on a CRRA class utility function – a power law function – because it is only compatible with balanced growth in which the economy grows at a constant rate, g, i.e. exponentially (Gabaix 2009):

$$c_t = c_0(1 + g)^t \text{ or } c_t = c_0 e^{gt}. \tag{2.19}$$

Is the exponential growth function in Eq. (2.19) pervasive in human behaviors or in the behaviors of an economic system? The answer is no. Figure 2.5 shows the growth path using Eq. (2.19), with a constant growth rate of 5% and the present value of consumption normalized to one.

To be more concrete, let c_0 be the current value of production of the economy. Then, a constant growth rate of 5% per annum implies that the size of the economy grows to 150 in a century. This is depicted as a solid black line in the figure. The growth of the economy's size by 15 000% is not in any way a reliable prediction of a future economy.

A more pervasive economic transition may be captured by making the growth rate depend upon an underlying parameter, Γ:

$$c_t = c_0(1 + g(\Gamma))^t \text{ or } c_t = c_0 e^{g(\Gamma)t}. \tag{2.20}$$

Let the growth rate be 5% as before at the initial period. Attributable to the underlying parameter, let it gradually fall in a linear way to 2% during the 100-year period. The size

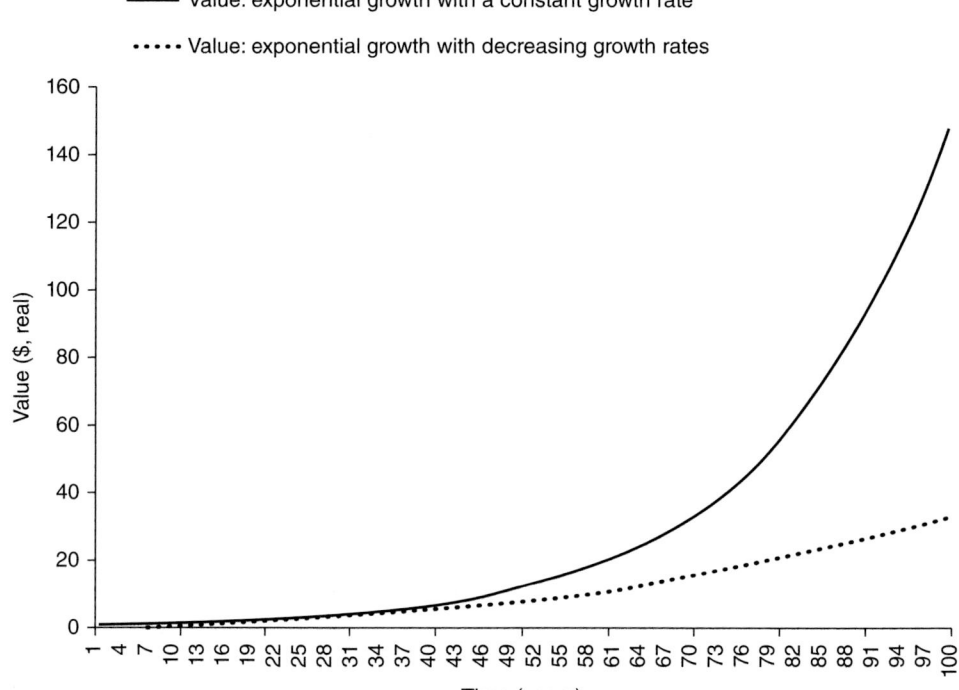

Figure 2.5 Exponential growth under a power law utility function.

of the economy can be projected using the declining growth rate, which is shown as a dashed line in Figure 2.5. With this reparameterization, the value of the economy grows from 1 to 32 in a century or by 3200%, which is by far a more reasonable prediction of the future economy faced with numerous constraints.

What this reparameterization reveals is that the power law distribution does not reflect in a comprehensive manner human behavioral patterns or behaviors of an economic system. The additional parameter in the above equation, Γ, has ample economic, psychological, behavioral meanings. For example, it may capture a decreasing-returns-to-scale technology, a diminishing marginal utility of an economic agent, limited lifespan of an individual, environmental degradation, or constraints in stocks of natural resources.

The additional parameter is also found in a biological system or a physical system. A virus–antibody relationship is one such example. When a virus enters the body, it multiplies rapidly without limits, like the fractal growth. However, its expansion is constrained by the formation of an antibody. A human behavior and a social system is an even more constrained system than a biological or physical system by numerous constraints, say, social antibodies.

That the fractal does not pervasively reflect individual or social behaviors begs the following question: Is the fractal universal? If not, what does the fractal describe? Below I have copied Michael Frame's response, without editing, to the above question. Michael Frame was one of the students of Mandelbrot and is now a Professor at Yale University where Mandelbrot had the tenure as Professor of Mathematics. He is also one of the last persons to see Mandelbrot before his death in 2010:

> Not nearly everything is fractal or self-similar. Fractals are useful descriptions for physical objects that are built by the same forces applied over many length scales, and for biological objects grown by the same instructions applied over many length scales. But of course not everything is built or grown in such a way (Frame 2018, personal communication).

2.5 Finding Order in Chaos

Throughout the historical development of the chaos theory, intellectual inquires have been as vigorous on finding an order in a reportedly chaotic system as those endeavors on searching for and conceptualizing a chaotic system (Feigenbaum 1978; Lanford 1982; Collet and Eckmann 1979, 1980). As a matter of fact, the chaos theory is now interpreted by many scholars to be equally a theory of order, that is, a theory of finding a hidden order in a chaotic system (Tziperman 2017).

In this direction, the pioneering works by Mitchell Feigenbaum are both revealing and illustrative. In the late 1970s, he was examining a bifurcation diagram of the population equation which is known as one of the representative cases of the chaos theory (Feigenbaum 1978). The population equation is the equation of growth of a biological, e.g. a species, population which takes the following normalized form (May 1976):

$$POP_{t+1} = \gamma \cdot POP_t(1 - POP_t) \tag{2.21}$$

In Eq. (2.21), the population size takes on the values from zero to one, with one being the maximum possible population and zero being the extinction of the species. The population of the species grows at the rate of γ, but declines owing to, inter alia, predators and food constraints.

The population bifurcation diagram by Robert May is the diagram of population sizes, POP_t, with varying growth rates, γ in Eq. (2.21). At a certain range of the growth rate, your choice of a particular growth rate leads to a certain population size, while, on this range, the higher the growth rate, the higher the resultant population size. For example, a growth rate of 2.5 leads to a population size of 0.6, while a growth rate of 2.9 leads to a population size of 0.655 172.

Surprisingly, when the growth rate goes out of the certain range, say, $\gamma = 3$, the population size in Eq. (2.21) was found to jump between two numbers: it takes on one number this year and another number the next year, whose pattern is repeated thenceforth. In other words, the population size bifurcates at the growth rate of 3. When the growth rate is increased further to a certain point, the population was found to bifurcate again. That is, it jumps from one to another among four population sizes. When the growth rate is raised again to a certain point, the population was observed to bifurcate once again, jumping from one number to another among eight different population sizes.

A population bifurcation diagram is shown in Figure 2.6 in which population sizes are calculated using growth rates that are in the range from 2.9 to 3.6, using Eq. (2.21). For each growth rate chosen, 2000 iterations were conducted using Eq. (2.21). The range of growth rates is selected by the author in a way that visualizes the bifurcation pattern

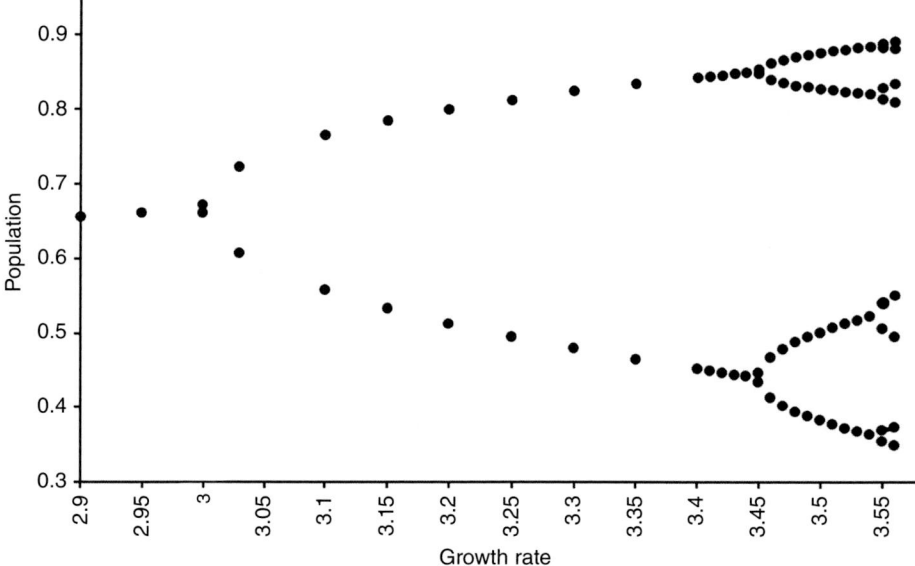

Figure 2.6 Population bifurcation.

clearly. The figure shows the population size bifurcates at the growth rate of 3, bifurcates the second time at the growth rate of 3.45, and bifurcates the third time at the growth rate of 3.55.

In Figure 2.6, it is notable that each successive population bifurcation comes at a faster pace. That is, the second bifurcation comes faster than the first bifurcation, the third bifurcation comes faster than the second bifurcation, and the fourth bifurcation comes faster than the third bifurcation. After repeated bifurcations, a chaos then arrives in which there is no stable population. That is, the population size jumps from one number to another without settling on any number.

Looking at a chaotic population bifurcation map similar to the one drawn in Figure 2.6, Feigenbaum wondered how fast each successive bifurcation point arrives. This inquiry eventually led to a ground-breaking discovery in the field of chaos theory: the successive bifurcations, although chaotic looking, occur at a constant rate. This constant number, which is 4.669 201 609…, has come to be known as the Feigenbaum constant (Feigenbaum 1978; Alligood et al. 1996).

In the language of the fractal theory, he discovered an exact scale of self-similarity in a fractal image, i.e. an exact scale at which a self-similar population diagram is repeated. Put differently, if we scale up (or down) the population diagram by 4.669 201 609, then a particular point of bifurcation will look exactly like the previous (next) point of bifurcation.

Feigenbaum continued his work to explore other nonlinear equations, e.g. a quadratic map and a Henon map, to see whether a scaling factor can be similarly determined for them. Much to his surprise, the scaling factors for these other nonlinear equations were exactly the same as that for the population bifurcation diagram which is a logistic map, i.e. 4.669 201 609… (Alligood et al. 1996). Not only was there a regularity in these more complex maps, but also Feigenbaum realized that the regularity appeared exactly the same way as that in a simpler map.

What he discovered was revolutionary in the field of chaos theory: a whole class of the set of mathematical equations which represents a chaotic system is behaving in the way that is predictable to some degree or acting with some regularity. More straightforwardly stated, there is an order in the set of equations which was previously thought of by many as a system of chaos.

Further, the Feigenbaum constant is applicable across a range of chaotic equations, and as such, it gives a researcher an analytical tool with which a chaotic system can be examined. For example, scientists can rely on a simple equation of chaos to predict the outcome of a more complex equation that contains a chaos.

To explain the derivation of the Feigenbaum constant, readers must understand beforehand the basics of the bifurcation theory (Poincaré 1880–1890; Kuznetsov 1998). Figure 2.6 is a bifurcation diagram of a logistic map, i.e. a logistic function. At the aforementioned bifurcation points of 3.0, 3.45, and 3.55, an infinitesimal change in the bifurcation parameter, that is, the population growth rate, leads to a sudden emergence of a qualitatively different solution for the system.

At the range of the bifurcation parameter between 1 and 3, it is said that the map has a period-one point. That is, it has one periodic point that the equation returns to after a large number of iterations. In the range of the bifurcation parameter between 3.0 and 3.45, it has period-two points or two periodic points that the equation returns to. In the range between 3.45 and 3.55, it has period-four points or four periodic points that the equation returns to. In the range above 3.55, it has period-eight points and above.

A period doubling refers to the transition from period-x points to period-$2x$ points. A period quadrupling refers to the transition from period-x points to period-$4x$ points. In the bifurcation theory, a bifurcation is a period doubling, quadrupling, octupling, and so on, which is followed by the onset of chaos (Rasband 1990).

Notice that a bifurcation also characterized the onset of a catastrophe, as shown in Figure 2.1 (Thom 1975; Zeeman 1977). In the system of fold catastrophe, for example, at the point of bifurcation emerges a qualitatively different system in which there is no stable equilibrium. In the chaos theory, at the point of bifurcation follows a chaos. This is another way to establish the theory of catastrophes.

The Feigenbaum constant is the limiting ratio of each bifurcation interval to the next interval of every period-doubling. More formally, let a one-parameter map be defined as follows, with a being the bifurcation parameter:

$$q_{i+1} = g(q_i|a). \tag{2.22}$$

With a_n being the parameter value of the nth bifurcation point, the Feigenbaum constant is defined by the following limit:

$$\delta = \lim_{n \to \infty} \frac{a_{n-1} - a_{n-2}}{a_n - a_{n-1}}. \tag{2.23}$$

Henceforth, the author will explain calculations of the Feigenbaum constant from several important mappings pertinent to the chaos theory. Let's start with a nonlinear mapping:

$$q_{i+1} = a - q_i^2. \tag{2.24}$$

Table 2.2 Calculating the Feigenbaum constant for a nonlinear map.

Number of bifurcation	Number of period (2^n)	Value of bifurcation parameter (a_n)	Feigenbaum ratio: $\dfrac{a_{n-1} - a_{n-2}}{a_n - a_{n-1}}$
1	2	0.75	—
2	4	1.25	—
3	8	1.368 098 9	4.233 7
4	16	1.394 046 2	4.551 5
5	32	1.399 631 2	4.645 8
6	64	1.400 828 6	4.663 9
7	128	1.401 085 3	4.668 2
8	256	1.401 140 2	4.668 9

Source: Alligood et al. (1996).

In Eq. (2.24), a is the bifurcation parameter and q is the variable of the system. The values of a for which a bifurcation occurs, that is, the period doubles, are denoted by a_1, a_2, a_3, etc. In Table 2.2, calculation of the Feigenbaum constant from the values of the bifurcation parameter is summarized for this nonlinear map (Alligood et al. 1996).

The parameter values at the first eight bifurcation points are, in order from the first bifurcation point, 0.75, 1.25, 1.368, 1.394, 1.399, 1.400, 1.4010, 1.4011, which are shown in the third column of the table. The Feigenbaum ratio in Eq. (2.23) is shown in the fourth column for each period doubling. This ratio is shown to be converging to the Feigenbaum constant which is the limit of this ratio. At the eighth bifurcation point, this ratio is 4.6689, truncated at the fifth decimal digits.

Now, let's consider for the calculation of the Feigenbaum constant the following logistic mapping, which is the population equation in Eq. (2.21), with a again being the bifurcation parameter:

$$q_{i+1} = a \cdot q_i (1 - q_i). \tag{2.25}$$

The bifurcation diagram of this logistic mapping is already shown in the population bifurcation diagram in Figure 2.6. Procedures for calculating the Feigenbaum constant for the logistic map are summarized in Table 2.3 (Alligood et al. 1996). As explained, successive bifurcations occur at the values of the bifurcation parameter $a = 3$, 3.449, 3.544, 3.564, and so on. These bifurcation points are again points of period doubling, quadrupling, octupling, and so on.

The Feigenbaum constant is approximated by the Feigenbaum ratio displayed in the last column, which has its value of 3.6692 at the seventh bifurcation point and 3.6694 at the eighth bifurcation point.

Now, the above-explained two nonlinear maps are simple systems of chaos. Can we find the Feigenbaum constant in a generalized system of chaos such as the Mandelbrot set? The third example for the calculation of the Feigenbaum constant is for the Mandelbrot set defined in Eqs. (2.10) and (2.11).

Table 2.3 Calculating the Feigenbaum constant for a logistic map.

Number of bifurcation	Number of period (2^n)	Value of bifurcation parameter (a_n)	Feigenbaum ratio: $\dfrac{a_{n-1} - a_{n-2}}{a_n - a_{n-1}}$
1	2	3	—
2	4	3.449 489 7	—
3	8	3.544 090 3	4.751 4
4	16	3.564 407 3	4.656 2
5	32	3.568 759 4	4.668 3
6	64	3.569 691 6	4.668 6
7	128	3.569 891 3	4.669 2
8	256	3.569 934 0	4.669 4

Source: Alligood et al. (1996).

For the Mandelbrot set shown in Figure 2.4, the Feigenbaum constant can be obtained via shrinking sizes of circles in the complex plane. Formally, the Feigenbaum constant is obtained by the limit of the ratio of the diameter of the nth circle and that of the $(n-1)^{th.}$ circle in the complex plane of the Mandelbrot set (Frame et al. 2017).

A couple of mathematical terms need to be defined before we proceed. The first circle on the Mandelbrot set which is attached to the big cardioid on the right is called the period-2 component, that is, period-2^1 component of the Mandelbrot set. The second circle attached to the first circle is called the period-4 component, that is, period-2^2 component. Similarly, successive circles are called thus.

The bifurcation parameter c_1 is the point of attachment on the real axis between the first circle and the big cardioid. In other words, it is the point of attachment between the period-2^1 component and the period-2^0 component. In the same way, c_2 is the point of attachment on the real axis between the period-2^2 component and the period-2^1 component. Generally, the bifurcation parameter c_n is the point of attachment on the real axis between the period-2^n component and the period-2^{n-1} component.

The diameter of the first circle is then the distance between the first point of attachment and the second point of attachment, i.e. $|c_2 - c_1|$. The diameter of the second circle is $|c_3 - c_2|$. The diameters of the succeeding circles are defined accordingly.

The Feigenbaum constant is the limit of the ratio of the diameters of the two successive circles in the complex plane (Frame et al. 2017):

$$\delta = \lim_{n \to \infty} \frac{|c_{n-1} - c_{n-2}|}{|c_n - c_{n-1}|}. \tag{2.26}$$

The procedure for calculating the Feigenbaum constant for the Mandelbrot set is explained in Table 2.4. The second column shows the periods and the third column the value of the bifurcation parameter. The fourth column lists the Feigenbaum ratio for each bifurcation point from the third bifurcation points onward. The sequence of the Feigenbaum ratio is numerically shown in the table to converge to the Feigenbaum constant whose value is $\delta = 4.669\,162\,24\ldots$ at the 10th bifurcation point.

Table 2.4 Calculating the Feigenbaum constant for the Mandelbrot set.

Number of bifurcation	Number of period (2^n)	Value of bifurcation parameter (c_n)	Feigenbaum ratio: $\frac{|c_{n-1}-c_{n-2}|}{|c_n-c_{n-1}|}$
1	2	−0.75	—
2	4	−1.25	—
3	8	−1.368 098 939 4 …	4.233 738 275 …
4	16	−1.394 046 156 6 …	4.551 506 949 …
5	32	−1.399 631 238 9 …	4.645 807 493 …
6	64	−1.400 828 742 4 …	4.663 938 185 …
7	128	−1.401 085 271 3 …	4.668 103 672 …
8	256	−1.401 140 214 699 …	4.668 966 942 …
9	512	−1.401 151 982 029 …	4.669 147 462 …
10	1024	−1.401 154 502 237 …	4.669 162 24 …
∞		−1.401 155 189 0 …	

Source: Frame et al. (2017).

2.6 Catastrophe Theory Applications

The catastrophe theory has experienced one of the most dramatic bubbles in intellectual history: it burst onto the academic scene in the early and mid-1970s, quickly became a fad among researchers for a short period of time, and then its credibility and applications faded suddenly after serious critiques were leveled against the theory (Rosser 2007). This was particularly the case in its applications to economic problems.

For the chaos theory, the author has already mentioned some of the applications to psychology, health sciences, financial economics, and atmospheric science in Section 2.3. A particularly relevant discussion and modeling to the readers of this book is also found in climate system dynamics (Pielke and Zheng 1994; Tziperman et al. 1995; Tziperman 1997). The discussion in this section is centered on the evaluations of the major applications of the catastrophe theory because richer experiences are found there, especially in the contexts of economics and policy.

The first application to economics was made by Zeeman through a model of stock market crashes, which was published in the first volume of the *Journal of Mathematical Economics* (Zeeman 1974). Zeeman modeled the stock market dynamics via interactions between fundamentalists who know true values of stocks and chartists who chase trends (Rosser 2007). Long after Zeeman's publication and the steam of the catastrophe theory dissipated, the events of stock market crashes have come to dominate world market trends since the late 1980s through a series of big crashes: the Black Monday crash in 1987, the collapse of the dot-com bubble in 2000, and the crash of 2008 that was started by subprime mortgage failures in the US (Shiller 2005).

In the modern economic theory of stock market crashes, which is much different from Zeeman's, psychological factors, such things as speculation and competition drive the formation of a bubble, defined as the size of deviation from the fundamental value, which eventually bursts when a mood suddenly shifts (Akerlof and Shiller 2009; Shiller 2014).

Early applications of the catastrophe theory to economics which appeared in similarly prestigious economics journals included discontinuous structural transformations in the general equilibrium theory (Rand 1976), an analysis of business cycles (Varian 1979), a model of bank failures (Ho and Saunders 1980), a model of inflationary hysteresis (Fischer and Jammernegg 1986), foreign currency speculation (George 1981), and particularly many applications in urban and regional sciences, including a cusp catastrophe model of urban density (Amson 1975).

Applications to ecological–economic systems were directed to the systems with a discontinuous change in a biological population. In particular, a collapse of a biological system to extinction was modeled as a result of human interactions, e.g. a multiple equilibria model of fishery dynamics, or as a result of the predator–prey dynamics, e.g. a fold catastrophe model of the Great Lakes' trout dynamics (Clark 1976; Walters 1986).

Catastrophe theory applications to economics and other disciplines quickly lost steam and the grounds for further growth (Rosser 2007). If you have followed the discussions in this chapter up to this point, it would not be difficult to understand why. Many critiques were written against the applications (see, for example, Zahler and Sussman 1977; Sussman and Zahler 1978).

The chief reason for their demise is that the potential function of the catastrophe theory which renders a system a catastrophe is too limited to be applied to social and behavioral systems. It is almost impossible to find an economic behavior or an economic system that can be pushed into a straight-jacket of one of the elementary catastrophes. The same is true of any ecological system in which strong and complex relationships exist with human activities.

Stated another way using the language in Section 2.2, catastrophe models are not persuasive, without substantive modifications, for applications to human behaviors, systems, or ecological dynamics where there are strong adaptive behavioral responses, or regulation mechanisms, or where a large number of agents with varied preferences are involved in the event considered.

2.7 Conclusion

This chapter provides a critical review of the catastrophe and chaos theories and their applications to economic and policy studies. The author attempts to elaborate all major concepts and essential mathematical formulations that pertain to this field of research, which include the catastrophe theory as defined by Thom and Zeeman, seven elementary catastrophe models, the chaos theory as defined by Lorenz, butterfly effect, Lorenz attractor, fractal theory, Mandelbrot set, power law function, bifurcation theory, singularity, and Feigenbaum constants.

The theories, mathematical models, and concepts introduced in this chapter are discussed early in the book, so that they can be appropriated as a key element in the presentations of the rest of the chapters. They will be frequently referred to in the descriptions of the theories of economics and policy interventions on catastrophic events. As mentioned before, after reading through the chapters on the economics and policy studies presented in Chapters 4–6, readers will be able to re-evaluate these scientific models and theories with additional insights gained from those chapters.

Throughout this chapter, the author emphasizes economic, behavioral, and policy implications of the theories and models of catastrophe or chaos. There are several clear messages that have emerged in this chapter for those researchers who are inclined to adopt these theories and models for their applied works.

First and foremost, regulation mechanisms or adaptive systems are by and large amiss in the catastrophe theory and models. Because of this, in the applications to behavioral or policy studies, one must be cautious in embedding assumptions in the existing models on market behaviors as well as policy factors and be mindful that omissions of such factors could severely bias policy outcomes (Lenton et al. 2008; Weitzman 2009; Nordhaus 2011; Seo 2013).

Second, researchers must heed Feigenbaum's and others' researches that discovered or revealed an underlying order in a previously thought chaotic system (Feigenbaum 1978). Certain behaviors or systems may appear to occur at random or to be chaotic, i.e. without any order. However, it is often the case that such behaviors or systems appear to be random or chaotic only until an incisive analysis reveals a hidden causality or mechanism that underlies the random behaviors.

Third, the theory of catastrophes and the theory of chaos have common features but also distinct features. Most notably, there is a similarity in that the onset of a catastrophe as well as that of a chaos in the bifurcation theory occurs suddenly at a certain value of the parameter of the concerned system.

On the other hand, a system in chaos does not necessarily lead to a catastrophic system. For instance, a fractal is very common in nature, i.e. on coasts, with clouds, trees, leaves, reefs, broccoli, etc., which are all perceived to be an orderly natural function or realization (Mandelbrot 1983; Frame et al. 2017).

A fractal pattern may also be present in financial markets and other economic behaviors, that is, anthropogenically driven systems (Mandelbrot 1963; Mandelbrot and Hudson 2004). A fractal may materialize as a power law function in modeling a certain human-driven system or behavior (Mandelbrot 1997; Taleb 2005; Gabaix 2009). However, it should be cautioned that there may lie an additional parameter that constrains a fractal expansion in a highly constrained system such as a human behavior or social system.

References

Alley, R.B., Marotzke, J., Nordhaus, W.D. et al. (2003). Abrupt climate change. *Science* 299 (5615): 2005–2010.

Alligood, K.T., Sauer, T.D., and Yorke, J.A. (1996). *Chaos: An Introduction to Dynamical Systems*. New York: Springer.

Amson, J.C. (1975). Catastrophe theory: a contribution to the study of urban systems? *Environment and Planning B* 2: 177–221.

Akerlof, G.A. and Shiller, R.J. (2009). *Animal Spirits: How Human Psychology Drives the Economy, and why it Matters for Global Capitalism*. New Jersey: Princeton University Press.

Arrow, K.J., Cropper, M.L., Eads, G.C. et al. (1996). Is there a role for benefit-cost analysis in environmental, health, and safety regulation? *Science* 272: 221–222.

Bergkamp L (2016) Decision theory and the doom scenario of climate catastrophe. Climate Etc. https://judithcurry.com/2016/09/11/decision-theory-and-the-doom-scenario-of-climate-catastrophe.

Bernoulli, D. (1954). Exposition of a new theory on the measurement of risk. *Econometrica: Journal of the Econometric Society* 22: 23–36.

Boccaletti, S., Grebogi, C., Lai, Y.-C. et al. (2000). The control of chaos: theory and applications. *Physics Reports* 329: 103–197.

Chapman, C.R. and Morrison, D. (1994). Impacts on the earth by asteroids and comets: assessing the hazard. *Nature* 367: 33–40.

Chen, Y.S. (2013). *The Primeval Flood Catastrophe: Origins and Early Development in Mesopotamian Traditions*. Oxford: Oxford University Press.

Clark, C.W. (1976). *Mathematical Bioeconomics*. New York: Wiley-Interscience.

Collet, P. and Eckmann, J.-P. (1979). Properties of continuous maps of the interval to itself. In: *Mathematical Problems in Theoretical Physics* (ed. K. Osterwalder). New York: Springer.

Collet, P. and Eckmann, J.-P. (1980). Iterated maps on the interval as dynamical systems. In: *Birkhäuser*. Boston, MA.

Cowell, E.B., Chalmers, R., Rouse, W.H.D. et al. (1895). *The Jataka; or, Stories of the Buddha's Former Births*. Cambridge, UK: Cambridge University Press.

Cristersson M (2017) Interactive Lorenz attractor & the Mandelbrot set. malinc.se.

Dar, A., Rujula, A.D., and Heinz, U. (1999). Will relativistic heavy-ion colliders destroy our planet? *Physics Letters B* 470: 142–148.

Denton, T.A., Diamond, G.A., Helfant, R.H. et al. (1990). Fascinating rhythm: a primer on chaos theory and its application to cardiology. *American Heart Journal* 120: 1419–1440.

Douady, A. (1986). Julia sets and the Mandelbrot set. In: *The Beauty of Fractals: Images of Complex Dynamical Systems* (ed. H.-O. Petgen and D.H. Richter). Berlin: Springer.

Edgar, G. (2004). *Classics on Fractals*. Boulder, CO: Westview Press.

Emanuel, K. (2011). *Edward Norton Lorenz (1917–2008): A Biographical Memoir by Kerry Emanuel*. Washington, DC: National Academy of Sciences.

Fischer, E.O. and Jammernegg, W. (1986). Empirical investigation of a catastrophe theory extension of the Phillips curve. *Review of Economics and Statistics* 68: 9–17.

Feigenbaum, M.J. (1978). Quantitative universality for a class of non-linear transformations. *Journal of Statistical Physics* 19: 25–52.

Fractal Foundation (2009). *Educators' Guide*. Albuquerque, NM: Fractal Foundation.

Frame M, Mandelbrot B, Neger N (2017) Fractal geometry. Yale University, New Haven. http://users.math.yale.edu/public_html/people/frame/fractals.

Gabaix, X. (2009). Power laws in economics and finance. *Annual Review of Economics* 1: 255–293.

George, D. (1981). Equilibrium and catastrophes. *Scottish Journal of Political Economy* 28: 43–61.

Gilmore, R. (1981). *Catastrophe Theory for Scientists and Engineers*. New York: Wiley.

Gleick, J. (1987). *Chaos: Making a New Science*. London: Penguin Books.

Gore, A. (2006). *An Inconvenient Truth: The Planetary Emergency of Global Warming and What we Can Do about it*. New York: Rodale Books.

Hanemann, W.M. (2000). Adaptation and its management. *Climatic Change* 45: 511–581.

Hawking S, Tegmark M, Russell S, Wilczek F (2014) Transcending complacency on superintelligent machines. *Huffington Post*. https://www.huffingtonpost.com/stephen-hawking/artificial-intelligence_b_5174265.html.

Hirsch, M.W., Smale, S., and Devaney, R.L. (2003). *Differential Equations, Dynamical Systems, and an Introduction to Chaos*, 2e. Boston, MA: Academic Press.

Ho, T. and Saunders, A. (1980). A catastrophe model of bank failure. *Journal of Finance* 35: 1189–1207.

Jaffe, R.L., Buszaa, W., Sandweiss, J., and Wilczek, F. (2000). Review of speculative disaster scenarios at RHIC. *Review of Modern Physics* 72: 1125–1140.

Koopmans, T.C. (1965). On the concept of optimal economic growth. *Academiae Scientiarum Scripta Varia* 28 (1): 1–75.

Kurzweil, R. (2005). *The Singularity Is near: When Humans Transcend Biology*. New York: Penguin.

Kuznetsov, Y.A. (1998). *Elements of Applied Bifurcation Theory*, 2e. New York: Springer.

Lanford, O.E. (1982). A computer-assisted proof of the Feigenbaum conjectures. *Bulletin of American Mathematical Society* 6: 427–434.

Lenton, T.M., Held, H., Kriegler, E. et al. (2008). Tipping elements in the earth's climate system. *Proceedings of the National Academy of Sciences of the United States of America* 105: 1786–1793.

Lombardi, F. (2000). Chaos theory, heart rate variability, and arrhythmic mortality. *Circulation* 101: 8–10.

Lorenz, E.N. (1963). Deterministic nonperiodic flow. *Journal of the Atmospheric Sciences* 20: 130–141.

Lorenz, E.N. (1969a). Atmospheric predictability as revealed by naturally occurring analogues. *Journal of the Atmospheric Sciences* 26: 636–646.

Lorenz, E.N. (1969b). Three approaches to atmospheric predictability. *Bulletin of the American Meteorological Society* 50: 345–349.

Lorenz, E.N. (1993). *The Essence of Chaos*. Seattle: University of Washington Press.

Lusk, J.L. and Tonsor, G.T. (2016). How meat demand elasticities vary with price, income, and product category. *Applied Economic Perspectives and Policy* 38: 673–711.

Mandelbrot, B. (1963). The variation of certain speculative prices. *Journal of Business* 36 (4): 394–419.

Mandelbrot, B. (1967). How long is the coast of Britain? *Science* 156 (3775): 636–638.

Mandelbrot, B. (1983). *The Fractal Geometry of Nature*. New York: Macmillan.

Mandelbrot, B. (1997). *Fractals and Scaling in Finance*. New York: Springer.

Mandelbrot, B. (2001). Scaling in financial prices: I. Tails and dependence. *Quantitative Finance* 1: 113–123.

Mandelbrot, B. (2004). *Fractals and Chaos*. Berlin: Springer.

Mandelbrot, B. and Hudson, R.L. (2004). *The (Mis)Behavior of Markets: A Fractal View of Risk, Ruin, and Reward*. New York: Basic Books.

Masterpasqua, F. and Perna, P.A. (ed.) (1997). *The Psychological Meaning of Chaos: Translating Theory into Practice*. APA Books, The American Psychological Association.

May, R.M. (1976). Simple mathematical models with very complicated dynamics. *Nature* 261 (5560): 459–467.

Mendelsohn, R. (2000). Efficient adaptation to climate change. *Climatic Change* 45: 583–600.

Mills, M.J., Toon, O.B., Turco, R.P. et al. (2008). Massive global ozone loss predicted following regional nuclear conflict. *Proceedings of the National Academy of Sciences of the United States of America* 105: 5307–5312.

National Academies of Sciences, Engineering, and Medicine (NASEM) (2017) Human genome editing: Science, ethics, and governance. National Academies Press, Washington, DC.

National Research Council (NRC) (2013) Abrupt impacts of climate change: Anticipating surprises. National Academies Press, Washington, DC.

National Research Council (NRC) (2015) Climate intervention: reflecting sunlight to cool Earth. National Academies Press, Washington, DC.

Nobel Prize (2012) The Nobel Prize in Physiology or Medicine 2012. Sir John B. Gurdon and Shinya Yamanaka. https://www.nobelprize.org/nobel_prizes/medicine/laureates/2012.

Nordhaus, W. (2008). *A Question of Balance: Weighing the Options on Global Warming Policies*. New Haven, CT: Yale University Press.

Nordhaus, W. (2011). The economics of tail events with an application to climate change. *Review of Environmental Economics and Policy* 5: 240–257.

Nordhaus, W. (2013). *The Climate Casino: Risk, Uncertainty, and Economics for a Warming World*. New Haven, CT: Yale University Press.

Olson, S. (2004). The genius of the unpredictable. *Yale Alumni Magazine* November–December 2004.

Pareto, V. (1896). *Cours d'Economie Politique*. Geneva: Droz.

Pascal, B. (1670). *Penseés* (trans. WF Trotter, 1910). London: Dent.

Pielke, R.A. and Zheng, X. (1994). Long-term variability of climate. *Journal of Atmospheric Science* 51: 155–159.

Poincaré H (1880–1890) Mémoire sur les Courbes Définies par les Équations Différentielles I–VI, Oeuvre I. Gauthier-Villars, Paris.

Posner, R.A. (2004). *Catastrophe: Risk and Response*. New York: Oxford University Press.

Rand, D. (1976). Threshold in Pareto sets. *Journal of Mathematical Economics* 3: 139–154.

Rasband, S.N. (1990). *Chaotic Dynamics of Nonlinear Systems*. New York: Wiley.

Rosser Jr., J.B. (2007). The rise and fall of catastrophe theory applications in economics: was the baby thrown out with bathwater? *Journal of Economic Dynamics and Control* 31: 3255–3280.

Schuster, E.F. (1984). Classification of probability laws by tail behavior. *Journal of the American Statistical Association* 79 (388): 936–939.

Seo, S.N. (2013). An essay on the impact of climate change on US agriculture: weather fluctuations, climatic shifts, and adaptation strategies. *Climatic Change* 121: 115–124.

Seo, S.N. (2015). Adaptation to global warming as an optimal transition process to a greenhouse world. *Economic Affairs* 35: 272–284.

Seo, S.N. (2016). *Microbehavioral Econometric Methods: Theories, Models, and Applications for the Study of Environmental and Natural Resources*. New York: Academic Press.

Seo, S.N. (2017). *The Behavioral Economics of Climate Change: Adaptation Behaviors, Global Public Goods, Breakthrough Technologies, and Policy-Making*. New York: Academic Press.

Shiller, R.J. (2005). *Irrational Exuberance*, 2e. Princeton, NJ: Princeton University Press.

Shiller, R.J. (2014). Speculative asset prices. *American Economic Review* 104 (6): 1486–1517.

Stewart, I. (2002). *Does God Play Dice? The New Mathematics of Chaos*, 2e. London: Wiley Blackwell.

Strogatz, S.H. (1994). *Nonlinear Dynamics and Chaos: With Applications to Physics, Biology, Chemistry, and Engineering*. Boston, MA: Addison-Wesley.

Sussman, H. and Zahler, R. (1978). A critique of applied catastrophe theory in applied behavioral sciences. *Behavioral Science* 23: 383–389.

Taleb NN (2005) Mandelbrot makes sense. Fat tails, asymmetric knowledge, and decision making: Nassim Nicholas Taleb's essay in honor of Benoit Mandelbrot's 80th birthday. *Wilmott Magazine* (2005): 51–59.

Taleb, N.N. (2007). *The Black Swan: The Impact of the Highly Improbable*. London: Penguin.

Thom, R. (1956). Les singularités des applications différentiables. *Annales Institute Fourier (Grenoble)* 6: 43–87.

Thom, R. (1972). *Stabilité Structurelle et Morphogenèse: Essai d'une Théorie Générale des Modèles*. New York: Benjamin.

Thom, R. (1975). *Structural Stability and Morphogenesis*. New York: Benjamin-Addison-Wesley.

Thom, R. (1977). What is catastrophe theory about? In: *Synergetics*, Springer Series in Synergetics, vol. 2 (ed. H. Haken). Berlin: Springer.

Thomson, J.A., Itskovitz-Eldor, J., Shapiro, S.S. et al. (1998). Embryonic stem cell lines derived from human blastocysts. *Science* 282 (5391): 1145–1147.

Trippi, R. and Chorafas, D.N. (1994). *Chaos Theory in the Financial Markets*. New York: McGraw-Hill.

Turco, R.P., Toon, O.B., Ackerman, T.P. et al. (1983). Nuclear winter: global consequences of multiple nuclear explosions. *Science* 222: 1283–1292.

Tziperman, E. (1997). Inherently unstable climate behavior due to weak thermohaline ocean circulation. *Nature* 386: 592–595.

Tziperman, E. (2017). *Chaos Theory: A Brief Introduction*. Boston, MA: Harvard University.

Tziperman, E., Cane, M.A., and Zebiak, S.E. (1995). Irregularity and locking to the seasonal cycle in an ENSO prediction model as explained by the quasi-periodicity route to chaos. *Journal of Atmospheric Science* 52 (3): 293–306.

United States Department of Energy (US DOE) (2008) Genomics and its impact on science and society: The Human Genome Project and beyond. US DOE, Washington, DC.

Varian, H.R. (1979). Catastrophe theory and the business cycle. *Economic Inquiry* 17 (1): 14–28.

Walters, C. (1986). *Adaptive Management of Renewable Resources*. New York: Macmillan.

Weitzman, M.L. (2009). On modeling and interpreting the economics of catastrophic climate change. *Review of Economics and Statistics* 91 (1): 1–19.

Zahler, R. and Sussman, H.J. (1977). Claims and accomplishments of applied catastrophe theory. *Nature* 269: 759–763.

Zeeman, E.C. (1974). On the unstable behavior of the stock exchanges. *Journal of Mathematical Economics* 1: 39–44.

Zeeman, E.C. (1977). *Catastrophe Theory – Selected Papers 1972–1977*. Reading, MA: Addison-Wesley.

Zeng, X., Pielke, R.A., and Eykholt, R. (1993). Chaos theory and its applications to the atmosphere. *Bulletin of the American Meteorological Association* 74: 631–644.

Zipf, G.K. (1949). *Human Behavior and the Principle of Least Effort*. Cambridge, MA: Addison-Wesley.

3

Philosophies, Ancient and Contemporary, of Catastrophes, Doomsdays, and Civilizational Collapses

3.1 Introduction

In the previous chapter, the author explained the core sciences and mathematical models of catastrophes through the catastrophe theory, chaos theory, fractal theory, and Feigenbaum constant, and their applications to social, economic, and ecological relationships. This chapter makes a sharp turn to follow up with the presentation of a set of philosophical traditions, ancient and modern, that were concerned with events of catastrophic consequences.

The presentation of this chapter is less mathematical than Chapter 2 but should certainly be more thought-provoking at times. These two chapters will, if successful, level the playing field for the introduction to the economics of catastrophic events and the empirical economic models of catastrophes, which are ensued in Chapters 4 and 5.

This chapter begins with two environmental classics that have gained prominence in the environmental fields by lucid but alarmist expositions of environmental catastrophes. The author revisits Rachel Carson's classic *Silent Spring*, and assesses her descriptions on then-future environmental disasters and catastrophes (Carson 1962).

The classic book provided impetus to the introduction of many US environmental regulations and federal agencies, including the Clean Air Act, Clean Water Act, and Toxic Substances Control Act, which will be discussed at length in Chapter 6 (US Congress 1978; US EPA 1977, 1990).

Another environmental classic that is reviewed belongs to the field of wildlife and wilderness conservation, initiated by Aldo Leopold through his influential book entitled *A Sand County Almanac*. Leopold was the first Professor of wildlife management and the founder of the Wilderness Society (TWS) (Leopold 1949; TWS 2017). Leopold's thoughts bear on the concept of the value of wildlife and wilderness and how human creativities and endeavors in the natural world should be valued.

In the context of catastrophe studies, his ideas have direct bearing on how to value ecological losses and changes as well as losses of human lives caused by catastrophic events, probably the most critical concept in the economics and policy of catastrophes (Mendelsohn and Olmstead 2009; Freeman et al. 2014). Should the value of a wild animal or an insect be the same as that of a human being? What is the value of human endeavors and innovations that alter natural landscapes? These are only a few of the philosophical questions related to Leopold's work.

In the sections that follow discussion on these environmental classics, we shift our focus to the archaeological literature that is concerned with collapses of past once-glorious civilizations – in particular, a series of rather recent studies that attribute past climatic and weather changes or ecological abuses to these civilizational collapses, e.g. the Mayan civilization (Diamond 2005; Gill et al. 2007; Kenneth et al. 2012; Drake 2012). These studies are often appropriated by various governments as a basis for strong climate actions or environmental regulations to prevent catastrophic global warming.

The archaeological literature on societal collapses, dubbed a "collapsiology," which is one of the major fields of archaeology, provides a more integrative description of these events in which a range of factors including race, internal conflicts, outsiders, ideologies, social hierarchy, diseases, and economic systems are intertwined in a societal collapse. We will go through some of these more balanced accounts of civilizational changes which do not even see them as a collapse (Schwartz and Nichols 2006; Middleton 2017).

After the archaeology of collapses, the chapter shifts its focus to the much older, indeed ancient, traditions of philosophies. One is from the western theology and probability theory tradition, which is known as Pascal's wager (Pascal 1670). The other is even more ancient, about 3000 years old. It is from the eastern spiritual/philosophical school in India that proposed randomness in occurrences/disappearances of things, life, and the universe (Tipitaka 2010).

The probability theory of Blasé Pascal is essentially a probability theory of infinity. In the theory of Pascal's wager, the assumption of infinity overshadows all empirical evidence and observations. In this corner problem, an extremely unlikely event completely overwhelms all highly likely events and outcomes due to the infinite value attached to an extremely unlikely outcome (Hájek 2012).

Pascal's probability bet sheds much light on today's environmentalism and environmentalists whose focus is solely on the extremely harmful consequences of human activities that might occur but with only an infinitesimal probability (Weitzman 2009). An extreme value attached to such an outcome of probability will overshadow all other reasonable considerations of likelihoods and distributions of possible events (Nordhaus 2011).

An even older classical tradition of intellect is a randomness theory of life which is deeply rooted in ancient Indian thoughts (Tipitaka 2010). In this school, a flower emerges out of nowhere at random. There is no need for a cause. There is no need for the cause-and-effect relationship to hold for the universe to function. As such, life ends at random, without any cause, without any warning. The end of life is, therefore, quite catastrophic, but so is the beginning of life.

Although one might think at first that such an extreme view of life is not acceptable to today's standards of scientists and philosophers, it shares many similarities with some of the most powerful scientific traditions today. To be more specific, at the foundation of quantum physics, there is a quantum whose behaviors are random, not deterministic, uncertain (Nobel Prize 1918; Feynman et al. 2005).

Before we delve into each of these literatures, the author is compelled to declare that the present chapter is not intended to be a comprehensive treatise on the philosophical traditions on catastrophes. The purpose of this chapter is to introduce a bundle of some of the most profound philosophical thoughts on catastrophic events, thereby helping readers to better evaluate the scientific theories of catastrophes presented in Chapter 2 and the economics to be presented afterwards. It is hoped that the review in this chapter will clarify similarities and differences between scientific models and philosophical traditions in conceptualizing catastrophes.

Table 3.1 A summary of topics covered.

Literature	Subjects	Primary sources
Environmental literature: environmentalism	Environmental catastrophe from toxic chemicals	R. Carson: *Silent Spring* (1962)
	Ecology of wilderness	A. Leopold: *A Sand County Almanac* (1949)
Archaeological literature: collapsiology	Climate doomsday studies	Gill et al. (2007), Kenneth et al. (2012), and others
	Archaeology of civilizational collapses	Middleton (2017) and others
Philosophical literature	Pascal's wager	Blaise Pascal: *Pensées* (1670)
	Indian school of randomness	C. sixth century (BCE), documented in Tipitaka (2010)
	Quantum physics[a]	Feynman et al. (2005) and others

a) In connection with the randomness theory.

A summary of the topics and literatures covered in this chapter is presented in Table 3.1. Broadly, there are three literatures covered and seven subjects. Primary sources and authors for each subject are noted in the table.

3.2 Environmental Catastrophes: *Silent Spring*

It is widely recognized that Rachel Carson's book entitled *Silent Spring* was instrumental in the establishment of early environmental regulations and the Environmental Protection Agency (EPA) in the US (Carson 1962). The US Clean Air Act, the Clean Water Act, and the Toxic Substances Control Act were all influenced by the strong public responses to Carson's book (US Congress 1978; US EPA 1977, 1990). The book is now recognized as one of the influential classics on environmental protection.

As a science writer in the 1960s, Carson warned against dire consequences of continued, unregulated uses of toxic chemicals such as pesticides and insecticides on human societies and ecosystems. What was told in the book by Carson in an imaginary setting was "a shadow of death" and "a spring without voices."

In the first chapter of the book entitled "A Fable for Tomorrow," she writes in a voice full of death, grim, and an evil spell about spring days in the heartland, perhaps, of the US (Carson 1962):

> Then a strange blight crept over the area and everything began to change. Some evil spell had settled on the community: mysterious maladies swept the flocks of chickens; the cattle and sheep sickened and died. Everywhere was a shadow of death.
> There was a strange stillness. The birds, for example – where had they gone? Many people spoke of them, puzzled and disturbed. The feeding stations in the

backyards were deserted. The few birds seen anywhere were moribund; they trembled violently and could not fly. It was a spring without voices.

On the farms the hens brooded, but no chicks hatched. The farmers complained that they were unable to raise any pigs, the litters were small, and the young survived only a few days.

A grim specter has crept upon us almost unnoticed, and this imagined tragedy may easily become a stark reality we all shall know.

Carson is concerned about the "irrecoverable" or "irreversible" lethal impacts of new chemicals which are produced at an astounding rate, 500 new chemicals each year then, on the air, land, rivers, and sea (Vogel and Roberts 2011). To survive, adaptability of men and animals is required to be beyond the limits of biological experiences; that is, they cannot adapt with current adaptive capacities.

In the chapter entitled "The Obligation to Endure," Carson elaborates on the irreversibility of consequences as well as the adjustments to be made by humans and biological systems which cannot be made on the scale of a man's life:

The most alarming of all man's assaults upon the environment is the contamination of air, earth, rivers, and sea with dangerous and even lethal materials. This pollution is for the most part irrecoverable; the chain of evil it initiates not only in the world that must support life but in living tissues is for the most part irreversible.

To adjust to these chemicals would require time on the scale that is nature's; it would require not merely the years of a man's life but the life of generations. And even this, were it by some miracle possible, would be futile, for the new chemicals come from our laboratories in an endless stream; almost five hundred annually find their way into actual use in the United States alone. The figure is staggering and its implications are not easily grasped—500 new chemicals to which the bodies of men and animals are required somehow to adapt each year, chemicals totally outside the limits of biologic experience.

What is the true cost of the uses of these chemicals on humanity? Despite Carson acknowledging mankind's successes in controlling infectious diseases such as smallpox, cholera, and plague, she foretells humanity's fate as an obsolete form of life on Earth caused by the devastating impacts of chemicals. In the chapter entitled "Human Price," she writes:

Only yesterday mankind lived in fear of the scourges of smallpox, cholera, and plague that once swept nations before them. Now our major concern is no longer with the disease organisms that once were omnipresent; sanitation, better living conditions, and new drugs have given us a high degree of control over infectious disease. Today we are concerned with a different kind of hazard that lurks in our environment . . . . 'We all live under the haunting fear that something may corrupt the environment to the point where man joins the dinosaurs as an obsolete form of life,' says Dr. David Price of the United States Public Health Service.

In the above, the true cost of toxic chemicals and environmental degradation is stated to be as large as the end of human civilizations, but the reading of the entirety of her book may confirm that an even larger catastrophe was in mind: the end of all life, both humans and animals, on the planet.

It is interesting to note that Carson states that we were no longer concerned with infectious diseases due to a high degree of control acquired by "sanitation, better living conditions, and new drugs." While the world nearly eradicated smallpox and cholera during the past century, it should be noted that other fatal infectious diseases have newly emerged to kill millions of people in the world.

Recent outbreaks of deadly infectious diseases include the Ebola virus disease in 2014–2015 that killed 11 310 people in Central Africa out of about 30 000 cases reported (WHO 2017). Another case is the Zika virus in Brazil in 2015 which can cause a birth defect of the child's brain called microcephaly – a baby born with a smaller head – when infected during pregnancy (WHO 2015).

Humanity's fate is imagined by Carson to depend upon the choice of one of the two roads. One is the road to extinction of the human race and the other is the road to our only chance to assure the preservation of the Earth. The other road comprises an extraordinary portfolio of alternatives to chemical control. In the chapter entitled "The Other Road," Carson writes:

> WE STAND NOW where two roads diverge. … The road we have long been traveling is deceptively easy, a smooth superhighway on which we progress with great speed, but at its end lies disaster. The other fork of the road—the one 'less traveled by'—offers our last, our only chance to reach a destination that assures the preservation of our earth.
> A truly extraordinary variety of alternatives to the chemical control of insects is available. Some are already in use and have achieved brilliant success. Others are in the stage of laboratory testing. Still others are little more than ideas in the minds of imaginative scientists, waiting for the opportunity to put them to the test.

In Carson's the road less traveled, the remedies for the ills of man and ecology must be a biological solution founded on the understanding of the biological organisms that scientists are invested in controlling. In other words, the lethal impacts of hazardous chemicals must be stopped only by manufacturing new chemicals that do not pose such lethal consequences.

This means that a host of alternative approaches to solving chemical problems is excluded from the set of remedies proposed by Carson, such as limiting rather than banning the uses of chemicals, making available alternative chemicals that achieve the same functions but do not cause lethal impacts, and levying a high tax on the use of a specific chemical (Tietenberg and Lewis 2014):

> All have this in common: they are biological solutions, based on understanding of the living organisms they seek to control, and of the whole fabric of life to which these organisms belong.

Interestingly, she seems to immediately contradict her own proposal of a biological-control solution by stating that the "control of nature" is a phrase that is conceived by arrogant minds in the Neanderthal age:

> As crude a weapon as the cave man's club, the chemical barrage has been hurled against the fabric of life—a fabric on the one hand delicate and destructible, on the other miraculously tough and resilient, and capable of striking back in unexpected ways. These extraordinary capacities of life have been ignored by the practitioners of chemical control who have brought to their task no 'high-minded orientation', no humility before the vast forces with which they tamper. The 'control of nature' is a phrase conceived in arrogance, born of the Neanderthal age of biology and philosophy, when it was supposed that nature exists for the convenience of man.

In any regulation, the reality is that an economic assessment is inevitable. Upon the passage of the Toxic Chemicals Control Act (1978) through the US Congress, the EPA went through a special review process from 1975 to 1989 to make decisions on canceling registrations of cancer-causing pesticides (US Congress 1978). A statistical analysis of the agency's decisions showed that the EPA in fact carefully weighed the risks against the benefits of regulating or allowing pesticides (Cropper et al. 1992).

The authors found that the higher the risks of a pesticide, the higher was the likelihood of cancelation of registration of the pesticide. The higher the benefits of a pesticide, the lower was the likelihood of cancelation of registration of the pesticide. This finding is one of the many researches that demonstrated an important and unavoidable role of the economic analysis of values and markets for regulating toxic chemicals.

From another perspective, Carson's vision of a biological solution has become materialized through the advances of green chemistry and green engineering. Green chemistry is a subfield of chemistry whose primary concern lies in the design of chemical products and processes for the purpose of reducing or eliminating the use and generation of hazardous substances (Anastas and Zimmerman 2012).

Since the publication of *Silent Spring* in 1962, the field of green chemistry has emerged to demonstrate that fundamental scientific knowledge and techniques can be utilized to protect human health as well as ecological resources. Key scientific advances in green chemistry such as catalysis, design of safer chemicals and environmentally benign solvents, and development of renewable feedstocks are most certainly some of what Carson had in mind (Anastas and Kirchhoff 2002).

Other notable works that warned humanity about environmental or ecological catastrophes in the early days of the environmental literature in the 1960s and the early 1970s include Ehrlich's *The Population Bomb* and the Club of Rome's *The Limits to Growth* (Ehrlich 1968; Meadows et al. 1972).

More than 50 years since the publication of *The Population Bomb* which warned of millions of deaths in the 1970s due to overpopulation and food shortages, industrialized societies today are grappling with a completely different reality in population growth: a serious decline in the birth rate of industrialized societies, called a population crash (IIASA 2007; Pearce 2011).

The global economy as well as the global community has also survived the Rome Club's dismal prediction that the world economy would crash at the threshold of the twenty-first century owing to the failure in ensuring sustainable growth. The resilience of the world economy owing to the complex ecology of markets and market participants was foretold by pre-eminent economists of the time, which turned out to be, over the course of time, proven true (Nordhaus 1992a,b).

3.3 Ecological Catastrophes: The Ultimate Value Is Wilderness

Aldo Leopold is widely regarded as the father of wilderness preservation across the environmental and ecological societies in the world. Today, environmental organizations such as the World Wildlife Fund (WWF) and TWS are engaged in wilderness and wildlife preservations with worldwide member participation. Leopold's publication *A Sand County Almanac* is regarded as a cornerstone in the literature of wilderness preservation (Leopold 1949).

A historical and enduring influence of this book cannot be understated in the broader literature of environmentalism – or environmental actions – in the West. The concepts, observations, outlooks, and critiques on many aspects of the modern economy and society put forth by Leopold hold powerful sway in the academic as well as nonacademic literature of environmentalism.

In the purview of this book, Leopold's visions and proposals offer a refreshing perspective on catastrophes and how humanity should get prepared for them, which will be elaborated in this section. In short, Leopold asks us what we should value in the event of a truly catastrophic event.

In the chapter entitled "Marshland Elegy," Leopold writes that the ultimate value lies in the wildness or wilderness of everything (Leopold 1949):

> Thus always does history, whether of marsh or market place, end in paradox. The ultimate value in these marshes is wildness, and the crane is wildness incarnate. But all conservation of wildness is self-defeating, for to cherish we must see and fondle, and when enough have seen and fondled, there is no wilderness left to cherish.

In this short paragraph, Leopold seems to be conflicted between "wildness incarnate" and "where there is no wilderness left to cherish." Could he choose one over the other? Or should he pick some from the former world and others from the latter world? This conflict was not clarified in the paragraph.

In another chapter entitled "Thinking Like a Mountain," he elaborates the ecological perspective or viewpoint in which one sees the natural systems and processes as if the mountain thinks about them (Leopold 1949):

> We reached the old wolf in time to watch a fierce green fire dying in her eyes. I realized then, and have known ever since, that there was something new to me in those eyes, something known only to her and to the mountain. I was young then, and full of trigger-itch; I thought that because fewer wolves meant more deer, that no wolves would mean hunters' paradise. But after seeing the green fire die, I sensed that neither the wolf nor the mountain agreed with such a view.

> I now suspect that just as a deer herd lives in mortal fear of its wolves, so does a mountain live in mortal fear of its deer. And perhaps with better cause, for while a buck pulled down by wolves can be replaced in two or three years, a range pulled down by too many deer may fail of replacement in as many decades.

Ignoring the ecology of the natural worlds or, in Leopold's language, failing to think like a mountain would provide a probable ultimate cause of numerous ecological catastrophes. Leopold argues that the Dust Bowl disaster in the 1930s, the most catastrophic ecological disaster in US history (Warwick 1980), was caused by the failure of learning to think like a mountain (Leopold 1949):

> So also with cows. The cowman who cleans his range of wolves does not realize that he is taking over the wolf's job of trimming the herd to fit the range. He has not learned to think like a mountain. Hence we have dustbowls, and rivers washing the future into the sea.

He then sums up his view of wilderness, borrowing from Henry Thoreau, in a pithy declaration (Thoreau 1854). It is proclaimed that the ultimate value lies in the wildness or wilderness, and, as such, the wild world is the salvation of the world (Leopold 1949):

> *Perhaps this is behind Thoreau's dictum: in wildness is the salvation of the world.*

In the chapter entitled "The Outlook," Leopold this time describes the land ethics in a more general framework, which is without doubt beyond the scope of the land as physically defined, based on value-loaded terms such as love, respect, and admiration. In the chapter, he makes a sharp separation of "economic value" from the value "in the philosophical sense":

> It is inconceivable to me that an ethical relation to land can exist without love, respect, and admiration for land, and a high regard for its value. By value, I of course mean something far broader than mere economic value; I mean value in the philosophical sense.
> Your true modern is separated from the land by many middlemen, and by innumerable physical gadgets. He has no vital relation to it; to him it is the space between cities on which crops grow. Turn him loose for a day on the land, and if the spot does not happen to be a golf links or a 'scenic' area, he is bored stiff.

In explaining land uses, he then sharply distinguishes "what is ethically and esthetically right" from "what is economically *expedient*." The distinction is undoubtedly made on what is moral (or ethical) versus what is immoral (or unethical). Leopold seems to characterize the former as "integrity, stability, and beauty" and the latter as "otherwise":

> The 'key-log' which must be moved to release the evolutionary process for an ethic is simply this: quit thinking about decent land-use as solely an economic problem. Examine each question in terms of what is ethically and esthetically right, as well as what is economically expedient. A thing is right when it tends to preserve the integrity, stability, and beauty of the biotic community. It is wrong when it tends otherwise.

In elaborating his land ethics by way of critiquing the economics of land uses, Leopold simply does not confront the reality that economic land uses can be also ethically conscious, esthetically cognizant, righteous, integral, and beautiful. In the ecology of markets and market participants, the complex web of transactions is built on the mutually agreeable contracts and terms, while the values of all things can be expressed and negotiated among the parties (Nordhaus 1992a).

Leopold then further sharpens his critiques of land use economics by calling it "the fallacy ... tied around our collective neck" by economic determinists. He states that land uses are determined not by economics, i.e. purse, but by "an innumerable host of actions and attitudes which is determined by land users' tastes and predilections" (Leopold 1949):

> It of course goes without saying that economic feasibility limits the tether of what can or cannot be done for land. It always has and it always will. The fallacy the economic determinists have tied around our collective neck, and which we now need to cast off is the belief that economics determines all land-use. This is simply not true. An innumerable host of actions and attitudes, comprising perhaps the bulk of all land relations, is determined by the land-users' tastes and predilections, rather than by his purse. The bulk of all land relations hinges on investments of time, forethought, skill, and faith rather than on investments of cash. As a land-user thinketh, so is he.

These statements are rather ignorant of the fact that the economics is primarily concerned about land users' tastes and predilections which manifest into a host of actions and attitudes. He does not seem to be aware that economics, i.e. market transactions, is determined by demand forces and supply sources at the same time. That is, what determines market transactions is not one person's tastes and predilections, but the push-and-pull repeated interactions between the expressions of the seller's tastes and predilections and those of the buyer's tastes and predilections (Nordhaus 1992a).

Before we move on to the next section, let me summarize the extreme ecological perspectives of Aldo Leopold, which will turn out to continue to provide refreshing alternative insights in the ensuing chapters that deal with the economics (Chapter 4), empirical models (Chapter 5), and policy-making (Chapter 6) of and on catastrophic events.

First, the view that the ultimate value lies in wildness or wilderness is an extreme view on the determination of values or how values are determined in society (Hanemann 1994; Maler and Vincent 2005; Mendelsohn and Olmstead 2009; Freeman et al. 2014). The view expresses attachment to – or love of – nature as it is or as it has been, but does not reflect the reality that the natural world must be managed by humans or animals to be more useful or friendlier. It underestimates the fact that humans' innovations on and with natural systems add value to the existing value of natural systems.

In the value system of Leopold, human endeavors such as creative works, redesigns, and utilizations of the Earth systems only deteriorate the value of nature as it is currently existent. Existence value is all there is according to Leopold, which is in contradiction to the reality in which the value is determined by all the aspects and works put into the object to be valued.

In Leopold's value system, a disappearance of the entire universe because of a black hole created by a natural sequence of events should not be considered a catastrophe at all: it is just a natural event (Jaffe et al. 2000; Ellis et al. 2008). The destruction of the Planet by collision with an exoplanet or an asteroid should not be viewed as a catastrophe at all (Chapman and Morrison 1994).

Let me give more familiar examples. Following Leopold's strict interpretation, the devastation caused by tropical cyclone Bhola in 1970 which killed as many as 280 000 people in Bangladesh should be interpreted as a natural event, not as a catastrophe (IMD 2015; JTWC 2017; Seo 2017). The devastation of the Haiti earthquake in 2010 that claimed 220 000 lives should not be viewed as a catastrophe to the extent that it is not caused by a human force (Utsu 2013; Swiss Re Institute 2017).

The second salient view of Leopold is that of an ecology-centric decision-making framework. However, if we admit that natural systems, or ecological systems as preferred by ecologists, must be managed or appropriated by humans to habituate there and provide enjoyments, it is not at all clear that the natural systems should be managed by the perspectives of a mountain, that is to say, by the perspectives of an ecological system or the ecology itself. The better way, and perhaps the right way, to manage them is through the perspectives of humanity.

The human race should manage the natural or ecological systems in a socially welfare-optimizing manner, considering the current and future generations of people that will live on this planet as well as the health of ecosystems that support life on this planet. The anthropogenic management must be a sustainable one, or put another way, it should take into consideration the size of the overall stock of the natural and ecological resources that must not dwindle away (Solow 1974, 1993).

One may be inclined to criticize this position as being human-centric, as that unique to the Anthropocene. Another may argue that humanity cannot treat animals and plants as resources to be appropriated as much as human race desires without limits. But, these critiques are by and large misguided in the contexts of catastrophic situations and events for the reasons explained shortly.

The fact that most natural resources and ecological services are owned by an individual or a group of individuals means that the care and compassionate relationships with natural resources and beings can be exercised even in the anthropocentric worldview. How much an individual or a community cares and extends compassionate heart toward natural beings ultimately depends on the preferences of the individuals – or borrowing Leopold's language, the tastes, predilections, devotion, and faith of the individual.

A situation may arise in which an individual human being should decide whether s/he should save another human being or a nonhuman animal in the event of an unprecedented catastrophe. Imagine the situation depicted in the Noah's ark problem introduced in Chapter 1. In this situation, an ecology-centric decision-making, e.g. saving a horse instead of another human being, would seem irrational.

A similar line of reasoning can be applied to numerous catastrophe situations to demonstrate that an ecology-centric decision framework is superseded by an anthropocentric decision framework in those situations. In the context of planet-ending catastrophes, Leopold's prominent ecological perspective seems unconvincing, to say the least.

3.4 Climate Doomsday Modelers

Recently, a new brand of scientists, researchers, and politicians has emerged whose primary message is an impending doomsday caused by climate change. This section describes some of this doomsday literature whose cause is fixed on climate change, or environmental degradation, or population explosion, or unsustainable economic development.

In the letter properly titled "What drives societal collapses?" published by the journal *Science*, Weiss and Bradley argued that past climatic changes brought about doomsdays for ancient civilizations, although the authors acknowledged that the corpus of archaeological literature indicated otherwise (Weiss and Bradley 2001):

> The archaeological and historical record is replete with evidence for prehistoric, ancient and pre-modern societal collapse. ... Each of these collapse episodes has been discussed intensively within the archaeological community, commonly leading to the conclusion that combinations of social, political, and economic factors were their root causes. That perspective is now changing with the accumulation of high-resolution paleoclimatic data .... These climatic events were ... highly disruptive, leading to societal collapse – an adaptive response to otherwise insurmountable stresses.

Other groups of researchers, many of whom are archaeologists, have gone further to obtain the paleoclimate data and other climate indicators and correlated them with past collapse episodes of once-prosperous monumental civilizations in human history, e.g. the Maya civilization and the Western Roman Empire.

The Maya civilization was a Mesoamerican civilization developed in the area that encompasses most parts of present-day southeastern Mexico, Guatemala, Belize, Honduras, and El Salvador. The Mayan civilization had existed from 2000 BCE to 1679 CE when the Mayans fell to the Spanish invaders.

Relying on the ice-core data extracted from Greenland and overlaying them onto the times of the Maya collapses, researchers argued that the collapse of the Mayan civilization was caused by climatic changes (Gill et al. 2007). The researchers argued that their ice-core data show that a minimum in solar insolation and a low in solar activity occurred during that time, which was accompanied by severe cold and dryness over Greenland. These hemispheric climatic conditions in turn were propitious for severe droughts in the Maya Lowlands, which resulted in the most severe drought of the past 7000 years. This devastated the Yucatan Peninsula, which was followed by the collapses of the Mayan cities in four phases.

Another group of researchers offered the evidence of the Belizean cavern's stalagmites – lumpy, rocky spires on cave floors. Stalagmites are formed by water and minerals dripping from above in the cave. Since they grow quicker in larger rainfall years, the authors argued that the stalagmites can give scientists a way to estimate historical precipitation levels in the times of the Mayan collapse (Kenneth et al. 2012).

Kenneth and coauthors reported that their analysis of the stalagmite record shows that the Mayan civilization suffered the longest dry-spell of the last 2000 years between 1020 CE and 1100 CE. The extreme droughts were followed by severe crop failures in Mayan urban centers, which then led to famine, mass migration, and death.

According to the authors, the Spanish conquistadors had no role in the end of the Mayan civilization. By the time they arrived in the late sixteenth century, Mayan populations had decreased by 90% and urban centers had been largely abandoned. Before the Spanish arrivals, farms had been deserted and reclaimed by forests.

Another study blames climate change and environmental degradation for the collapse of the Mycenaean civilization, the first Greek civilization which existed in the

Late Bronze Age in Greece (Drake 2012). Relying on the data of various climate indicators such as oxygen-isotope speleothems, stable carbon isotopes, alkenone-derived sea surface temperatures, and changes in warm-species dinocysts and foraminifera in the Mediterranean, the author reported a sharp increase in northern hemisphere temperatures that preceded the collapse of Mycenaean palatial centers and a sharp decrease in temperatures that occurred during their abandonment. Thus, he concluded that the collapse of the Mycenaean civilization was caused by the dramatic climate changes.

Besides these studies, in the article entitled "Five civilizations that climate change may have doomed," the following five collapses are cited (Le Page 2012): the Mycenae, the first Greek civilization (between 1600 and 1100 BCE), the Moche in northern Peru (between 200 CE and 600 CE), Chichen Itza in the Mayan civilization (between 200 CE and the thirteenth century CE), the Western Roman Empire (between the third century and fifth century CE) (Buntgen et al. 2011), and the Thirty Years War or the General Crisis in Europe (between 1618 and 1648 CE) (Zhang et al. 2011).

The most frequently told ecological collapse story in the western world is the collapse of the Easter Island civilization. Easter Island is located in a Polynesian region of the Pacific Ocean and is a very remote territory of Chile. It is one of the most remote human-inhabited islands in the world, with a population of about 20 000 at a historical peak. Europeans found the Easter Islanders in the eighteenth century.

The eco-disaster story of Easter Island goes like this (Diamond 2005): the Easter Islanders, to support their ever-growing island population, had cut down all the palm trees, which were the only trees that survived on this remote Pacific island. They had to clear the palms for farmlands, but also to move their famed *moai* statues.

According to the story told in the frame of an ecological disaster, it did not dawn on the Easter Islanders that chopping trees for farmland would deteriorate the island's ecosystem, resulting in reductions in food production. Further, they were not cognizant that they were in fact, by chopping down trees, getting rid of the means and resources for escaping the island in the event of a catastrophe by building boats from the palm trees (Diamond 2005).

The story goes on: when the Europeans found them in the eighteenth century, the population of the island had already decreased to 2000 from about 15 000 a century earlier. This is equivalent to saying that the Europeans had nothing to do with the collapse of the Easter Island civilization (Middleton 2017). It died because of the unbelievable ignorance of the Easter Islanders of the ecological sustainability of their own island, not because of the Europeans. Thus, the story is told.

A widely publicized theory or argument of societal collapses through the mass media and popular books is deeply rooted in the concepts of environmental degradation, ecological sustainability, overpopulation, and climate changes. Some or all of these factors are claimed to have played a dominant role in such past collapses (Diamond 2005).

Past and future collapses attributed to the mistreatment of the environment and sustainability cited by Diamond are the Greenland Norse, the Easter Islanders, the Polynesians of Pitcairn Island, the Anasazi of southwestern North America, the Maya of Central America, the genocide of Rwanda, the ecological failure of Haiti, the problems facing the developing nation China, and the environmental problems facing Australia.

The thesis that societies have collapsed or will collapse owing to human behaviors that destroy their natural environments or overpopulation that exceeds nature's carrying capacity is broadly refuted by archaeologists who specialize in these historical events (McAnany and Yoffee 2009; Middleton 2017). They argue that these purportedly collapsed societies, in the face of such societal crises, showed rather high resilience and bounced back, albeit with a different form.

They also refute the notion that Euro-American colonial triumphs had little to do with the "collapses" in these societies and only environmental–geographical–climate factors were to blame for such events. They find that the complex relationships among race, political labels of societal "success" and "failure," and economic systems played a prime role in the historical events of collapses.

The review of climate doomsday models in this section is narrowly focused on the studies from the archaeological literature. To be sure, there is a broader literature of climate science and policy which warns or foretells a climate-change-induced global catastrophe through various physical routes and feedbacks, e.g. a hockey-stick-like rise in global temperature, abrupt climate changes, a reversal in the global ocean thermohaline circulation, ocean euxinia (an anoxic and sulfidic ocean), the melting of polar icesheets, and tipping elements in the global climate system (Mann et al. 1999; Broecker 1997; Meyer and Kump 2008; Oppenheimer and Alley 2005; Gore 2006; Lenton et al. 2008; NRC 2013).

Some of this literature even goes as far as forecasting the end of all civilizations on the planet or mass extinction of life on Earth by as much as 50% of all species on the planet (Posner 2004; Weitzman 2009; Wagner and Weitzman 2015; Kolbert 2014). This scientific literature of climate change will be presented in Chapters 4, 5, and 6 to the extent that they are pertinent to the discussion of the economic theory of life-ending or universe-ending catastrophes.

3.5 Collapsiology: The Archaeology of Civilizational Collapses

A study of civilizational collapses has long fascinated archaeologists and historians. Unlike the studies cited in Section 3.4, that attribute climatic shifts and subsequent crop failures to once-glorious societies' collapses, many archaeologists see a civilization's collapse as a by far more complex process, with many factors, mostly nongeophysical, intertwined. The collapse of a civilization is a highly complex operation, oversimplification of which can seriously distort what really happened in those times (Wheeler 1966).

In Middleton's *Understanding Collapse*, some of the well-established definitions by archaeologists of a society or civilization's collapse are summarized, which, the present author thinks, contain insightful as well as practical concepts pertinent to the discussion of this book (Middleton 2017).

According to Middleton, a societal collapse can be defined as "an abrupt political change and reduction in social complexity that has knock-on effects throughout society, visible to archaeologists in the material culture," as is the case, for example, in Renfrew's *Approaches to Social Archeology* and Tainter's *The Collapse of Complex Societies* (Renfrew 1984; Tainter 1990).

This definition is practical in the sense that it permits a quantitative analysis of a collapse event, if we understand the social complexity in the above definition as a quantitative concept that can be measured by the number of parts of a society or the number of levels in the hierarchy of the society.

This definition can be applied to an empirical study of civilizational collapses by specifying social complexity as a set of empirical components. Schwartz and Nichols in the book entitled *After Collapse*, for example, adopts the following set of empirical criteria for identifying a societal collapse (Schwartz and Nichols 2006):

i) the fragmentation of states into smaller political entities;
ii) the partial abandonment or complete desertion of urban centers, along with the loss or depletion of their centralizing functions;
iii) the breakdown of regional economic systems;
iv) and the failure of civilizational ideologies.

Looked from the perspective of the archaeology of collapses, Diamond's and others' attributions of environmental degradation or climate change are only narrowly fixated on an ecological/environmental crisis of a society (Diamond 2005). However, a civilization's collapse does not occur solely from an ecological crisis. There must have been a whole set of interwoven disasters, most of which nonecological factors, that actually brought about the collapse of the once-dominant civilization.

As described in Section 3.4, some may be inclined to frame the collapse of the Mycenaean civilization in the Late Bronze Age in Greece as a climate doomsday story (Drake 2012). But, while the archaeological evidence that supports such a climate doomsday story is not at all unequivocal, whatever available evidence also points rather to multiple nonclimate factors that involve race, politics, and economic systems.

According to Middleton who specializes in Mycenaean civilization, there is no evidence that the population had been so high in Greece in the Late Bronze Age that it could not be supported by its resources. Further, there is very little evidence from paleoclimatic data to support the argument that there was a mega drought around 1200 BCE that could have destabilized the civilization. Further, no paleoclimatic evidence of such a mega drought exists in the parts of Greece that collapsed (Middleton 2017).

The ecological-collapse story of the Easter Islanders told by ecologists is also far from right (Diamond 2005). Archaeological evidence shows that the real threats that the Easter Islanders faced were not any climatic shift, nor ecological destruction of their natural environments, nor even their unsustainable ways of living (Middleton 2017).

The real threat to their survival and way of life was indeed the arrival of outsiders, i.e. the Europeans. The disaster and collapse, if we interpret that there was any such collapse at all, of the Easter Island were brought about by the outsiders who with them came animals, diseases, and Christianity. The local culture was targeted and destroyed by the Christian outsiders. The island's population dropped drastically by slave trading by the Europeans, through which many locals were kidnapped to work in Peru or in the mines in the Chincha Islands (Middleton 2017).

In collapsiology, that is, the study of civilizational collapses, a sharper definition of a collapse is needed with reference to what exactly collapsed. Most often, scholars, e.g. those cited in Section 3.4, refer to the collapse of a civilization. However, a civilization is a flexible as well as an intangible concept which refers broadly to many sets of traditions. As such, it is not correct in most circumstances to say that a civilization collapsed (Middleton 2017).

A more accurate way to state societal collapses is that states or nations collapsed. This is because states or nations are tangible units, which are identifiable and quantifiable.

A fascinating conclusion that arises from the change in the perspective is that the past civilizations, described in Section 3.4 as collapsed ones, should be seen to have never collapsed. What collapsed was the states, that is, the states of Mayan civilization, of the Mycenaean civilization, of the Easter Island civilization.

Middleton in *Understanding Collapse* elaborates the importance of this distinction by way of the historical changes of the Mayan civilization. The Maya was a world of kings and cities, with a set of ideologies and traditions, interconnected and spread across a large land area with distinct geographies and ecological systems. The Mayan cities and lands were divided and governed by super-states which, with wide influence, arose and fell through the years. There were about 60–70 independent states, whose fortunes had fluctuated.

During the so-called Terminal Classic period, one of the Mayan time periods between 750 and 1050 CE, a series of state collapses had occurred in the Mayan world, but it is not possible to pinpoint a single civilizational collapse. Similarly, specific parts of the Maya culture collapsed, as in the case of the divine-king ideology, but it is impossible to proclaim the disappearance of the Mayan civilization itself.

Archaeological remains indicate that there were cases of an individual rapid collapse of a state. For example, at the battle at Cancuén in the Pasion river region of the southern Maya lowlands which was waged against the unknown outsiders, the Mayan state was unprepared and the battle was lost, with kings and rulers found massacred and buried in cisterns with their regalia.

The Mayan civilization, notwithstanding, did not disappear through these collapses of independent states. Even in the sixteenth century when the Spanish arrived in the Mayan world, archaeological artifacts tell us that it continued to be a complex society: there were cities, trades occurred, books were compiled, new cities were founded, and cities and urban centers were periodically abandoned as the result of infighting, politics, famine, and plagues (Middleton 2017).

It was the Spanish armies that destroyed the last independent Mayan kingdom, the kingdom of the Itzas, in 1697 (Middleton 2017). The Spanish, as Christians, destroyed the pagan "idols" and captured and baptized the last king. The Spanish Christians were determined to destroy the ancient Mayan culture: thousands of books were burnt, with only four of them remaining today, and Mayan families were broken up, with children sent for re-education.

Even so, the Mayan civilization has survived through the times of these great challenges until today, with millions of their descendants still living in Central America. From the archaeologist's point of view based on archaeological evidence, the Mayan civilization never collapsed (Middleton 2017):

> The idea of a collapse of Maya civilisation seems just wrong .... Via many individual collapses, Classic Maya society transformed through the Terminal Classic and into the Postclassic – a development that is hardly surprising when compared with the changing map of Europe across any five-century period. Maya society continued to change with the arrival of the Spanish, and through the colonial and modern eras. If we value the Maya's so-called Classic period more than their culture at other times, this is our choice – but it is one that should be recognised and questioned.

3.6 Pascal's Wager: A Statistics of Infinity of Value

The review of environmentalism, climate doomsday studies, and civilizational collapses reveals how the concepts of catastrophes are embedded and appropriated in empirical studies of various issues. The next two sections deal more directly with the conceptualizations of a catastrophe in a philosophical context.

Blaise Pascal (1623–1662) was a French philosopher, mathematician, and physicist whose legacy and fame are encapsulated in his influential book entitled *Pensées* (Pascal 1670). The book describes Pascal's wager in which the consequence of infinity is put into a statistical analysis. It has been widely interpreted that Pascal's wager explicated the consequence in a statistical analysis of an event that is extremely unlikely to occur but which causes the infinite loss if it were to occur.

In the past few decades, as scientists, economists, and philosophers have pondered over the slim possibility of a truly catastrophic event in which nearly all is lost, Pascal's statistical bet has been revived from the tomb of statistics but in a barely identifiable form. As explained in Chapter 1, a truly catastrophic event may end all life or even the universe, which we may interpret as an infinity of loss.

Predictions are available by numerous scientists that each one of the following events has a small probability of ending all life or the universe: destruction of the Earth from an asteroid collision, disappearance of the universe into a black hole created by strangelets born from the Large Hadron Collider experiments by the European Organization for Nuclear Research (CERN), the moment of singularity in artificial intelligence (AI), nuclear winters created from nuclear explosions and wars, and an end of all civilizations on the planet by global warming (Turco et al. 1983; Chapman and Morrison 1994; Mills et al. 2008; Dar et al. 1999; Jaffe et al. 2000; Kurzweil 2005; Lenton et al. 2008; Weitzman 2009; Hawking et al. 2014).

The statistical apparatus of Pascal is quite simple and we may be able to trace its origin to the Indian classical texts, one of which is presented in Section 3.7 in relation to modern quantum physics. It may not be as obvious to modern scholars to discern whether Pascal's true intention in setting up his probability argument was indeed approximately close to what has been interpreted by others in the past.

Pascal puts his statistical apparatus into a form of a gamble between belief in God's existence and nonbelief in God's existence (Hájek 2012):

> "God is, or He is not." But to which side shall we incline? Reason can decide nothing here. There is an infinite chaos which separated us. A game is being played at the extremity of this infinite distance where heads or tails will turn up …. Which will you choose then? Let us see. Since you must choose, let us see which interests you least. You have two things to lose, the true and the good; and two things to stake, your reason and your will, your knowledge and your happiness; and your nature has two things to shun, error and misery. Your reason is no more shocked in choosing one rather than the other, since you must of necessity choose.

In Pascal's game apparatus, a gambler's choice is determined by the magnitude of happiness that comes with choosing one of the two alternatives, in which you gain infinity of happiness if one's choice in God's existence turned out to be truthful and rewarded.

3.6 Pascal's Wager: A Statistics of Infinity of Value

If the gambler's choice of God's existence turned out to be false, then Pascal formulates, most likely erroneously, that there is nothing lost from a false choice (Hájek 2012):

> But your happiness? Let us weigh the gain and the loss in wagering that God is .... If you gain, you gain all; if you lose, you lose nothing. Wager, then, without hesitation that He is.

Pascal formulates his concept of infinity by way of an infinite number of lives to be had by believing in the existence of God. In this game to be played, there is an equal chance of a wrong or right choice in this life. Therefore, your total welfare is $\left(\frac{1}{2} * \text{one life's happiness}\right)$ in this life. However, if you have two lives, your welfare in believing in God is, according to Pascal,

$$\prod_{2}^{God} = \left(\frac{1}{2} * \text{one life's happiness}\right) + (1 * \text{one life's happiness}). \tag{3.1}$$

If you have n lives, your payoff from believing in God is then

$$\prod_{n}^{God} = \left(\frac{1}{2} * \text{one life's happiness}\right) + (n-1) * (1 * \text{one life's happiness})^{n-1}. \tag{3.2}$$

On the other hand, your payoff from not believing in God in the game is, according to Pascal who thinks that there is no life at all after this life if you do not believe in God or, to put it differently, there is no God:

$$\prod_{n}^{Non\_God} = \left(\frac{1}{2} * \text{one life's happiness}\right). \tag{3.3}$$

The concept of infinity, i.e. the "eternity of life and happiness," is arrived through the following reasoning:

$$\lim_{n \to \infty} \prod_{n}^{God} = +\infty. \tag{3.4}$$

It should be noted that multiple variations to the above set of equations are possible to match with variations in Pascal's arguments. The contents of Eq. (3.1) to Eq. (3.4) are elaborated in French verbally by Pascal as follows (Hájek 2012):

> Let us see. Since there is an equal risk of gain and of loss, if you had only to gain two lives, instead of one, you might still wager. But if there were three lives to gain, you would have to play (since you are under the necessity of playing), and you would be imprudent, when you are forced to play, not to chance your life to gain three at a game where there is an equal risk of loss and gain. But there is an eternity of life and happiness.

He stakes his entire argument on the division of "infinity" versus "the finite." If there is only one life, then it doesn't make any difference whether you choose to believe or not to choose to believe. In the following single paragraph, he repeats the word "infinity" by as many as nine times, which ends with the suggestion of giving all:

But there is an eternity of life and happiness. And this being so, if there were an infinity of chances, of which one only would be for you, you would still be right in wagering one to win two, and you would act stupidly, being obliged to play, by refusing to stake one life against three at a game in which out of an infinity of chances there is one for you, if there were an infinity of an infinitely happy life to gain. But there is here an infinity of an infinitely happy life to gain, a chance of gain against a finite number of chances of loss, and what you stake is finite. It is all divided; wherever the infinite is and there is not an infinity of chances of loss against that of gain, there is no time to hesitate, you must give all.

What Pascal's wager is trying to get at is the concept of infinity, placed against the finite, applied to the statistical apparatus. Although his intention seems – at the least it is interpreted certainly this way – to be proving the existence of God, Pascal's gambling framework, as we examined above, does not achieve this goal. In fact, it has little to do with proof of the existence of God, which is also by now widely recognized in the philosophical and theological literature.

Still, Pascal's wager is somewhat pertinent in terms of the concept of infinity in statistical and probability analyses. Pascal's bet can be interpreted to have some bearing on a statistical analysis in which great uncertainty dominates our understanding of a catastrophic event, therefore the variance in the occurrence of an event cannot be defined, and so it is unbounded or infinite.

The situations where the variance of an event is not defined, unbounded, or infinite were expounded by both serious statisticians and popular authors. Among the serious scientists are catastrophe theorists. The catastrophe theories have such cases where the variance cannot be defined or unbounded (Thom 1975; Zeeman 1977). You have no control at all over which values the control variable will take in the catastrophe models, upon which the system falls into a catastrophe.

In a chaotic system such as the Lorenz attractor, it is extremely difficult, if possible at all, to predict the movement of the chaotic system precisely: it does not have a steady state, nor does it have a cyclical pattern. Every time you look at the system, it takes on a different point in a completely different loop. A tiny error or perturbation to the system attributed to an uncontrollable circumstance will push it to a strange location in a completely different loop (Lorenz 1963, 1969a,b).

However, as elucidated in Chapter 2, a chaotic system isn't necessarily a disorderly system (Feigenbaum 1978; Mandelbrot 1983, 2004). This is to say that a chaotic system can be bounded in various ways so that it can have empirical limits in unpredictability or chaotic behaviors. For example, a weather system is very difficult to predict, as Lorenz delved into this and proclaimed, but can still be approximated with a certain level of precision (Zeng et al. 1993; Tziperman et al. 1995).

In the popular literature, an event that is highly unlikely but surprises greatly when it does occur was referred to as a black swan event (Taleb 2007). Before its occurrence, a black swan event is unexpected as well as unpredictable. That is, we do not expect a swan that is black. What we expect is a white-appearance swan with differences in many attributes other than its color.

According to the author, a black swan event is described succinctly to have the following characteristic (Taleb 2007):

> First, it is an outlier, as it lies outside the realm of regular expectations, because nothing in the past can convincingly point to its possibility.

In the applied statistics literature, an event in which there is infinity of uncertainty with regard to its realization was referred to as a fat-tail event in the statistical applications of the economics of climate change (Weitzman 2009). An infinity of uncertainty gives rise to a fat-tail distribution of a statistical event (Schuster 1984).

In the fat-tail distribution of climate change, it is argued that a truly catastrophic climate outcome, e.g. a 20° C increase in temperature by the year 2100, has a fat-tail probability of occurrence. That is, it is not possible to reduce the probability of a truly "strange" event to zero through research and investments, however hard scientists may attempt to do so. The author gives a detailed analysis and critique of this conclusion in Chapter 4.

Finally, it is rather notable that while Pascal's wager relied on the concept of infinity of happiness, its applications and modifications in the scientific and popular literature are only concentrated about the infinity of destruction, say, by climate change, chaos, catastrophe, doomsday, etc.

3.7 Randomness in the Indian School of Thoughts

Looking east, we find a treasure trove of thinkers and philosophies that preached the concepts and consequences of catastrophes, from which the author will introduce one that seems to be most spectacular, as spectacular as the catastrophe theory, but at the same time whose legacy has turned out to be surprisingly enduring, far more enduring than the catastrophe theory.

According to the Buddhist sutras, i.e. the collection of Buddha's words, compiled in the Tipitaka (2010), there were 16 unique philosophical views, other than the Buddha's teachings, that were popular in the Indian philosophical traditions at the Buddha's time, probably all of which have survived to this day and are upheld by some at the present time as well.

Among the 16 philosophical views, of interest to the economics of catastrophes is the theory of randomness or the theory of occurrences of all things by chance. The theory of random occurrences is a very critical concept in catastrophe theories because it refutes, among other things, the theory of a cause and effect, causality, or causal relationships.

That is, it refutes in the most profound manner that a particular event occurs in dependence upon a set of causes and conditions. According to the randomness theorists, even the universe occurs at random and disappears at random as well. The proponents of the theory of randomness were referred to as sages in the Buddha's time.

In the sutra named "All embracing net of views," the Buddha elaborates the theory of random occurrences in the section entitled the "Doctrines of Fortuitous Origination" (Adhiccasamuppannavāda) as follows:

> There are, bhikkhus, some recluses and brahmins, who are fortuitous originationists, and who on two grounds proclaim the self and the world to originate fortuitously. And owing to what, with reference to what, do these honorable recluses and brahmins proclaim their views?

He explains the theory of fortuitous origination, that is, the origination by pure chance, in terms of the self and the world, but it is applicable to all material and mental objects and phenomena because every such object has the conception of self-nature according to the ancient Buddhist and Indian philosophical traditions (Tsongkhapa 2001).

He explains the two types of the fortuitous originationists. The first one is from the perceptionist, i.e. the arguments based on perceptual experiences:

> There are, bhikkhus, certain gods called 'non-percipient beings'. When perception arises in them, those gods pass away from that plane. Now, bhikkhus, this comes to pass, that a certain being, after passing away from that plane, takes rebirth in this world. Having come to this world, he goes forth from home to homelessness. When he has gone forth, by means of ardor, endeavor, application, diligence, and right reflection, he attains to such a degree of mental concentration that with his mind thus concentrated he recollects the arising of perception, but nothing previous to that. He speaks thus: 'The self and the world originate fortuitously. What is the reason? Because previously I did not exist, but now I am. Not having been, I sprang into being.' This, bhikkhus, is the first case.

The second type of the fortuitous originationists is the rationalist's or the investigator's case which is "hammered by reason" and through his own "flight of thoughts":

> In the second case, owing to what, with reference to what, are some honorable recluses and brahmins fortuitous originationists, proclaiming the self and the world to originate fortuitously?
> Herein, bhikkhus, a certain recluse or a brahmin is a rationalist, an investigator. He declares his view – hammered out by reason, deduced from his investigations, following his own flight of thought – thus: 'The self and the world originate fortuitously.'

The Buddha says that there is no other possibility, i.e. there are only two grounds for the theory of fortuitous origination:

> It is on these two grounds, bhikkhus, that those recluses and brahmins who are fortuitous originationists proclaim the self and the world to originate fortuitously. Whatever recluses or brahmins there may be who proclaim the self and the world to originate fortuitously, all of them do so on these two grounds or on a certain one of them. Outside of these there is none.

Having presented the theory of random occurrences or fortuitous originations, the author has a feeling that many readers will not grasp immediately the significance of the randomness theorists. As such, its enduring influences through time and pertinence especially in the context of the catastrophes need some elaborations.

The critical importance of the view of randomness in the ancient Indian philosophical traditions emerges strongly in relation to the modern scientific theories of catastrophe as well as chaos. In the catastrophe theory expounded by Thom and others, the catastrophe occurs randomly or by chance. In the potential functions of the catastrophe models, there is no mechanism to move the needle from one point to another. It just moves on its own and falls upon the tipping point (Thom 1975).

In the extreme interpretation of the chaos theory, a trajectory of a concerned variable such as a weather variable cannot be predicted. A miniscule change in a conditioning variable or an environmental condition that occurs by chance, or by error, or fortuitously, or randomly triggers the system into a whole new trajectory unidentifiable from the original trajectory that was predicted by, for example, a weather model (Lorenz 1969a,b).

It should be made clear that this Indian philosophical view is not entirely a nonscientific viewpoint, at least as far as the way the science is defined today. Modern physics is defined, if not dominated, by quantum physics. Quantum physics is the physics at the level of a quantum, the smallest material unit or subatomic particle (Nobel Prize 1918). Quantum physicists argue that there is no material center for the universe as we know and experience it, neither a center for an individual being or mind.

In a quantum, unlike an atom, there is no nucleus, i.e. a central entity, around which electrons circle. Further, in quantum physics, there is no duality between a particle and a wave. A photon, i.e. a quantum of light, is at the same time a particle and a wave, or it is neither a particle nor a wave (Feynman et al. 2005).

According to Niels Bohr, classical physics or Newtonian physics is based on the principle of separated properties which quantum physics refutes (Faye 2014). The principle of separated properties holds that two objects or systems separated in time and space have each independent inherent states or properties.

Based on this principle of independent states of objects or systems, classical physicists can argue that the physical world can be described accurately by a physicist independently of the measurement apparatus or the measurer. In quantum physics, by contrast, a physical object or phenomenon is interpreted to have no definitive characteristics until someone attempts to describe it. Similarly, a physical object or phenomenon is interpreted to have no definitive characteristics independent of the describer.

Another theory of quantum physics that is particularly pertinent to catastrophe studies is the uncertainty principle or indeterminacy rule (Heisenberg 1927). Heisenberg showed that, in the quantum physics world, a researcher cannot measure both the speed of a particle and its position precisely at the same time. This uncertainty principle is applicable to any pair of complementary variables with regard to the properties of a particle.

In Heisenberg's article, the uncertainty principle is stated as follows: if a researcher measures the momentum of a particle precisely, he cannot measure the position of the particle precisely; if he measures the position accurately, he cannot measure the momentum accurately. At the quantum level where the electron is spread out like a sea wave, if the momentum of a particle is well defined, there is no well-defined position of the particle, and vice versa.

Some quantum physicists go even further to refute the classical principle of causality which states that every event has a cause (Faye 2014). That is, there is an implication in quantum physics that interactions between and among quanta occur at random or probabilistically. At the smallest particle, i.e. quantum, level, there is no causality rule that governs their behaviors.

More precisely stated, when a quantum hits another atom, a scientist cannot predict with precision in which direction it will bounce off. Stated differently, "the electron waves are waves of probability;" as such, a quantum's direction is both unpredictable and uncontrollable (Weinberg 2017). Quantum interactions occur at random; that is, at the smallest particle level there is no cause-and-effect relationship that defines their interactions. In the words of quantum physicists, "the moment-to-moment determinism of the laws of physics seems to be lost" in the world of quantum physics (Feynman et al. 2005; Weinberg 2017).

The meaning and future of quantum physics continue to be debated among physicists and philosophers alike. The author concludes this section with quotations by two of the most well-known physicists today with regard to the status of quantum physics (Weinberg 2017):

> Quantum mechanics is impressive. But, an inner voice tells me that it is not yet the real thing (Einstein (1926) as cited in Pais (1982)).
> I think I can safely state that no one understands quantum mechanics (Feynman 1967).

These quotations should apply, with even stronger force, to the fortuitous originationists in ancient India whose views are even more extreme and much broader in scope than modern quantum physicists in expounding the world of things and beings as well as their originations and dissolutions.

3.8 The Road to the Economics of Catastrophes

This concludes an introductory review of the philosophical traditions, broadly defined, which expounded on or are concerned with catastrophic events. As stated in the beginning, it is a selected review which does not intend to be a comprehensive coverage of the entire range of philosophical traditions on the topic. Notwithstanding, the selection of reviewed traditions was carefully done to be representative of the most direct and the most influential works on catastrophes, which have received much attention in relation to modern-day events as well as enduring legacies.

Further, the review of this chapter, unlike other similar reviews available, was carefully planned to cover a broad spectrum of intellectual works: environmental and ecological catastrophes, climate doomsday studies, archaeology of civilizational collapses, a western philosophical tradition, an Indian school of thoughts, and some aspects of quantum physics.

Throughout this chapter, the author highlights many similarities of the philosophical traditions' conceptualizations with the scientific theories of catastrophes and chaos explained in Chapter 2. It may even be quite novel to argue that the "modern" sciences of catastrophes are, to a large extent, mathematical and statistical formalizations of the old catastrophe philosophies.

It is hoped that the sciences and mathematical models of catastrophes and chaos presented in Chapter 2 and the philosophical works of catastrophes reviewed in this chapter have paved the way for the presentation of the economics of catastrophes in the ensuing two chapters. The economics chapters will delineate, aiming for an integrative framework of a large variety of disciplines and components, how behavioral, market, and policy forces are in play in dealing with numerous catastrophic risks.

The economics of catastrophic events will be presented both from a theoretical perspective and from an empirical perspective. Chapter 5 will give the reader an opportunity to have a closer look at and assess global datasets on numerous catastrophe events and their consequences, of which hurricane, cyclone, typhoon, and earthquake data are central and presented at a regional scale for each specific event.

The original sources of these catastrophe data are of particular concern to this book and will be presented in detail, including the International Best Track Archive for tropical cyclone data and the Specialized Regional Meteorological Center (SRMC) reports of individual cyclone events in different ocean basins for about five decades (NCEI 2016; IMD 2015; ABOM 2017; JMA 2017).

References

Anastas, P.T. and Kirchhoff, M.M. (2002). Origins, current status, and future challenges of green chemistry. *Accounts of Chemical Research* 2002: 686–694.

Anastas, P.T. and Zimmerman, J.B. (ed.) (2012). *Innovations in Green Chemistry and Green Engineering*. Berlin: Springer.

Australian Bureau of Meteorology (ABOM) (2017) Individual cyclone reports. ABOM, Australian Government, Canberra. www.bom.gov.au/cyclone/history/index.shtml (accessed 22 May 2018).

Broecker, W.S. (1997). Thermohaline circulation, the Achilles heel of our climate system: will man-made CO_2 upset the current balance? *Science* 278: 1582–1588.

Büntgen, U., Tegel, W., Nicolussi, K. et al. (2011). 2500 years of European climate variability and human susceptibility. *Science* 331: 578–582.

Carson, R. (1962). *Silent Spring*. Boston, MA: Houghton Mifflin.

Chapman, C.R. and Morrison, D. (1994). Impacts on the earth by asteroids and comets: assessing the hazard. *Nature* 367: 33–40.

Cropper, M.L., Evans, W.N., Berardi, S.J. et al. (1992). The determinants of pesticide regulation: a statistical analysis of EPA decision making. *Journal of Political Economy* 100: 175–197.

Dar, A., Rujula, A.D., and Heinz, U. (1999). Will relativistic heavy-ion colliders destroy our planet? *Physics Letters B* 470: 142–148.

Diamond, J. (2005). *Collapse: How Societies Choose to Fail or Succeed*. New York: Viking Press.

Drake, B.L. (2012). The influence of climatic change on the Late Bronze Age collapse and the Greek Dark Ages. *Journal of Archaeological Science* 39: 1862–1870.

Ehrlich, P.R. (1968). *The Population Bomb*. Oakland, CA: Sierra Club/Ballantine Books.

Ellis, J., Giudice, G., Mangano, M. et al. (2008). Review of the safety of LHC collisions. *Journal of Physics G: Nuclear and Particle Physics* 35 (11).

Faye J (2014) Copenhagen interpretation of quantum mechanics. Stanford Encyclopedia of Philosophy Archive, Fall 2014 Edition. https://plato.stanford.edu/archives/fall2014/entries/qm-copenhagen.

Feigenbaum, M.J. (1978). Quantitative universality for a class of non-linear transformations. *Journal of Statistical Physics* 19: 25–52.

Feynman, R. (1967). *The Character of Physical Law*. Cambridge, MA: MIT Press.

Feynman, R.P., Leighton, R.B., and Sands, M. (2005). *The Feynman Lectures on Physics: The Definitive and Extended Edition*, 2e. Boston, MA: Addison Wesley.

Freeman, A.M. III, Herriges, J.A., and Cling, C.L. (2014). *The Measurements of Environmental and Resource Values: Theory and Practice*. New York: RFF Press.

Gill, R.B., Mayewski, P.A., Nyberg, J. et al. (2007). Drought and the Maya collapse. *Ancient Mesoamerica* 18: 283–302.

Gore, A. (2006). *An Inconvenient Truth: The Planetary Emergency of Global Warming and What we Can Do about It*. New York: Rodale Books.

Hájek, A. (2012). *Pascal's Wager*. Stanford Encyclopedia of Philosophy https://plato.stanford.edu/entries/pascal-wager.

Hanemann, W.M. (1994). Valuing the environment through contingent valuation. *Journal of Economic Perspectives* 8: 19–43.

Hawking S, Tegmark M, Russell S, Wilczek F (2014) Transcending complacency on superintelligent machines. *Huffington Post*. https://www.huffingtonpost.com/stephen-hawking/artificial-intelligence_b_5174265.html.

Heisenberg, W. (1927). Über den anschaulichen Inhalt der quantentheoretischen Kinematik und Mechanik. *Zeitschrift für Physik* 43 (3–4): 172–198.

International Institute of Applied Systems Analysis (IIASA) World Population Program (2007) Probabilistic projections by 13 world regions, forecast period 2000–2100, 2001 revision. IIASA, Vienna, Austria. www.iiasa.ac.at/Research/POP/proj01 (accessed 22 May 2018).

India Meteorological Department (IMD) (2015) Report on Cyclonic Disturbances in North Indian Ocean from 1990 to 2012. Regional Specialised Meteorological Centre—Tropical Cyclones, New Delhi.

Jaffe, R.L., Buszaa, W., Sandweiss, J., and Wilczek, F. (2000). Review of speculative disaster scenarios at RHIC. *Review of Modern Physics* 72: 1125–1140.

Japan Meteorological Agency (JMA) (2017). *RSMC Best Track Data*, 1951–2017. Tokyo: JMA.

Joint Typhoon Warning Center (JTWC) (2017) Annual Tropical Cyclone Reports. JTWC, Guam, Mariana Islands. http://www.usno.navy.mil/jtwc/annual-tropical-cyclone-reports.

Kennett, D.J., Breitenbach, S.F.M., and Aquino, V.V. (2012). Development and disintegration of Maya political systems in response to climate change. *Science* 338: 788–791.

Kolbert, E. (2014). *The Sixth Extinction: An Unnatural History*. New York: Henry Holt and Company.

Kurzweil, R. (2005). *The Singularity Is Near*. New York: Penguin.

Lenton, T.M., Held, H., Kriegler, E. et al. (2008). Tipping elements in the earth's climate system. *Proceedings of the National Academy of Sciences of the United States of America* 105: 1786–1793.

Leopold, A. (1949). *A Sand County Almanac: And Sketches Here and There*. Oxford: Oxford University Press.

Le Page M (2012) Five civilisations that climate change may have doomed. New Scientist 3 August. https://www.newscientist.com/gallery/climate-collapse.

Lorenz, E.N. (1963). Deterministic nonperiodic flow. *Journal of the Atmospheric Sciences* 20: 130–141.

Lorenz, E.N. (1969a). Atmospheric predictability as revealed by naturally occurring analogues. *Journal of the Atmospheric Sciences* 26: 636–646.

Lorenz, E.N. (1969b). Three approaches to atmospheric predictability. *Bulletin of the American Meteorological Society* 50: 345–349.

Maler, K.-G. and Vincent, J.R. (2005). *The Handbook of Environmental Economics (Vol. 2): Valuing Environmental Changes*. Amsterdam: North Holland.

Mandelbrot, B.B. (1983). *The Fractal Geometry of Nature*. New York: Macmillan.

Mandelbrot, B.B. (2004). *Fractals and Chaos*. Berlin: Springer.

Mann, M.E., Bradley, R.S., and Hughes, M.K. (1999). Northern hemisphere temperatures during the past millennium: inferences, uncertainties, and limitations. *Geophysical Research Letters* 26: 759–762.

McAnany, P.A. and Yoffee, N. (ed.) (2009). *Questioning Collapse: Human Resilience, Ecological Vulnerability, and the Aftermath of Empire*. Cambridge, UK: Cambridge University Press.

Meadows, D.H., Meadows, D.L., Randers, J., and Behrens, W.H. (1972). *The Limits to Growth*. New York: Universe Books.

Mendelsohn, R. and Olmstead, S. (2009). The economic valuation of environmental amenities and disamenities: methods and applications. *Annual Review of Resources* 34: 325–347.

Meyer, K.M. and Kump, L.R. (2008). Oceanic euxinia in earth history: causes and consequences. *Annual Review of Earth and Planetary Sciences* 36: 251–288.

Middleton, G.D. (2017). *Understanding Collapse: Ancient History and Modern Myths*. Cambridge, UK: Cambridge University Press.

Mills, M.J., Toon, O.B., Turco, R.P. et al. (2008). Massive global ozone loss predicted following regional nuclear conflict. *Proceedings of the National Academy of Sciences of the United States of America* 105: 5307–5312.

National Centers for Environmental Information (NCEI) (2016) International Best Track Archive for Climate Stewardship. https://www.ncdc.noaa.gov/ibtracs.

National Research Council (NRC) (2013) Abrupt impacts of climate change: anticipating surprises. National Academies Press, Washington, DC.

Nobel Prize (1918) The Nobel Prize in Physics 1918. Max Karl Ernst Ludwig Planck. http://www.nobelprize.org/nobel_prizes/physics/laureates/1918.

Nordhaus, W.D. (1992a). The ecology of markets. *Proceedings of the National Academy of Sciences of the United States of America* 89: 843–850.

Nordhaus, W.D. (1992b). Lethal model 2: the limits of growth revisited. *Brookings Papers on Economic Activity* 1992 (2): 1–59.

Nordhaus, W. (2011). The economics of tail events with an application to climate change. *Review of Environmental Economics and Policy* 5: 240–257.

Oppenheimer, M. and Alley, R.B. (2005). Ice sheets, global warming, and Article 2 of the UNFCCC. *Climatic Change* 68: 257–267.

Pais, A. (1982). *Subtle Is the Lord: The Science and the Life of Albert Einstein*. Oxford, UK: Oxford University Press.

Pascal, B. (1670). *Penseés* (trans. WF Trotter, 1910). London: Dent.

Pearce, F. (2011). *The Coming Population Crash: And our Planet's Surprising Future*. Boston, MA: Beacon Press.

Posner, R.A. (2004). *Catastrophe: Risk and Response*. New York: Oxford University Press.

Renfrew, C. (1984). *Approaches to Social Archeology*. Cambridge, MA: Harvard University Press.

Schuster, E.F. (1984). Classification of probability laws by tail behavior. *Journal of the American Statistical Association* 79 (388): 936–939.

Schwartz, G.M. and Nichols, J.J. (ed.) (2006). *After Collapse: The Regeneration of Complex Societies*. Tucson, AZ: University of Arizona Press.

Seo, S.N. (2017). Measuring policy benefits of the cyclone shelter program in the North Indian Ocean: protection from intense winds or high storm surges? *Climate Change Economics* 8 (4): 1–18. doi: 10.1142/S2010007817500117.

Solow, R.M. (1974). The economics of resources or the resources of economics. *American Economic Review* 64: 1–14.

Solow, R.M. (1993). An almost ideal step toward sustainability. *Resources Policy* 19: 162–172.

Swiss Re Institute (2017). *Natural Catastrophes and Man-made Disasters in 2016: A Year of Widespread Damages*. Zurich, Switzerland: Swiss Re.

Tainter, J.A. (1990). *The Collapse of Complex Societies*. Cambridge, UK: Cambridge University Press.

Taleb, N.N. (2007). *The Black Swan: The Impact of the Highly Improbable*. London: Penguin.

The Wilderness Society (TWS) (2017) Aldo Leopold. http://wilderness.org/bios/founders/aldo-leopold.

Thom, R. (1975). *Structural Stability and Morphogenesis*. New York: Benjamin-Addison-Wesley.

Thoreau, H.D. (1854). *Walden: Or, Life in the Woods*. Boston: Ticknor and Fields.

Tietenberg, T. and Lewis, L. (2014). *Environmental and Natural Resource Economics*. New York: Prentice Hall.

Tipitaka (2010) Brahmajāla Sutta: The All-embracing Net of Views. Digha Nikaya. Translated by Bhikkhu Bodhi. www.accesstoinsight.org (accessed 22 May 2018).

Tsongkhapa, J. (2001). *The Great Treatise on the Stages of the Path to Enlightenment: Volume 1, Volume 2, Volume 3*. Ithaca, New York: Snow Lion Publication.

Turco, R.P., Toon, O.B., Ackerman, T.P. et al. (1983). Nuclear winter: global consequences of multiple nuclear explosions. *Science* 222: 1283–1292.

Tziperman, E., Cane, M.A., and Zebiak, S.E. (1995). Irregularity and locking to the seasonal cycle in an ENSO prediction model as explained by the quasi-periodicity route to chaos. *Journal of Atmospheric Science* 52 (3): 293–306.

United States Congress (1978). *Toxic Substances Control Act of 1978*. Washington, DC: US Congress.

United States Environmental Protection Agency (US EPA) (1977). *The Clean Air Act Amendments of 1977*. Washington, DC: US EPA.

United States Environmental Protection Agency (US EPA) (1990) The Clean Air Act Amendments of 1990. US EPA, Washington, DC.

Utsu T (2013) Catalog of damaging earthquakes in the world (through 2013). International Institute of Seismology and Earthquake Engineering, Tsukuba, Japan. https://iisee.kenken.go.jp/utsu/index_eng.html.

Vogel, S.A. and Roberts, J.A. (2011). Why the Toxic Substances Control Act needs an overhaul, and how to strengthen oversight of chemicals in the interim. *Health Affairs* 30: 898–905.

Wagner, G. and Weitzman, M. (2015). *Climate Shock: The Economic Consequences of a Hotter Planet*. Princeton, NJ: Princeton University Press.

Warrick, R.A. (1980). Drought in the Great Plains: a case study of research on climate and society in the USA. In: *Climatic Constraints and Human Activities. IIASA Proceedings*

Series, IIASA Proceedings Series 10 (ed. J. Ausubel and A.K. Biswas), 10. New York: Pergamon Press.

Weinberg, S. (2017). The trouble with quantum mechanics. *New York Review of Books* 64 (1).

Weiss, H. and Bradley, R.S. (2001). What causes societal collapses? *Science* 291: 609–610.

Weitzman, M.L. (2009). On modeling and interpreting the economics of catastrophic climate change. *Review of Economics and Statistics* 91 (1): 1–19.

Wheeler, M. (1966). *Civilizations of the Indus Valley and Beyond*. New York: McGraw-Hill.

World Health Organization (WHO) (2015). *Zika Virus Outbreaks in the Americas. Weekly Epidemiological Record*. Rome: WHO.

World Health Organization (WHO) (2017) Ebola outbreak 2014–2015. WHO, Rome. http://www.who.int/csr/disease/ebola/en.

Zeeman, E.C. (1977). *Catastrophe Theory: Selected Papers 1972–1977*. Reading, MA: Addison-Wesley.

Zeng, X., Pielke, R.A., and Eykholt, R. (1993). Chaos theory and its applications to the atmosphere. *Bulletin of the American Meteorological Association* 74: 631–644.

Zhang, D.D., Lee, H.F., Wang, C. et al. (2011). The causality analysis of climate change and large-scale human crisis. *Proceedings of the National Academy of Sciences of the United States of America* 108: 17296–17301.

4

Economics of Catastrophic Events: Theory

4.1 Introduction

From this chapter on, the author turns the book's focus away from the science and philosophical perspectives on catastrophes to the economics of and policy decisions on global-scale catastrophes. The latter lie at the heart of the endeavors made through this book, but will be explained by benefiting from the discussions provided in the former, i.e. Chapters 2 and 3.

This chapter provides a critical review of the economic theories on catastrophic events, including financial derivatives for catastrophes and governmental policy alternatives. Chapter 5 provides a series of empirical economic analyses conducted on the globally gathered empirical economic and scientific data on catastrophic events. Chapter 6 reviews and evaluates historical developments of policies and programs implemented in the range of catastrophic events introduced throughout this book, with in-depth discussions of the relevant US policies and global agreements.

The economics field of "truly big," say humanity-scale, catastrophes has received scant attention until the end of the first decade of the twenty-first century (Cropper 1976; Parson 2007). During this pre-catastrophe economics period, there were economic works on big events such as global biodiversity loss, economics of a great war, and sustainability and limits of growth, but the scale of damage in these events was thought to be limited to a particular region or a particular set of species (Meadows et al. 1972; Weitzman 1998; Nordhaus 2002).

A rigorous economic analysis of the "truly big" catastrophic event appeared emphatically with reference to a potential catastrophic consequence of unmitigated global warming by Weitzman (Weitzman 2009, 2011) and a series of analyses on fat-tail events in the context of global warming by William Nordhaus (Nordhaus 2008, 2011, 2012, 2013). The concern of these authors was a truly global catastrophe, more specifically, the end of human civilization as we know it, or the end of all life on the planet.

At around the threshold between the twentieth century and the twenty-first century, concerns about the possibility of a truly big catastrophe at a global scale that can literally destroy humanity itself were voiced by concerned scientists and policy professionals (Posner 2004; Parson 2007). A nuclear winter engendered by explosions of a large number of nuclear bombs may have the impact of killing most, if not all, of humanity (Turco et al. 1983; Mills et al. 2008). A collision of a large asteroid with Earth may permanently destroy the planet (Chapman and Morrison 1994; NRC 2010). Experiments in the Large Hadron Collider (LHC) to create the initial stages of the universe by the European

Natural and Man-made Catastrophes – Theories, Economics, and Policy Designs, First Edition. S. Niggol Seo.
© 2019 John Wiley & Sons Ltd. Published 2019 by John Wiley & Sons Ltd.

Organization for Nuclear Research (CERN) may create strangelets and subsequently a black hole that can fold the universe and all beings in it (Dar et al. 1999; Jaffe et al. 2000). There may come soon enough a time when the intelligence of artificial intelligence (AI) robots surpasses that of humans, killing humanity uncontrollably (Kurzweil 2005; Hawking et al. 2014).

Any of these events would be truly catastrophic in the sense that it would be nearly irreversible if any of these were indeed to strike and unfold as predicted by the concerned scientists. As such, humans have almost no experience at all of any of these events. The closest events of the truly big catastrophes, in terms of the scale of destruction, that humans have experienced are natural catastrophes, i.e. exceedingly catastrophic events of nature recorded in human history. These historical events were either an earthquake, or a hurricane, or a coupled event such as an earthquake–tsunami, each event killing hundreds of thousands of people and causing billions of dollars of economic losses (Utsu 2013; Swiss Re Institute 2017; Seo and Bakkensen 2017).

Figure 4.1 shows the number of human fatalities from a natural catastrophic event since 1970 through 2016 (Swiss Re Institute 2017). In this figure, drawn from a reinsurance company's research, a natural catastrophe was defined to be a natural event with total economic losses, insured losses, and number of fatalities in excess of the following threshold values: the number of people dead or missing exceeding 20, total economic losses exceeding US$99 million, and insured losses exceeding US$49.5 million.

The number of fatalities exceeded 100 000 six times during this period, three of which were earthquakes and the other three were tropical cyclones. Two salient characteristics of natural catastrophic events can be learned from the figure. The first is that each one of the six major catastrophic events results in a surprisingly large number of fatalities. In other words, while most natural disasters resulted in a loss of lives in the magnitude

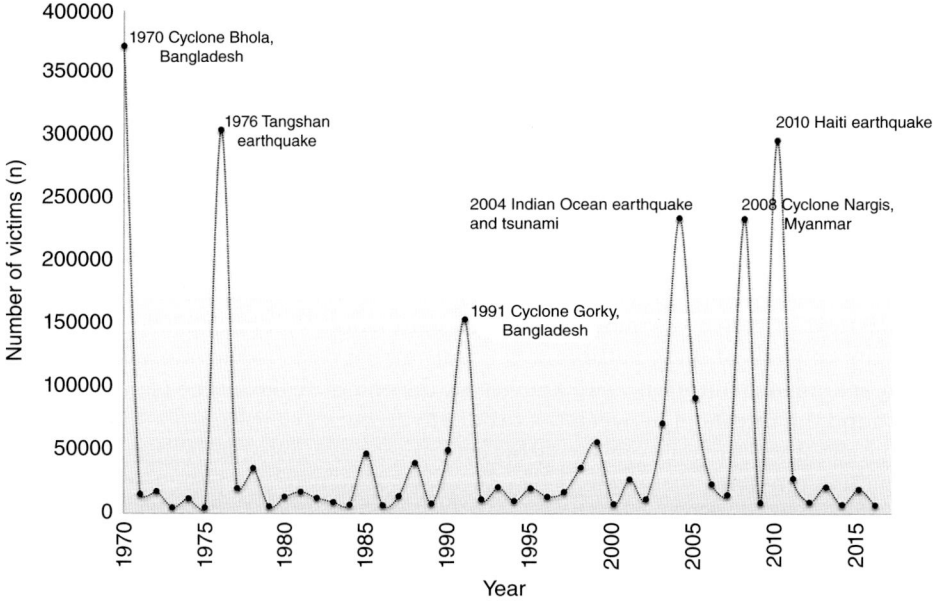

Figure 4.1 Number of victims from natural catastrophes since 1970. *Source*: Swiss Re Institute (2017).

of several hundreds, the number of lost lives increases more than 10 times in the six major catastrophe events. The six major events are a tail event (Mandelbrot 1997; Nordhaus 2011).

The second characteristic is that an occurrence of an exceptional natural catastrophe event is difficult to predict in advance. None of the six catastrophes in Figure 4.1 was predicted with precision well before the event. This may be especially true in the case of earthquake occurrences. The theory of plate tectonics indicates that it is near impossible to predict with accuracy a particular earthquake in advance (Kiger and Russell 1996). In the case of a hurricane, predictions of a genesis and the tracks of a hurricane are equally difficult to be forecast in advance, but a generated hurricane often takes several days before it reaches a landmass (Emanuel 2008; Seo 2015a).

In comparison with the natural catastrophes shown in Figure 4.1, man-made disasters do not show the two salient features of natural catastrophes. The historical data of the numbers of victims from man-made disasters show that they do not exceed 10 000 even in the most disastrous cases. Further, an analysis of the Swiss Re catastrophe data reveals that the standard deviation of the number of victims from man-made disasters was 1903 persons, which can be contrasted to the standard deviation of the number of victims from natural catastrophes which is 88 116 persons (Swiss Re Institute 2017). The large variance in the number of fatalities from natural catastrophes is a prominent characteristic of truly big catastrophic events.

Man-made disasters are diverse, including environmental disasters, airplane crashes, maritime disasters, wildfires, building collapses, a war, terrorist attacks, and mass killing. Of the major environmental disasters during this time period are nuclear accidents such as the Three Mile Island accident in Pennsylvania in 1979, the Chernobyl nuclear disaster in the former Soviet Union (presently Ukraine) in 1986, and the Fukushima nuclear disaster in Japan in 2011 (NEI 2016).

The economics of catastrophic events, described in this chapter, addresses the two prominent features of catastrophic studies. First, this chapter answers the essential question of how an individual and a society should cope with potential catastrophes. The author explains at length the financial derivatives that have been developed to help individuals deal with a low-probability–high-damage event: insurance, futures, options, and catastrophe bonds. Second, this chapter elaborates both the concepts that underlie the financial derivates and the historical market/public data on their applications to natural and man-made catastrophe events (Shiller 2004; Fabozzi et al. 2009; Barrieu and Albertini 2009).

The economics of catastrophic events will also elucidate the conundrum of a "truly big" humanity-ending catastrophe which presents a non-negligible probability of the end of human civilization on Earth, e.g. global warming, asteroid collision, strangelets, and AI. For this type of event, there is often a threshold clearly defined, which is also referred to as a tipping point, or singularity, or a bifurcation point in the catastrophe literature (Thom 1975; Kurzweil 2005). In addition, each of these humanity-scale events may exhibit a unique probability distribution in which the tail probability refuses to die out, namely, a fat-tail distribution (Schuster 1984; Gabaix 2009).

Economists' debates are centered on how humanity should manage these humanity-ending catastrophes in a socially optimal manner (Weitzman 2009; Nordhaus 2011; Yohe and Tol 2012). Should the global community prevent a potential occurrence of one of these events at all cost, even though the likelihood of an event occurring actually is very small? Should the global community manage this type of event in a globally optimal way by balancing the benefit and cost from an array of policy actions, taking into account the probabilities of various possibilities?

The author will evaluate existing policy proposals such as a globally optimal policy intervention and a generalized precautionary principle for policy interventions in the event of a humanity-scale catastrophe (Seo 2017a). In evaluating the policy proposals, alternative parameterizations of existing policy models, especially the dismal theorem, will be attempted, from which an opposite policy recommendation is shown to be possible.

4.2 Defining Catastrophic Events: Thresholds

What is or is not a catastrophe? The data shown in Figure 4.1 are constructed on the basis of a specific definition on what a catastrophe is. More specifically, an event with more than 20 human deaths or US$99 million loss is classified as a catastrophic event. That is, there were predetermined cut-off points or thresholds used by the reinsurance company's researchers (Swiss Re Institute 2017).

A threshold is one of the fundamental concepts in the field of catastrophe research. A threshold is a particular point in the range of values that a random variable can take, before and after which behaviors of a system or an outcome variable are starkly different. With reference to the catastrophe theory, it is the value of the potential function at which point the system suddenly turns into an unstable system (Thom 1975).

In climate change policy negotiations at the global level, a temperature ceiling placed at a 2° C increase from the pre-industrial level is one of the most important benchmarks for climate policy-makings (UNFCCC 2010, 2015). The rationale for the support of this benchmark by scientists has been that the global damage from a beyond 2° C increase in global temperature will be catastrophic as well as irreversible, according to the supporters. The temperature ceiling policy proposal is rooted on the assumption of a threshold in the climate system, also called a tipping point by climate scientists (Broecker 1997; Lenton et al. 2008).

In a slightly different form, a threshold can be defined as a critical point in the range of values of a random variable beyond which a fatality (death) is certain to occur. In toxicology where this definition was advanced, a threshold is a critical level of a toxic or hazardous chemical that is highly likely to result in the death of an individual who consumes the chemical (Lutz et al. 2014).

An application of this definition of a threshold to policy issues is found in environmental policy areas. In the US Clean Air Act, six criteria pollutants were determined and air quality standards for each of these pollutants were determined and implemented. In excess of the concentrations specified as the standards for each pollutant, protection of public health is assessed by the environmental agency to be threatened (US EPA 1977, 1990).

The six criteria pollutants are ground-level ozone, particulate matter (PM), sulfur dioxide, nitrogen oxides, lead, and carbon monoxide (US EPA 1990, 2010). The thresholds for these pollutants are called the National Ambient Air Quality Standards (NAAQS) in the Clean Air Act. For example, the NAAQS for ozone (O_3) is set at 0.070 ppm (parts per million), calculated from an 8-hour average concentration, while the NAAQS for sulfur dioxide is 75 ppb (parts per billion) per 1-hour daily maximum.

A more drastic application of the toxicology definition of a threshold is found in the US Toxic Substances Control Act (TSCA) (US Congress 1978). The TSCA gives the US Environmental Protection Agency (EPA) administrator an authority to ban a certain chemical based on her/his judgment of "an unreasonable risk of injury to health or the environment."

More specifically, Section 4.6 of the TSCA gives the administrator of the EPA the authority – if s/he concludes that there is a reasonable basis to judge that a chemical substance or mixture, during any stage of production, distribution, use, or disposal of it, presents an unreasonable risk of injury to human health or the environment – to prohibit the manufacturing, processing, or distribution in commerce of such substance or mixture (Vogel and Roberts 2011).

To continue with the conceptualizations of a threshold, let X be a non-negative random variable whose value is determined by the magnitude of the damage from a natural or man-made event, say, a tropical cyclone or an earthquake. Let the distribution of the random variable be parameterized by its mean value and variability: mean μ and variance σ^2.

Then, a high-damage outcome as well as a catastrophic outcome in the realizations of the random variable, i.e. the natural event concerned such as an earthquake, can be conceptualized by a $\tau - sigma$ event as follows:

$$\mathbb{C} = \{x | x - \mu \geq \tau\sigma\}. \tag{4.1}$$

In simple terms, a one-sigma event is a realization of an earthquake in which the realized damage is greater than the sum of the mean and the one standard deviation of the distribution of X; a two-sigma event is a realization in which the realized damage is greater than the sum of the mean and the two standard deviations of the distribution of X.

The larger the τ, the more harmful a $\tau - sigma$ event becomes. How large should it be to be considered a catastrophic earthquake event? To formalize the answer to this question, let's assume that the random variable, i.e. the magnitude of damage, is a function of an underlying variable:

$$X = f(Z). \tag{4.2}$$

In the case of hurricane damage, the underlying variable can be a variable of hurricane intensity such as maximum wind speeds (MWS) or minimum central pressure (MCP) (Emanuel 2008). In the case of earthquake damage, it can be a variable of earthquake intensity such as a Richter scale or a moment scale of magnitude (Richter 1958).

For the specification of the function in Eq. (4.2), let's consider the following two situations. One is where the damage increases linearly in response to the underlying variable:

$$X = \alpha + \beta Z. \tag{4.3}$$

In the linear damage function of Eq. (4.3), τ as the threshold for a catastrophic event is hard to define because the rate of increase in the magnitude of damages remains the same over the whole range of values of the underlying indicator.

In this situation, it is more appropriate to define a ω-percent event to refer to the severity of an event. To be more concrete, a three-sigma event in the normal distribution of the random variable can be stated as a 1% event. That is, the event would occur less than 1% of the time of the particular type of natural catastrophe, assuming the particular distribution.

In the second situation, on the other hand, the damage function is assumed to be a combination of two separate functions. In the first phase of the underlying indicator, the damage function takes on the linear form as shown in Eq. (4.3). In the second phase, the damage function takes on a nonlinear functional form in which the damage increases in an exponential way:

$$X = \begin{cases} \alpha + \beta Z, \forall Z \leq z_c; \\ e^{\gamma Z + \omega}, \forall Z \geq z_c. \end{cases} \quad (4.4)$$

In Eq. (4.4), z_c is the critical value at which point the damage function takes on another form which is an exponential function. z_c is often relied upon as a threshold value for policy-making purposes. In our preceding discussions, it fits well in the threshold of the 2° C temperature increase in climate change policy negotiations. The toxicological definition of a threshold can be expressed with an appropriate modification of Eq. (4.4).

4.3 Defining Catastrophic Events: Tail Distributions

The threshold value of z_c is a key concept for an analysis of catastrophic events as well as policy interventions. Particularly, when the value of Z can be controlled by humans, the threshold value is the most important indicator for policy-making purposes. For example, when the amount of a toxic chemical applied to a subject is varied by an experimenter, e.g. a doctor, the threshold value shall not be crossed. In the case of the phenomenon of global warming, a 2° C ceiling may or may not be a meaningful threshold when it comes to the human controllability of the threshold.

In natural catastrophes such as an earthquake, hurricane, and asteroid collision, however, the value of Z, i.e. the realized values of the random variable, is by and large uncontrollable by mankind. That is, an earthquake moment scale or a hurricane MCP is considered to be uncontrollable. In a random event such as this, a probability distribution of the random variable provides further indispensable information on the characteristics of the event, which is essential for policy-makers.

Continuing with the set-up used in Section 4.2, let X, the amount of economic damage from one of the above-mentioned natural events, be normally distributed with the two parameters as before: the mean μ and the standard deviation σ. The Gaussian (or normal) distribution is characterized by a bell-shaped probability density function (PDF), where the probability of realization is highest at the center of the distribution and falls, resembling the shape of the outer surface of a bell, as the realized outcome is further at the tails of the distribution.

In the Gaussian distribution, which is a symmetric distribution, the tail probability distribution, that is, $\tau - sigma$ events, is well defined (Nordhaus 2011):

$$P\{z | z - \mu \geq \sigma\} < \frac{0.33}{2};$$

$$P\{z | z - \mu \geq 2\sigma\} < \frac{0.05}{2};$$

$$P\{z|z - \mu \geq 3\sigma\} < \frac{0.003}{2};$$

$$P\{z|z - \mu \geq 4\sigma\} < \frac{0.0001}{2}. \tag{4.5}$$

What Eq. (4.5) shows is that, in the Gaussian distribution, a four-sigma event occurs extremely rarely, while a three-sigma event occurs very rarely. A four-sigma event, if it actually occurs, will be a total surprise or a shock to the observers. Specifically, a four-sigma hurricane is an event that would occur once in 20 000 hurricanes according to Eq. (4.5), assuming a Gaussian distribution of hurricane events. Assuming 10 hurricane events per year in the North Atlantic, it is a hurricane event that would occur only one time in 2000 years. Similarly, a three-sigma hurricane would occur once in 1000 hurricanes or once in 100 years.

By way of the tail distribution we can state the conundrum of dealing with a catastrophic event, either natural or man-made: if we are able to clearly define the critical threshold value, z_c, for a natural or man-made event, should we be concerned about it if the probability of z_c occurrence is tiny, say, a once-in-5000-years event? If yes, how much should the potentially affected be concerned and prepared by investing valuable resources?

The answer depends on, among other considerations, the type of the tail distribution (Mandelbrot 1997; Weitzman 2009; Nordhaus 2011). A probability distribution of a random variable can be classified into three types by its tail distribution, given that the tail distribution is clearly defined (Schuster 1984). The Gaussian distribution with a bell-shape PDF is classified as a medium-tail distribution. A thin-tail distribution has a PDF with a finite upper limit, meaning that the probability of occurrence is zero beyond the upper limit. A thin-tail distribution includes a uniform distribution and a triangular distribution.

A triangular distribution takes on a triangular form, and therefore has upper and lower limits. A triangular distribution is used when it is impossible for the random variable to take on the values beyond the two limits. With $a < c < b$, a triangular distribution can be defined by the following PDF (Casella and Berger 2001):

$$f(z) = \begin{cases} 0, z < a; \\ \dfrac{2(z - a)}{(b - a)(c - a)}, a \leq z < c; \\ \dfrac{2(b - z)}{(b - a)(b - c)}, c \leq z \leq b; \\ 0, z > b. \end{cases} \tag{4.6}$$

If the probability distribution of an event follows a thin-tail distribution such as the triangular distribution specified in Eq. (4.6), a threshold value, i.e. a tipping point, is not of any concern as long as it lies outside the upper and lower limits. This is because there is no possibility of the occurrence of z_c.

In a medium-tail distribution, for example, the normal distribution, the threshold value z_c should be of concern to the potentially affected by the random event, but the threshold value should be of no grave concern at the national level such as the US federal government. The reason is that the probability of occurrence quickly approaches zero the rarer the outcome is in the normal distribution. Put differently, nearly all realizations of the random event will fall within the bounds that are expected or commonly prepared by the affected. Imagine a once-in-1 000 000-years hurricane event. Of course, this does not mean that the US federal government should not be concerned about hurricane events altogether.

Since it may not be immediately clear, let me corroborate further through an illustrative example: a housefire in a city. The annual number of housefires in the city of New York shows a well-defined normal distribution with a mean and a standard deviation. Every year, housefires occur μ_1 times, with a standard deviation of σ_1. As such, 99.99% of the time, the number of annual housefires remains within three standard deviations, i.e. in the range of $(\mu_1 - 3\sigma_1, \mu_1 + 3\sigma_1)$, according to the tail probabilities shown in Eq. (4.5).

The medium-tail distribution of the random event makes it possible for a financial company to offer a housefire insurance to potential victims of housefires in New York City, from which the company gains profits. For a potential victim, it is welfare-improving to purchase a housefire insurance to avoid a large loss in the event of a devastating housefire. The housefire insurance improves the Pareto welfare of the society (Fabozzi et al. 2009).

The example of housefire events and the availability of a housefire insurance provide an illustrative situation in which there is no need for a national policy intervention to cope with a natural catastrophe event. In addition to a basket of insurance products offered in the market, there are other financial as well as nonfinancial options that help potential victims deal with various potential catastrophe events, including options, futures, and catastrophe bonds. These financial derivatives are explained in later sections of this chapter.

The third class of the tail distribution is a fat-tail, also called a heavy-tail, distribution. A fat-tail distribution is defined against a thin-tail distribution and a medium-tail distribution. Unlike the other two types of a tail probability distribution, the probability of a severe tail event does not fall to zero even in a very long tail event. The probabilities of tail events remain substantial, however long the tails are extended.

The fat-tail distribution is also called a power law distribution or Zipf's law distribution after the influential work by Zipf who defined the power law (Zipf 1949). It is also called a Pareto–Levy–Mandelbrot (PLM) distribution after Wilfred Pareto who encountered empirically a power law in income distribution and Mandelbrot who expounded the critical meaning of the power law extensively (Pareto 1896; Mandelbrot 1963, 1997, 2001).

More formally, a fat-tail distribution can be defined by the following power law PDF or the countercumulative distribution function (Mandelbrot 2001; Nordhaus 2011):

$$P[Z > z] \sim z^{-\varphi}, \text{ or } P[Z = z] = kz^{-(1+\varphi)}. \tag{4.7}$$

The PLM distribution has a power law tail which is defined by a shape parameter, φ, and a constant. The shape parameter takes on non-negative values and defines the shape of the power law tail of the PLM distribution. The larger the φ, the closer the tail of the Pareto distribution to the normal (Gaussian) distribution. The closer the φ to zero, the fatter the tail of the Pareto distribution becomes.

From the power law distribution in Eq. (4.7), a fat-tail distribution is shown in Figure 4.2, assuming specific values for the shape parameter and the constant from the PDF derived in Eq. (4.7), against a medium-tail distribution. Both tail distributions are

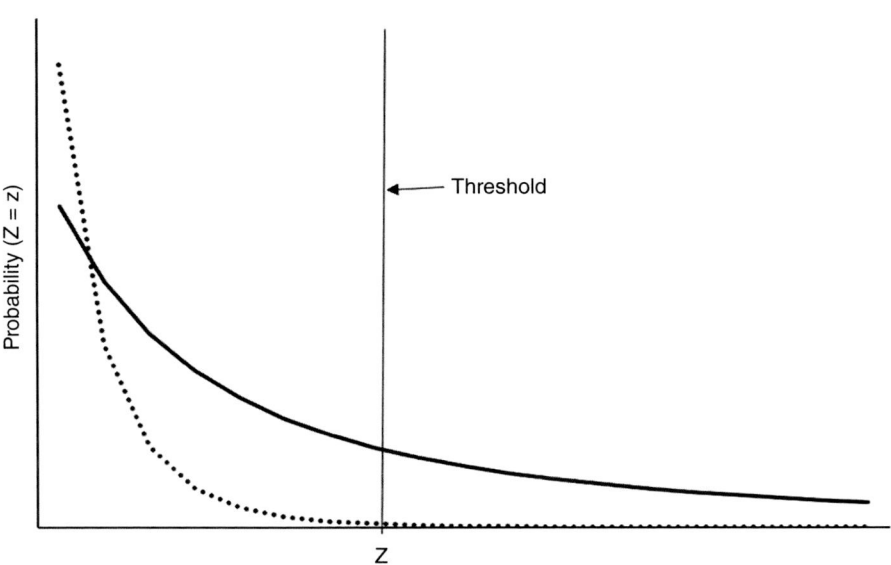

Figure 4.2 Pareto–Levy–Mandelbrot distribution.

drawn by varying the value of the shape parameter in Eq. (4.7). The vertical line is a hypothetical threshold value, approximately the four-sigma point of the medium-tail distribution.

In the medium-tail distribution in the figure, the probability of occurrence of an event falls rapidly to approximate zero at the threshold value, i.e. the four-sigma point. By contrast, a salient feature of the fat-tail distribution is that the probability of a four-sigma event remains significant. Further, the probability of even a longer tail event, e.g. a 10-sigma event, remains substantial.

How does a fat-tail distribution come about? Put differently, why is there a fat-tail distribution in a certain natural (or man-made) event? As explained in depth in Chapter 2, Benoit Mandelbrot made an influential discovery of a fractal whose distribution can be explained by a power law. He argued that certain speculative prices, i.e. cotton prices, and prices of financial indexes exhibit a power law (Mandelbrot 1963, 1997; Fama 1963).

Others argued that the fat-tail distribution can manifest due to an extremely large uncertainty in the random event (Taleb 2007; Weitzman 2009). In a fat-tail event, it is not possible to exclude the possibility of even an extremely severe tail event such as a 10-sigma event in Figure 4.2. That is, we cannot treat a 10-sigma event as virtually nonexistent or assign a vanishingly small probability to it. As shown in Figure 4.2, decision-makers are not certain about a four-sigma event much more than about a 10-sigma event if it follows a power law distribution.

The power law distribution is characterized by scale invariance, which is explained in detail in Chapter 2 (Mandelbrot 2001). Stated differently, the scale invariance means that the elasticity of the curve that has a power law distribution in Figure 4.2 remains invariant regardless of at which point it is measured. The scale invariance is expressed by the exponent of the power function in Eq. (4.7), i.e. $-\varphi$.

A fat-tail distribution may arise from a catastrophic event with an exceptionally large uncertainty, which means that it is not applicable to a host of catastrophic events. For the fat-tail distribution to be applicable, an exceptionally large uncertainty should not be reducible by further research, extensive funding, technological advances, or any other means to understand, get prepared for, and adapt to the fat-tail event concerned.

With regard to the fat-tail distribution, three policy-relevant questions can be posited with regard to catastrophe events. One is whether there is a truly fat-tail event with an exceptionally large uncertainty which cannot be reduced by any means or actions possible. In other words, is it a hypothetical distribution or an empirical distribution when it comes to catastrophes? The author examines relevant experiences in catastrophe studies to address this question in the empirical chapters of this book, Chapters 5–7.

The author offers three economic examples in Chapter 2, while discussing the power law distribution in the context of the fractal theory, where scale invariance can be inadequate: the Bernoulli utility function, empirical price elasticities of demand, and the constant economic growth rate. For many economic behaviors besides these examples, scale variance is expected to appear.

The second policy question is whether the financial derivatives designed specifically to help individuals cope with a host of catastrophic events and disasters, to be explained at length shortly, such as insurance, options, futures, bonds, and other derivatives, are still an effective instrument that an individual or a community can rely upon to address the fat-tail problems.

It has been already explained that traditional financial options and strategies can be a way to improve the welfare of both the insureds and the insurer, and thereby the social welfare, in a host of situations of catastrophes. Does this conclusion hold also for a fat-tail catastrophe or a very fat-tail catastrophe? Section 4.4 gives an example in which a financial derivative is purposefully designed to deal with a fatter-tail catastrophe.

The third policy-relevant question is whether policy interventions in a fat-tail event by a national government or an international organization should be based upon a benefit-cost analysis, which is a standard analytical tool that should be conducted before policy interventions in most, if not all, policy issues (Arrow et al. 1996a; Nordhaus 2008). If not, how fat should the fat-tail be for it to be abandoned?

An alternative policy approach rooted in the concept of a critical threshold – a tipping point – has been promoted by some researchers who are concerned with a truly big unstoppable catastrophe (Weitzman 2009). They argue that in a fat-tail event, uncertainty is so large that it makes the standard benefit-cost analysis useless. They instead support a precautionary principle in policy interventions in such events, under which a threshold critical value of a policy variable is prevented at all cost. The author will come back to the approach of the precautionary principle in later sections of this chapter.

4.4 Insurance and Catastrophic Coverage

What is the Pareto optimal strategy – or more loosely, a socially optimal strategy – to cope with a host of catastrophic events, and even a fat-tail catastrophic event? This is one of the central questions of the economics of catastrophes and disasters. The author will begin by describing market and financial instruments that have been devised historically and that are available in the market to hedge individuals against a range of catastrophic events. Later in this chapter the author will elaborate policy questions and options on catastrophic events that can be relied upon if at all needed.

The answer to the above question depends on the type of the catastrophe that one is concerned with. The answer would vary across a locally catastrophic event, a nationally catastrophic event, and a globally catastrophic event. The answer would also hinge on whether people have previous experiences of the catastrophic event of concern, e.g. a hurricane catastrophe, or have never experienced it before, e.g. a global warming catastrophe. The answer may be different with a fat-tail catastrophe from that given to a thin-tail or medium-tail catastrophe event.

The earliest financial security that was devised to hedge an individual against a catastrophic event is insurance (Shiller 2004). In the medieval era, people had already come to terms with the concept of insurance as a way to cope with a catastrophic but largely uncontrollable event. The concept of risk pooling was already understood by people in ancient Rome and medieval times. A form of insurance contract, e.g. death insurance or housefire insurance, existed in ancient times.

A formal expression of risk pooling and insurance first appeared in the seventeenth century based on the widespread understanding, by then, of the concept of probability. Shiller quotes an anonymous letter sent to Count Oldenburg written in 1609 which describes a housefire insurance as follows (Shiller 2004):

> Why don't we start a fund, in which people pay 1% of the value of their home every year into the fund, and then we will use the fund to replace the house if there's a fire? ... no doubt that it would be fully proved, if a calculation were made of the number of houses consumed by fire within a certain space in the course of 30 years, that the loss would not amount, by a good deal, to the sum that would be collected in that time.

This idea is more formally expressed today. Let X_i be the random variable of a housefire which takes on either zero or one. Let P_i be the probability of the housefire event occurring. This is known as the Bernoulli trial in probability theory. The mean of this Bernoulli distribution is P_i and the variance is $P_i(1 - P_i)$ (Casella and Berger 2001).

When this Bernoulli trial is repeated independently across a large number of similar households (n), then the sample average $\overline{X_n} = ((X_1 + \ldots + X_n)/n)$ converges in probability to P_i as n becomes large. This is the law of large numbers in probability. Then, the number of housefire converges in probability to nP_i.

For example, if there are 1 000 000 houses with similar conditions and the probability of housefire is 0.01% for each household each year, the average number of housefires in a year in the city is almost certainly 100. By the law of large numbers, it is extremely rare to have 110 housefires in the city in a year. This is the fundamental concept of risk pooling in insurance (Cutler and Zeckhauser 2004).

Given the almost certain number of housefires that is expected, an insurance company can offer a housefire insurance contract to all households in the city. The insured will pay a premium and, in exchange, the insurer will pay a claim should a specified contingency occur such as a housefire. The housefire insurance improves the welfare of both the insureds and the insurer, and, thereby, the Pareto efficiency of the society (Cutler and Zeckhauser 2004).

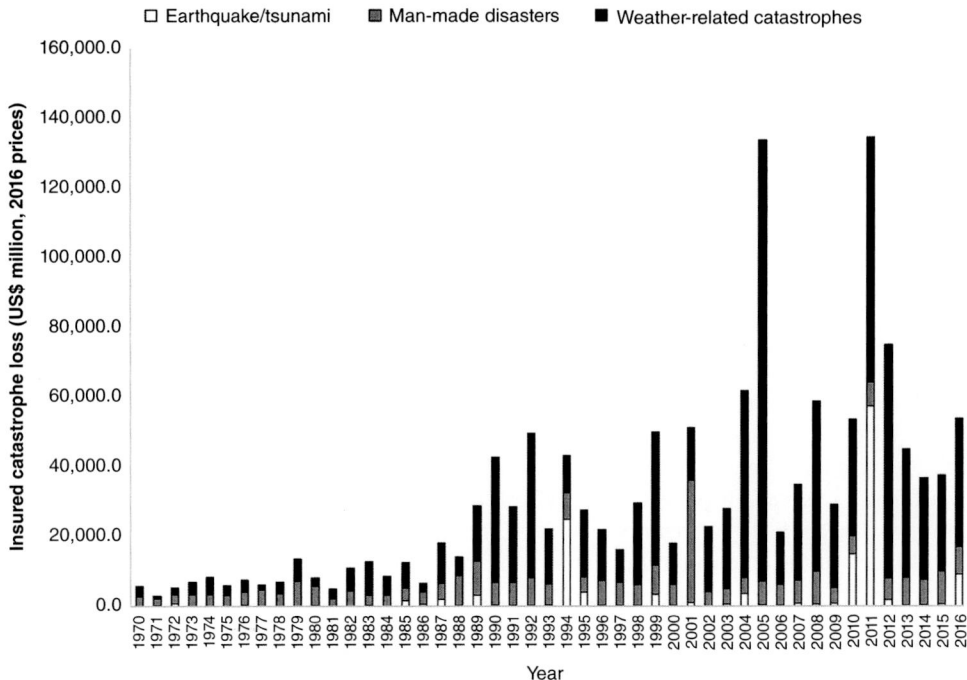

Figure 4.3 Annual insured catastrophe losses, globally. *Source*: Swiss Re Institute (2017).

Globally, the number of insurances offered and the magnitude of insured losses for natural and man-made catastrophic events have increased markedly since 1970. Figure 4.3 shows the magnitudes of annual insured catastrophe losses since 1970 from data provided by Swiss Re (Swiss Re Capital Markets 2017). In the early 1970s, the magnitude of insured losses amounted to several billion US dollars. By the 2010s, it amounted to the range of US$40 billion to US$140 billion. In the figure, all dollars are in 2016 prices.

In Figure 4.3, the amount of insured losses is broken down into three types: earthquake/tsunami, man-made, and weather-related. Of the three types of insured losses, the magnitude of insured losses from man-made disasters shows a steady increase over this time period, although at a much smaller growth rate than that from weather-related catastrophes.

The magnitude of insured losses from earthquakes/tsunamis exhibits a highly volatile pattern over this period. The magnitude of annual insured losses from earthquakes/tsunamis is small across the time period, with exceptions in four years: 2011, 1994, 2010, and 2016.

The insured losses from weather-related catastrophes reveal both of these patterns: a steady long-term growth and a highly volatile fluctuation. The increase in insured losses from weather-related catastrophes is remarkable. At the same time, there were long-tail events. The magnitude of annual insured losses amounted to over US$100 billion in 2005, the year of Hurricane Katrina, and US$70 billion in 2011, the year of Hurricane Sandy.

Table 4.1 Insured losses from catastrophes by world region in 2016.

Region	Number of catastrophes (n)	Victims (n)	Insured loss (US$ billion)	Total loss (US$ billion)	Insured (%)
North America	66	1 005	30.4	59.5	51.1
Latin America and Caribbean	22	1 009	1.4	6.4	21.9
Europe	51	1 509	7.5	15.5	48.4
Africa	44	1 761	1.7	3	56.7
Asia	128	5 309	8.8	83	10.6
Oceania/Australia	7	52	3.4	6.4	53.1
Seas/Space	9	253	0.5	0.8	62.5
World	327	10 898	54	175	30.9

Source: Swiss Re Institute (2017).

The insured losses account for about 30% of the total economic loss from catastrophic events, globally. As shown in Table 4.1, the total losses amounted to US$175 billion, of which US$54 billion was insured losses in 2016. Of the total losses, Asia accounted for US$83 billion and North America accounted for about US$60 billion (Swiss Re Institute 2017).

There is a stark difference between Asia and North America with regard to the percentage of insured losses. In North America, 51% of the total losses was insured. On the other hand, only 10% of the total losses in Asia was insured losses. Notably, almost 57% of the total losses from catastrophic events was insured losses in Africa, owing to foreign aids and international organizations.

To further explain a detailed structure that underlies an insurance contract for various catastrophe events, the author will begin here by describing a crop insurance and subsidy program in the US and later in this chapter come back to a hurricane-related insurance as well as catastrophe bonds for, e.g. earthquakes, cyclones, and tsunamis.

Agricultural insurance policies have played an important role in managing weather-related disasters and catastrophes, such as a severe drought, in agriculture in many world regions. In the US, the federal crop insurance program is a multiperil insurance policy and is the most widely purchased insurance policy for insuring against natural disasters and catastrophes in the country (Barry et al. 2000; Sumner and Zulauf 2012).

The modern crop insurance program in the US dates back to the Crop Insurance Reform Act of 1994 and the creation of the Risk Management Agency (RMA) in the following year (Sumner and Zulauf 2012). The RMA was created to administer the Federal Crop Insurance Corporation (FCIC) established in 1938 as a political response to severe droughts and related natural disasters, including the Dust Bowl during the 1930s (Warrick et al. 1975).

The federal crop insurance program in the US has grown remarkably since the early 1990s. As summarized in Table 4.2, the number of acres insured by the federal program grew from 101 million acres in 1990 to 295 million acres in 2013, which amounted to 83%

Table 4.2 Growth of the US federal crop insurance program.

Year	Total policies sold (million)	Buy-up policies sold (million)	Net acres insured (million)	Total liability (US$ billion)	Total premium (US$ billion)	Total indemnity (US$ billion)
1990	0.89	0.89	101.4	12.83	0.84	0.97
1995	2.03	0.86	220.5	23.73	1.54	1.57
2000	1.32	1.01	206.5	34.44	2.54	2.59
2005	1.19	1.05	245.9	44.26	3.95	2.37
2006	1.15	1.03	242.2	49.92	4.58	3.5
2007	1.14	1.02	271.6	67.34	6.56	3.55
2008	1.15	1.03	272.3	89.9	9.85	8.68
2009	1.17	1.08	264.8	79.57	8.95	5.23
2010	1.14	1.06	256.2	78.1	7.59	4.25
2011	1.15	1.07	265.3	114.07	11.95	10.71
2012			283	117.15	11.11	
2013			296	117.12	11.74	

Source: Sumner and Zulauf (2012).

of the total US crop acreage. The magnitude of total premium has increased by about 10-fold during this time period, and so has that of total indemnity or that of liability. In 2011, the total premium amounted to US$12 billion and the total liability to US$114 billion (Sumner and Zulauf 2012).

The number of crops separately insured by the federal program has increased from about 50 crops to over 300 crops during this period. Four major crops produced in the US – corn, cotton, soybeans, and wheat – account for more than two-thirds of the insured acres. Other than the major crops, insured items include a variety of fruit trees, nursery crops, pasture, rangeland, forage, and livestock (Shields 2013).

The US crop insurance is a multiperil insurance, meaning that a single insurance policy insures a policy-holder against a multitude of natural perils that may trigger low yields or low revenue, including severe drought, heavy and intense rainfall, heat stress, frost, fire, earthquakes, insects, and volcanic eruption (Sumner and Zulauf 2012; Wright 2014).

The crop insurance policies in the US are sold and serviced by 16 private insurance companies. A producer can purchase a crop insurance policy from one of these private companies and should pay premiums. The premiums are subsidized by the federal government. In the catastrophic coverage to be explained shortly, 100% of the premiums is subsidized. As the coverage level rises through a purchase of buy-up policies, the insurance premium increases and the federal government subsidy rate declines (Shields 2013).

A crop insurance policy purchased from one of the insurance companies guarantees the insured an indemnity payment when a specified contingency occurs, after subtracting insurance deductible. A policy can be written based on either loss of yields or loss of revenues. As of 2013, revenue-based policies accounted for 75% and yield-based policies for 25% of total active policies (Sumner and Zulauf 2012).

As mentioned above, a crop insurance policy with catastrophic coverage gives the insurance holder a federal subsidy that is equal to 100% of the premium. The catastrophic coverage has the following structure. In a yield-based insurance with catastrophic coverage, a producer is insured against crop losses in excess of 50% of her/his "normal" yield at 55% of the market price of the crop when the crop loss exceeds the selected loss threshold. It is thus called a 50/55 coverage.

In the crop insurance policy, the "normal" yield of the crop is specified based on the producer's actual production history. The producer is required to submit to the United States Department of Agriculture (USDA) actual annual crop yields for the last 4–10 years, from which a normal yield is calculated as the average of the annual yields. In addition, the market price for the crop is specified based on estimated market conditions (Shields 2013).

The catastrophic (CAT) coverage is the most basic policy. The coverage levels that are higher than the CAT coverage level (the 50/55 coverage), called a buy-up coverage, can be purchased. As shown in Table 4.2, nearly all farmers purchased buy-up policies even at larger premium payments because of additional protection provided.

A producer can buy-up the CAT coverage to any coverage level between 50/100 and 75/100. In the 75/100 coverage, the producer is insured against the loss in excess of 75% of the normal yield at 100% of the estimated market price of the crop. In limited areas, the 85/100 coverage is also available.

For buy-up coverages, a portion of the insurance premium should be paid by the policy-holder, while the rest is subsidized by the federal government. The higher the coverage level, the higher the insurance premium of the policy and the lower the subsidy rate by the federal government. The subsidy rate for the 60% coverage insurance is 64%, that for the 75% coverage insurance is 55%, and that for the 85% coverage is 38% (Shields 2013).

Besides the policies based on the actual production history of an individual producer which account for more than 90% of yield-based policies, there are other types of policies. An area-wide index is one such policy. In the area-wide index policy, an indemnity payment is triggered by an area-wide, e.g. the US county level, yield shortfall instead of an individual producer's yield shortfall (Halcrow 1949).

The theory behind an area-wide crop insurance is that although an individual producer's yield of a crop is controlled to a large extent by the producer's decisions, an area-wide county-level yield of the grain is by and large outside the control of the individual producer. Halcrow and other researchers suggested that with an area-wide-index-based insurance policy, a producer's purchase of an insurance policy is less motivated by perverse incentives such as adverse selection and the behaviors of the insured producer are less influenced by moral hazards (Skees et al. 1997; Miranda and Glauber 1997; Mahul 1999).

A growing concern with regard to crop insurance is that the cost to the federal government for the program has risen sharply, as shown in Figure 4.4 (USDA 2017). It has grown from US$2 billion in 2000 to US$14 billion in 2012. The steep increase is largely due to the increase in the federal crop subsidy, which has risen from about US$1 billion to about US$7 billion at the end of the period in the figure.

The steep increase in the government cost of the program begs the question: Can the crop insurance program work without the federal government subsidy or with much reduced subsidy rates by the federal government? With regard to the fundamental

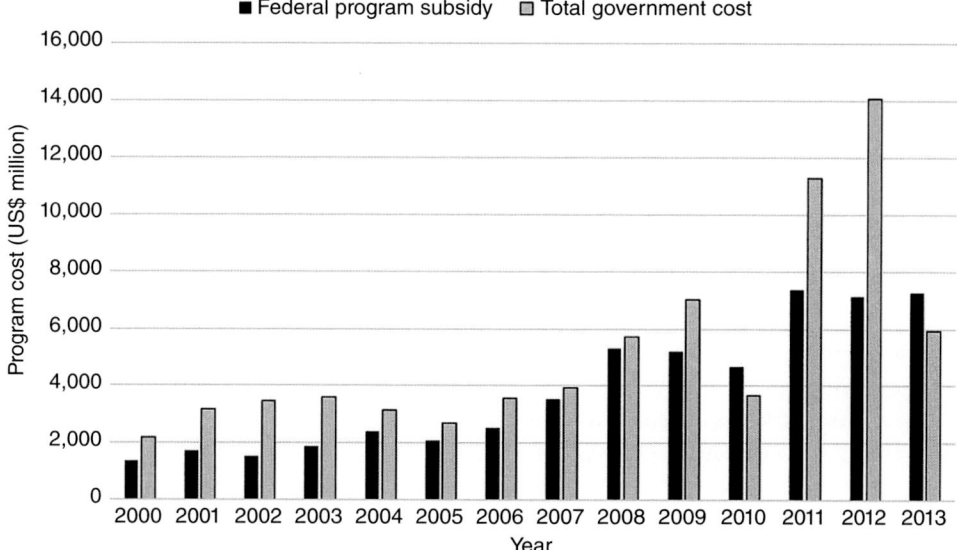

Figure 4.4 The government cost of federal crop insurance. *Source*: USDA (2017).

inquiry of this book which is how a society should devise ways to cope with catastrophe events, the answer to the above question is most likely positive. In other words, a crop insurance program without a federal subsidy can be designed in a way to still be an effective instrument for minimizing the damage from natural weather events, even though it would not provide as much political support for farming communities as the current federal crop insurance program does at present.

4.5 Options for a Catastrophic Event

One of the fundamental aspects of a natural catastrophe, i.e. a long-tail event, is that it shocks a "normal" economy or system. A severe drought, for example, may cause a severe drop in a farm's primary crop yield to the level below the subsistence level. In an agriculture-based economy in Sub-Saharan Africa, such a natural disaster tends to seriously hurt rural communities and ends up causing a large number of human deaths, especially of children and women (World Bank 2008).

Nonetheless, even the most vulnerable farming populations in the world have learned to cope with unanticipated natural disasters, even without the agricultural technological revolutions during the past half century (Evenson and Gollin 2003). They are well informed of and experienced with numerous risks that threaten the farms. They have devised various coping strategies, such as diversification of a farm portfolio, e.g. into a mixed portfolio of crops, animals, and forests, or drought-resilient species of animals, in preparation for natural changes in the climate and weather (Seo and Mendelsohn 2008; Seo 2010a,b, 2012, 2016a). They save good-years' incomes and harvests in preparation for a severe yield shortfall in a bad year which is mostly uncontrollable by them (Udry 1995).

Storage of harvested grains is another form of farm savings. Stored grains by a farm can be used for the farm household's consumption in a severe drought year or sold at a later time of the year when the market supply is low. By storing harvested grains during the harvest period and selling them in low-supply periods caused by, e.g. severe droughts, a farmer can avoid selling them at very low prices due to an oversupply during the harvest period. Storage is, on the other hand, costly due to the need for a building and air control inside the storage building.

A market instrument that formalizes the storage of grains during harvest time and sales at later times is called a futures market, but a forward contract was widely in use in rice-based East Asian countries even in medieval times (Working 1953, 1962). A producer can sell grain at a futures market instead of selling at a spot market immediately upon harvest of the grain. A farm producer sells a futures contract which is purchased by a consumer.

The futures market is possible because a futures contract can benefit both a producer and a consumer, i.e. it is Pareto welfare improving. A producer gains because s/he can avoid selling the grain at lower prices during the time of high supply and a consumer gains because s/he can avoid buying the grain at higher prices during the time of low supply (Barry et al. 2000; Fabozzi et al. 2009).

A futures contract contains the following elements. A seller of the contract agrees to deliver one unit of a specific grain (x) to a buyer of the contract at a specific price (F_x) on a designated future date. Let's say that the contract is for one bushel of maize to be delivered 12 months later.

At what price would the two traders settle for the contract? During the contract period, the following three costs are incurred: storage cost, interest rate, and dividends. The cost of storage is expressed as c_x as a percentage of the spot market price (P_x). The market interest rate (r) is lost by the seller by putting off the sale for one year. The dividend is any income earned from holding the grain during the contract period, d_x, again expressed as a percentage of the spot market price.

With an inter-temporal no arbitrage condition, the futures contract price is settled as follows, assuming no dividends from holding the grain (Fabozzi et al. 2009):

$$F_x = P_x + P_x \cdot (r + c_x) \tag{4.8}$$

Although formalized first in agricultural grain markets, a futures contract is widely used across the financial sector (Shiller 2004). A large number of different types of futures contracts are traded, among other places, at the Chicago Mercantile Exchange (CME Group 2017). Although it serves to help farmers to hedge against adverse natural shocks, a futures contract is most suited to the risk in a market with predictable or cyclical fluctuations, e.g. maize yields.

For the difficult-to-predict events or the events with high uncertainty, an option contract can be relied upon by an individual producer or consumer to directly address the risk of high volatility. An option contract gives the purchaser the right to sell or buy at a specified price of an underlying asset in the event of a catastrophe (Fabozzi et al. 2009).

An option contract that specifies the right of a purchaser for the option to buy an asset is called a call option. An option contract that specifies the right of a purchaser for the option to sell an asset is called a put option.

An option contract is also distinguished by the exercise date. Depending on the times when an option can be exercised, there are two types of an option contract: the American option and the European option. For the American option, the right for the option can be exercised only on the date specified in the contract. For the European option, the right for the option can be exercised on any date before the specified date in the contract.

A natural gas call option is traded in the following manner. If an investor expects with some certainty that there is a natural gas reservoir under a certain land area, s/he may buy a call option for the land. S/he would purchase a call option instead of purchasing the land directly since s/he may decide not to exercise the option at the exercise date if the market price of land stands below the strike price specified in the call option contract. If the market price of the land is higher than the strike price specified in the call option, then s/he will exercise the call option and purchase the land at the strike price which is lower than the market price at the exercise date.

By purchasing the natural gas option at a certain strike price, the investor can gain from the option contract. By applying a similar reasoning, it can be shown that the seller of the option can gain from the option trading. Again, this is to say that the option contract is Pareto welfare improving (Fudenberg and Tirole 1991; Mas-Colell et al. 1995).

What should be the price of the call option, say, for natural gas? The price is determined principally by the size of the risk in the land value, i.e. the volatility in the value of the asset that is traded. Given the current market value of the land and the strike price of the land specified in the option contract, the higher the risk, the higher the price of the call option. The lower the risk, the lower the price of the call option.

Let us look at the Black and Sholes option pricing model, the most basic of all models (Black and Sholes 1973). Let T be the time to exercise the option, σ^2 the variance of one-time price change of the asset, ρ_{rf} the risk-free interest rate, and Φ the standard normal cumulative distribution function (CDF). Let's further denote the current stock price of the asset by P_{stock} and the strike (exercise) price by P_{strike}. Then, the fair market value (price) of the call option, P_{call}, is determined by the following formula (Black and Sholes 1973):

$$P_{call} = P_{stock}\Phi(q_1) - P_{strike}\Phi(q_2). \tag{4.9}$$

Given the strike price and the two normal CDFs, the higher the current stock price, the higher the price of the option. Given the current stock price and the two normal CDFs, the higher the strike price, the lower the price of the option. The two quantities inside the parentheses capture the size of the risk:

$$q_1 = \frac{\ln\left(\frac{P_{stock}}{P_{strike}}\right) + \rho_{rf} T + \sigma^2 \left(\frac{T}{2}\right)}{\sigma\sqrt{T}}, \tag{4.10a}$$

$$q_2 = \frac{\ln\left(\frac{P_{stock}}{P_{strike}}\right) + \rho_{rf} T - \sigma^2 \left(\frac{T}{2}\right)}{\sigma\sqrt{T}}. \tag{4.10b}$$

The higher the risk-free interest rate, the larger the q_1 risk. The longer the time to exercise, the larger the q_1 risk. The larger the variance of one-time price changes, the higher the q_1 risk. The higher the q_1 risk, the higher the price of the call option.

For the q_2 component of the risk, the lower the risk-free interest rate, the lower the q_2 risk. The larger the variance of one-time price changes, the lower the q_2 risk. The lower the q_2 risk, the higher the price of the call option.

A put option contact, mentioned earlier, gives the right to the contract holder to sell an underlying asset at a strike price at or before the exercise date. For example, a property owner may purchase a put option for the sale of the property at the specified value, namely, the strike price. S/he would exercise the put option to minimize the loss from a precipitous fall in the property value if s/he should expect it to happen in the near future, owing to, for example, an economy-wide shock, such as the subprime mortgage crisis in 2007 and 2008 in the US (Shiller 2009).

An investor in a financial or commodity asset is concerned about a price bubble of the asset and a sudden burst of the bubble (Shiller 2005). S/he may purchase a relevant option for the asset in order to minimize the loss of value from the sudden price fall. At present, many types of options are traded in the commodity and financial markets, including crude oil options, natural gas options, corn options, gold options, and S&P 500 options (CME Group 2017).

When applied to natural catastrophic events, a manufacturer of corn-based products may purchase a corn call option in order to minimize the loss from a corn price hike caused by a severe drought in a certain year. When a severe drought or other natural perils were to occur, changes in corn prices and corn products' prices would be reduced because of the trades of corn options, thereby reducing the impacts on the economy and consumers.

An option can be applied to many catastrophe situations where there is no option contract available yet. As a hypothetical situation, consider a land owner who is concerned about the risk of being hit by a large asteroid that can devastate the rural county to which s/he belongs (NRC 2010). S/he may sign a put option contract for the right to sell the land at a strike price at a designated date. An investor would be interested in selling the option because s/he can purchase, if on sale, the land at a lower price if there were to be no asteroid strike. The land owner would be relieved of the fear of asteroid strikes, knowing that s/he can sell the land at the strike price any time before the exercise date.

For another situation, consider a coastal city, e.g. the Fukushima prefecture in Japan that suffered the catastrophic event of the earthquakes–tsunami–nuclear accident in 2011 or a coastal city along the US west coast (WNA 2017). Imagine that there is 80% likelihood that the city is hit by a major earthquake or tsunami in the next 10 years, based on the past history and scientific projections. A house owner in the city may purchase a put option for the right to sell the property at a strike price in three years.

If the forecasted earthquake or tsunami does occur within the three years, the house owner will be able to avoid a near total loss of the house value by exercising the put option at the exercise date. If the forecasted event does not occur, on the other hand, s/he would decide not to exercise the option.

To conclude, the key message of this section is that an option contract offers a promising strategy for the potential victims of a long-tail and fat-tail catastrophe for minimizing a truly catastrophic loss, that is, besides the situations of a host of medium-tail catastrophe events. Nonetheless, an option contract would not be a savior in an extremely fat-tail event, which the author called a humanity-ending or a universe-ending catastrophe while other authors called it the end of all civilizations on the planet.

In the extremely fat-tail humanity-ending catastrophe, there would be practically no demand for any option contract possibly devised because there would be no world left to live in after the catastrophe. More specifically, would someone purchase a put option for the right to sell the property at a strike price on 31 December 2020 and pay the price for the option because s/he suspects that the world will end before that date?

4.6 Catastrophe Bonds

A catastrophe bond, more commonly called a CAT bond, is still a new financial derivative which was developed during the mid-1990s in response to a dramatic failure of traditional insurances in protecting victims as well as disaster insurers against an exceptionally large damage event, specifically, Hurricane Andrew. The CAT bond market grew markedly after the Hurricane Katrina disaster in 2005 (Barrieu and Albertini 2009).

It is one of the insurance-linked securities (ILS) and the best known of them. The CAT bond offers another way to spread the risk of an unprecedented catastrophe, which is not covered by a traditional insurance, across the investors of the bond. As elaborated in this section, the CAT bond is evolving as a financial derivative that is designed to address a tail risk.

In 1992, Hurricane Andrew hit southern Florida, causing the then-largest damage in US history, equivalent to US$25 billion. Because it was a severe tail event at that time, multiple insurance companies went bankrupt as they did not have sufficient funds to pay the indemnity claims. These bankruptcies led insurers and reinsurers to seek a novel way to insure against a low-probability exceptional-damage catastrophic event (WSJ 2016).

The concept of reinsurance lies at the heart of the CAT bonds. Reinsurance refers to an insurance for an insurance contract (Borch 1962). That is, reinsurance is the insurance that is purchased by an insurance company from a reinsurance company to cover the claims for an extraordinary catastrophe event, that is, a long-tail event (WSJ 2016).

More broadly, it is an insurance-linked security. When insurers cannot write a complete insurance contract that specifies both a full list of contingencies and a full bundle of state-contingent payments for various reasons, including incomplete knowledge, high uncertainty, a high degree of heterogeneity among the insureds, or a prohibitively high cost (Froot 1999; Niehaus 2002), the CAT bond can be designed as an insurance-linked security to improve the welfare of the concerned parties by way of offering insurance against a catastrophe event with such above-mentioned situations through a reinsurance vehicle (Lakdawalla and Zanjani 2012).

Let's consider the following insurance company for hurricane events or earthquake events. Assume that the insurer has sufficient reserves to pay the claims up to US$1 billion but would not be able to cover the claims that exceed the threshold of US$1 billion in the realization of a long-tail event. In order to avoid a bankruptcy, the insurance company has the incentive to purchase a reinsurance for the long-tail event.

The insurance company would have a stronger incentive to purchase the CAT bonds if such long-tail events or such exceptionally large-damage events are observed to be occurring more frequently than before or if such events are predicted to occur more frequently in the future. An examination of the US historical hurricane datasets shows that high-damage events may have indeed increased since the late 2000s (Swiss Re Institute 2017). For example, total payments paid to policy-holders of the National Flood Insurance Program (NFIP) after Hurricane Katrina in 2005 amounted to US$17 billion, while Superstorm Sandy in 2012 and Hurricane Harvey, Irma, and Maria in 2017 caused exceptionally large indemnity payments (King 2013). The Hurricane Harvey Relief Bill alone in 2017 amounted to an aid package of US$15.3 billion (Kaplan 2017). On top of this, many hurricane scientists project that tropical cyclones will get more intense and intense hurricanes will strike more frequently due to a warming planet by the latter half of this century (Emanuel 2005, 2008, 2013; Kossin et al. 2013).

Another fundamental concept of CAT bonds is a trigger point. The definition of the trigger point cannot be left out in the CAT bond contract. It is the threshold point based upon which a legal judgment is made on whether a covered catastrophe has occurred or not. There are more than four methods presently used to define the trigger point, which will be explained shortly (Barrieu and Albertini 2009).

Without the trigger point crossed, the CAT bond works like a commercial bond between the investors and the issuer of the bond (Fabozzi et al. 2009). If the defined catastrophe occurs before the maturity date of the bond, i.e. the trigger point is crossed, then the CAT bond defaults, in which case a portion of the principal paid by the investors will begin to be drawn to pay the claims, and eventually all of the principal will be gradually used up for covering the indemnities.

As explained above, the issuer of the CAT bond – the reinsurer – sells the bond to cover a slice of the total risk involved in a natural or man-made event. For example, it can be designed to cover the indemnities exceeding US$1 billion but below US$1.5 billion from, say, a tropical cyclone event. In this example, the size of the CAT bond is US$500 million.

In this example, US$1 billion is the trigger point, also called the attachment point, while US$1.5 billion is called the exhaustion point. At the attachment point, a fraction of the principal must begin to be attached to cover indemnity claims, and all of the principal is exhausted to cover the claims at the exhaustion point. Beyond the exhaustion point, investors and the bond issuer bear no further liabilities.

To explain further, the author needs to elaborate the features of the mechanics of the CAT bond. First, the basic transaction of the CAT bond is as follows, which is similar to that of a commercial bond: an investor purchases the CAT bond from a bond issuer, through a Special Purpose Vehicle (SPV) which is licensed to sell the CAT bonds, with a payment of the principal equal to the face value of the bond. The investor receives periodic interest payments, usually quarterly, from the bond issuer. The maturity of the CAT bond is typically three years but can be in the range of one to five years (Edesess 2014).

Second, the issuer of the CAT bond is usually a reinsurer, but other entities have begun to issue them, such as an insurer, a government entity, pension funds, a corporation, and a nongovernmental organization (WSJ 2016). Mexico is the only national sovereign who issued the CAT bond in 2006 to hedge against earthquake risks in Mexico City. The World Bank issued the first CAT bond in 2014 linked to natural disaster risks in 16 Caribbean countries (World Bank 2014). The value of outstanding CAT bonds has increased steeply since the late 2000s from around US$20 billion in 2007 to over US$70 billion in 2015.

Third, the sponsor of the CAT bond is the insurance company, that is, the hurricane insurance company in our example. It is also called a cedent. It pays premiums in return for the coverage of the claims in the event of a defined catastrophe.

Fourth, who are the investors in CAT bonds? The CAT bond investors are principally institutional investors such as pension funds, endowment funds, and hedge funds. When it comes to a geographical decomposition of the investors, 59% of investors were from the US, 25% from Europe, and 11% from Bermuda by the end of 2014 (Swiss Re Capital Markets 2017).

The fifth detail of the mechanics of the CAT bond is where the principal should be invested. The principal paid for the CAT bond by the investors is deposited in very safe securities, usually a US Treasury money market fund. In this way, the funds are kept in safe reserves which can be relied upon to cover the claims in the event that a specified catastrophe is triggered (Barrieu and Albertini 2009).

However, the practice is not legally required. Before the mid-2000s, for example, CAT bond issuers entered into an interest rate swap with an investment bank counterparty in order to offer CAT bond investors a floating interest rate plus a fixed premium. In 2008, four CAT bond issuers defaulted because one of the investment bank counterparties, Lehman Brothers, went into bankruptcy. Since then, the CAT bond principal is invested in very safe assets (Barrieu and Albertini 2009; Edesess 2014).

Sixth, another complex aspect in designing a CAT bond is, as noted previously, how to define a trigger point. There are four basic types of trigger: the indemnity trigger, industry loss trigger, parametric trigger, and modeled trigger (Barrieu and Albertini 2009). By the end of 2013, 43% of CAT bonds relied on the indemnity trigger and 38% on the industry loss trigger. Only 11% of CAT bonds were based on the parametric trigger (Swiss Re Capital Markets 2017).

In the indemnity trigger, the attachment point and the exhaustion point are defined by actual claims or losses of the CAT bond issuer. Continuing with the above example of a US$1 billion attachment point and US$1.5 billion exhaustion point, the attachment point of US$1 billion is reached when the claims that must be paid to the claimants by the issuer pass US$1 billion in total, at which point the loss of the principal is triggered. The exhaustion point is defined in the same way by the actual claims made in total.

By contrast, the industry loss trigger relies on the index estimates of industry losses, i.e. not on an individual issuer's losses, on the insured event, after the event has occurred. The estimates of the index are made by an independent third-party service, e.g. the PCS (Property Claims Service) in the US.

The parametric trigger is defined by the value of a predefined parameter in the legal judgment of a trigger point. For instance, the trigger point may be set at 120 km h^{-1} in MWS, or a category 5 hurricane in the Saffir–Simpson hurricane scale, or an earthquake exceeding a magnitude of 7.0 in the earthquake moment scale. The modeled trigger, like the indemnity trigger, is based on the indemnity claims, but the claims are estimated or projected by an independent modeling company (Edesess 2014).

With the fundamental concepts, major elements, and the mechanics of the CAT bond grasped, readers might wonder what makes the CAT bond more attractive than other market securities to the investors and the issuers, which is a fundamental question regarding the future success of the financial derivative.

Let's first consider it from the standpoint of the investors. The primary attractiveness to the investors is that the risk of the CAT bond is largely uncorrelated with the other risks that investors must take into account in investing in other assets: equity (stock) market risk, credit risk, interest rate risk, exchange rate risk, government policy change risk, etc. In plain words, an occurrence of a natural catastrophe trigger point is almost uncorrelated with the risks due to abrupt changes in the economy and government policies such as stock market shocks, interest rate changes, debts, and foreign trades.

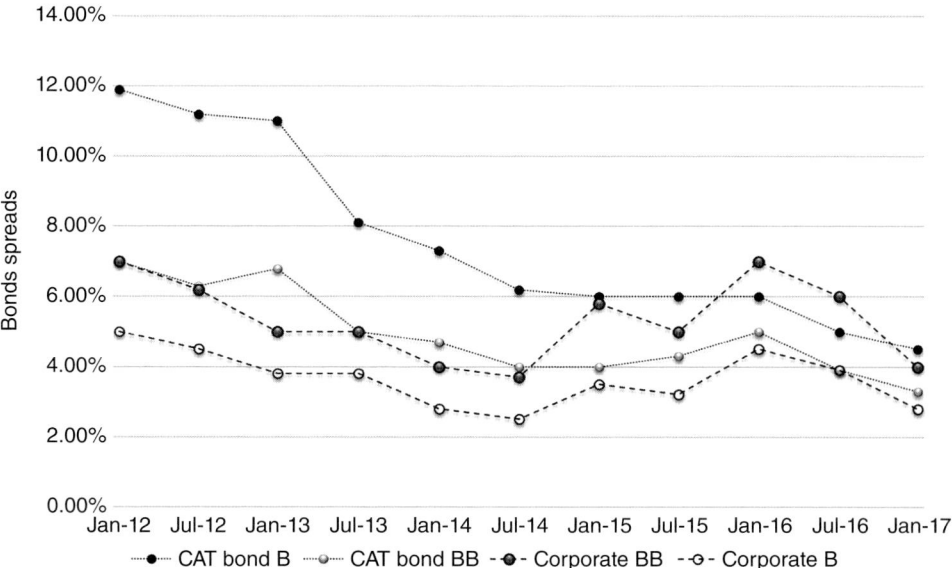

Figure 4.5 Spreads for CAT bonds versus high-yield corporate bonds. *Source*: Swiss Re Institute (2017).

From the practical viewpoint, however, the major attraction of the CAT bond to the investors has been the high interest rates earned from the principal investment. CAT bond issuers have offered to the investors the base interest rate on a US monetary market fund plus a fixed premium. The high interest rates earned by investors are captured by the historical CAT bond spreads (Figure 4.5).

The CAT bond spread is the interest rate of the CAT bond in excess of the interest rate earned by the risk-free US Treasury bond. In Figure 4.5, spreads of the two CAT bonds with different ratings, B and BB, are shown for the period from 2012 to 2017 (Swiss Re Capital Markets 2017). By January 2012, the bond spread for the CAT bond with the B rating was 12% and that for the CAT bond with the BB rating was 7%. By January 2017, the bond spread for the former fell to about 5% and that for the latter fell to about 3.2%.

The figure also shows the spreads of comparably rated high-yield corporate bonds. The BofA Merrill Lynch High Yield Option Adjusted Spread is drawn for the B and BB ratings. By January 2012, the corporate B bond spread was 5% and the corporate BB bond spread was 7%. By January 2017, the former fell to about 2.8% and the latter fell to about 4%. Although the difference between the CAT bond spread and the corporate bond spread has fallen during this specific time period, the CAT bond B still earned a higher interest by 2% than the high-yield corporate bonds with the same rating.

For the investors of CAT bonds, there is the risk of a credit cliff. That is, investors can lose the entire principal rapidly when a defined catastrophe is triggered. Computer modeling plays a crucial role in assessing the risk in CAT bonds as well as in assessing pricing, yields, and ratings of CAT bonds. The results of computer modeling are highly sensitive to the data used and other factors (Barrieu and Albertini 2009).

Next, consider the perspective of the CAT bond issuers. To the CAT bond issuers, why are CAT bonds attractive? First, issuing CAT bonds reduces the issuer's reserve requirement and it also increases the issuer's insurance protection. Second, as was the case for investors, there is negligible credit risk in the issue of CAT bonds. Further, the issuer can design a CAT bond for various tranches of the total risk involved in a natural or man-made catastrophe event (Edesess 2014).

Would the CAT bond become a financial derivative of choice for the range of catastrophe events surveyed in this book? Up to this point, its application has been rather limited to US hurricanes and earthquakes, European wind, Japanese earthquakes, and typhoons. Securitizations of other catastrophic risks are possible, e.g. terrorism risk. However, the success of such a securitization would depend on whether people have faith in the computer models of the catastrophe concerned.

As of December 2016, the distribution of outstanding CAT bonds – that is, bonds issued but not matured or redeemed – across the range of perils is shown in Figure 4.6. The vast majority of the outstanding CAT bonds are issued for peak perils, i.e. natural catastrophes or low-probability high-impact events, such as US wind and US earthquakes. About 62% of them are issued for US wind – in the events of hurricanes and tornadoes – and about 51% for US earthquakes (Swiss Re Capital Markets 2017). For US earthquakes, 40% of the outstanding CAT bonds are held for California earthquakes.

In summary, the CAT bond is another financial derivative that can be relied upon for dealing with a fat-tail catastrophe, like catastrophe options explained in Section 4.5. A CAT bond can be issued for a tranche of the risk that lies in a long tail of the distribution. Unlike the catastrophe options, will there be sufficient demands for the CAT bonds designed for an extremely fat-tail event?

More specifically, can the CAT bond be issued for the trigger point of US$5 trillion, about 10% of the world's gross domestic product (GDP), and the exhaustion point of

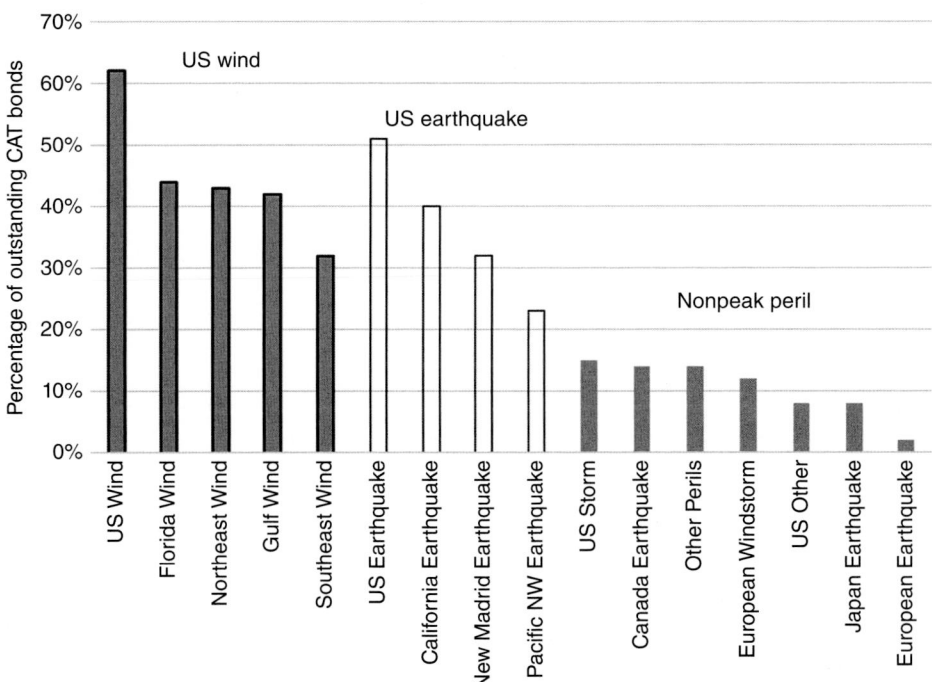

Figure 4.6 Outstanding CAT bonds by peril (as of December 2016). *Source*: Swiss Re Institute (2017).

US$10 trillion, about 20% of the world GDP, for a natural catastrophe event, say, an asteroid strike? It would certainly be difficult for a single reinsurance company to issue the asteroid CAT bond of such a large size. However, it may become of interest to an international organization such as the United Nations to issue the asteroid bond.

The demand for the asteroid bond can be high for several reasons. First, confidence in a computer modeling of an asteroid risk is high, which leads potential investors to assess the risk reliably. Second, given the extremely low-probability of an actual global-scale asteroid collision, the asteroid bond is almost like a commercial bond, only with a higher interest rate.

However, the incentive to issue the asteroid bond by, for example, the United Nations, would be very weak if the organization is certain that there will be negligible risk of a large-size asteroid hitting Earth. The CAT bond issuer would find it difficult to find the sponsors/cedents of the asteroid bond.

Let's consider this time a global warming CAT bond. There is a clear trigger point in the climate literature, including the 2° C increase in global average temperature. Given the high concern at the United Nations Framework Convention on Climate Change (UNFCCC), it may have strong incentive to issue the global warming CAT bond (UNFCCC 2015).

The demand is expected to be high, if we can trust the zeal expressed by the parties of the conference in the UNFCCC meetings, as well as the public concern on the 2° C ceiling which is evidently demonstrated by climate activists. In particular, the size of the adaptation fund called the Green Climate Fund (GCF) is expected to be US$100 billion per annum (UNFCCC 2010). Further, there is high confidence placed on the climate prediction models by many people, including the climate activists.

Considering these examples, we can conclude that the CAT bond as a financial derivative may be further developed, in tandem with the scientific modeling advances in the future, to be an alternative strategy for some of the catastrophes surveyed in this book.

4.7 Pareto Optimality in Policy Interventions

Should the government or an international organization intervene to minimize and manage the risk posed by a catastrophe, especially global-scale catastrophes? Through Sections 4.4–4.6, the author described market and financial instruments that have been devised to help individuals and communities deal with the risks of catastrophic events such as insurance, futures, options, and CAT bonds.

In addition to the market-institutional responses, behavioral responses of individuals who are potentially affected increase the resilience and decrease the risk from a catastrophic event. Buying an earthquake-resistant house is one way an individual can cope with a catastrophic earthquake event. At times of a hurricane strike, an individual can reduce the damage by taking simple actions such as staying indoors, staying in a cellar, or sealing up the doors and windows (Seo 2017a,b).

Beyond the institutional and behavioral responses, historical technological developments have played a major role in reducing the damage from catastrophic events, both natural and man-made catastrophes but especially for the former. In the case of hurricane risks, advances in hurricane path projection technology as well as in satellite monitoring of tropical cyclones have made remarkable strides during the past half century to become instrumental strategies for reducing drastically the number of cyclone fatalities (Seo 2015a; Seo and Bakkensen 2017).

Similarly, for the earthquake catastrophes, an enhanced earthquake risk mapping and zoning as well as advanced technologies to build earthquake-resistant buildings may have contributed greatly to reducing the number of fatalities from a severe earthquake event (NRC 1997). For example, one of the strongest earthquakes ever recorded in human history that triggered a tsunami and a nuclear meltdown in the northeast of Japan in 2011 resulted in a handful of human fatalities, while an earthquake in Haiti in 2010 killed several hundred thousand people owing to fragile houses and buildings (WNA 2017).

Are financial securities, behavioral responses, and technological capacities sufficient as responses to catastrophic events? In other words, are they Pareto optimal responses or the best social responses in the face of global challenges (von Neumann and Morgenstern 1947; Mas-Colell et al. 1995)? The answer may be "yes" in many catastrophe situations, including tropical cyclones and earthquakes, but the answer is most likely "no" in many other situations.

The author summarized in the previous sections that an extremely fat-tail event cannot be addressed optimally through financial securities and derivatives alone. More broadly, these situations in the "no" category involve the goods and services that are publicly (jointly) owned or consumed (Samuelson 1954, 1955). As noted in the CAT bonds section, a global-scale catastrophe cannot be handled efficiently by the issue of CAT bonds owing to the characteristics of the global-scale catastrophe as a global public good, more precisely in this case, a global public bad.

A public good is a good that is jointly consumed by the public, for which reason it was originally called a jointly consumed commodity by Samuelson (1954). A classic example is national defense. Once it is provided, every citizen of the nation benefits from the provision of the good. It has the characteristic of being nonrivalrous in consumption. That is, one person's consumption does not decrease the amount available for the others' consumption. Further, it has the characteristic of being nonexcludable. That is, it is not possible or extremely expensive to exclude someone from the consumption of the public good once provided (Mas-Colell et al. 1995).

A free market economy fails to provide the public good efficiently, that is, in a Pareto efficient manner (Samuelson 1954). For a public good, owing to the aforementioned two characteristics, individuals have an incentive to free ride. That is, an individual will wait for the others to pay for the provision of the public good and enjoy the good once it is provided without paying one's share of the cost once it is provided (Buchanan 1968).

The optimal – Pareto efficient – provision of the public good must be brought about by government interventions through taxation and public expenditure (Samuelson and Nordhaus 2009). In the case of national defense, the government should tax the citizens and provide the national defense system through government spending. Without such interventions, private security companies will not succeed in providing national defense due primarily to the salient characteristics of the national defense system, i.e. nonrival and nonexcludable.

A similar conclusion can be drawn for the financial securities, behavioral responses, and technological advances, discussed hitherto, as a response to catastrophic events. When the good that must be provided to minimize the damage of catastrophic events, for example, the purchase of CAT bonds discussed above, has the characteristics of the public good, it will not be optimally provided in the *laissez faire* economy.

Of the categories of public goods, a global public good is one in which the publicness of the good is extended to the scale of the entire globe (Nordhaus 2006). A global public good is consumed jointly by the "citizens" of the Earth. Since there is no global government like a national government, however, there is no political entity that can fix the incentive problems existent in the provision of a global public good.

In the context of the catastrophic risks surveyed in this book, ozone depletion in the stratosphere is a case of a global public good. The stratospheric ozone layer protects humans from ultraviolet B (UVB) radiation from the Sun which causes skin cancer and other fatal diseases. The ozone layer has been depleted by the releases into the atmosphere of cooling agents such as chlorofluorocarbons (CFCs) in air conditioners and refrigerators (Molina and Rowland 1974).

Since the sky or atmosphere is freely available to any individual or country, emissions of CFCs are not abated because it is costly to do so. No individual or country is willing to provide the abatement of the emissions of the ozone-depleting substances, a global public good. Even if a single country were to ban CFCs as cooling agents, it won't stop the depletion of the ozone layer in the world.

A global policy response to the ozone-depleting chemicals has been swift. The Montreal Protocol on Substances that Deplete the Ozone Layer was signed in 1987 as an international treaty, through which countries agreed to phase out CFCs, the main chemical that depletes the ozone layer (UNEP 1987). Subsequently, replacement coolants such as hydrofluorocarbons (HFCs) and hydrochlorofluorocarbons (HCFCs) were added to the chemicals to be phased out (UNEP 2016).

Another global public good that has received even more intense global policy attention is global warming. Releases of greenhouse gases from numerous anthropogenic activities form a greenhouse-like blanket in the Earth's atmosphere, thereby increasing the Earth's temperature. A globally warmed world would disrupt natural ecosystems as well as anthropogenic economic activities (IPCC 1990).

Since the atmosphere is freely available to any individual or country, no individual or country is willing to reduce the emissions of Earth-heating gases such as carbon dioxide at the expense of the country. Even if a single country were to eliminate the emissions of greenhouse gases, it would not make the slightest dent to the Earth's temperature (Nordhaus 1994). Therefore, an effective global policy intervention should make all countries shoulder responsibilities in abating greenhouse gas emissions, which is what the Paris Agreements attempted to achieve (UNFCCC 2015).

A Pareto efficient policy response at the global level should take the form of actions that are harmonized across the globe (Nordhaus 2008). It should aim to achieve the maximization of global welfare, taking into considerations long-term realizations of global warming and the impacts of such changes. However, high uncertainties with regard to many aspects of global warming and policy-making may hinder an analytical approach to a Pareto optimal global warming policy.

In the following, the author describes a generalized framework of a global policy intervention in response to a global catastrophic event. The description is based on the literature of global warming policy elucidated by William Nordhaus, but any global catastrophe policy based on the concept of Pareto optimality would have a similar framework and features (Koopmans 1965; Nordhaus 2008).

An optimal global warming policy is designed in a way that maximizes the global social welfare function, W, which is the discounted sum of population-weighted ($L(t)$) per capita consumption, $c(t)$ (Nordhaus 2013):

$$W = \sum_{t=1}^{T_{max}} U[c(t), L(t)]R(t). \quad (4.11)$$

$R(t)$ is the discount factor which is defined by the pure rate of time preference, ρ, which assigns weights to the utilities of different future generations (Arrow et al. 1996b; Nordhaus 2007):

$$R(t) = (1 + \rho)^{-t}. \quad (4.12)$$

The social utility at one point in time is represented by a constant elasticity utility function times the size of population:

$$U[c(t), L(t)] = L(t)\frac{c(t)^{1-\alpha}}{1-\alpha}. \quad (4.13)$$

The α is the constant elasticity of the marginal utility of consumption. It is interpreted as an aversion parameter to intergenerational inequality. It represents diminishing social valuations of consumptions of different generations. If this parameter is close to zero, consumptions of different generations are close substitutes, which reflects low aversion to intergenerational inequality. If this parameter is large, consumptions of different generations are not close substitutes, which reflects high aversion to intergenerational inequality (Nordhaus 1992).

With changes in global temperature over the twenty-first century, consumptions of different future generations will be affected. Let the global damage function be written as a function of global atmospheric average temperature, T_{at}, in a percentage loss of global production or consumption. For example, a quadratic functional form of the global warming damage function can be estimated from empirical research (Mendelsohn et al. 2006; Mendelsohn and Williams 2007; Tol 2009):

$$\Omega_t = d_1 T_{at}(t) + d_2 T_{at}(t)^2. \quad (4.14)$$

The changes in global temperature are determined by the physical interactive relationships between the amount of emissions of greenhouse gases and climate variables, given the climate system of the planet. This can be captured by a physical model such as a three-reservoir model of atmosphere, ocean surface, and deep ocean (Nordhaus and Boyer 2000).

Given the physical relationships, the global temperature can be controlled by altering the rate of control of greenhouse gases. A high rate of control of greenhouse gases will lower the emissions of them into the atmosphere, which will slow down the rate of change in global average temperature. The lower the rate of control, the smaller the impact of the control policy on the rate of global warming.

With no abatement costs incurred, the global community would set the control rate of emissions as high as possible, e.g. a 100% control rate, because the higher the rate of control, the smaller the damage from global warming, taking as given the temperature-damage relationship in Eq. (4.14). However, controlling emissions, i.e. cutting greenhouse gas emissions from numerous economic activities, is costly.

The abatement cost function should be estimated through either an examination of engineering details of greenhouse gas abatements or a statistical application to relevant empirical data, which will yield a nonlinear functional relationship (Nordhaus 1992).

With $\mu(t)$ a rate of control, the abatement cost function may take the following nonlinear form:

$$\Lambda(t) = v_t \mu(t)^\eta. \tag{4.15}$$

A Pareto optimal global warming policy entails that the cost of abatement incurred is equated to the global warming damage avoided for one ton of carbon (tC)-equivalent emissions at the equilibrium. In the optimal policy, the benefit and the cost of abating one ton of emissions are weighed against each other and balanced.

A unique feature in the benefit-cost analysis of global warming policies is that the damage from global warming unfolds over a long period of time in the future, i.e. a century or multiple centuries, while the cost of abating one ton of emissions today is incurred at the time of abatement. Let's assume that one tC-equivalent greenhouse gases emitted today causes a stream of dollar damages of Ω_t' annually and that abatement of one ton of greenhouse gases incurs a one-time cost Λ_t'.

In the simplest case in which market interest rate is γ_t and the constant decay rate of emissions δ, the optimality condition can be written as follows in an intuitive form (Falk and Mendelsohn 1993):

$$\frac{\Omega_t'}{\gamma_t + \delta} = \Lambda_t'. \tag{4.16}$$

A fundamental benefit-cost principle for global warming policies expressed in Eq. (4.16) gives the trajectory of control rates of emissions over the next century and beyond, which also gives the trajectory of carbon (tax) prices to be set by the global policy-maker for the entire time period over which global warming unfolds (Nordhaus 1994).

A Pareto-efficient trajectory of carbon price through the end of the twenty-first century is shown in Figure 4.7 (Seo 2017a). A Pareto optimal intervention calls for the imposition of carbon tax at 35$/ tC in 2015, which rises to 145$/tC in 2055 and further to 360$/tC in 2105, all expressed in 2005 US dollars.

A Pareto optimal policy prescription for the global warming catastrophe is so simple and yet very powerful at the same time. The simplicity of the carbon tax trajectory, however, belies the complexity of physical and economic changes that support the trajectory. There is a great deal of uncertainty regarding physical changes in the climate system, changes in ecosystems, and changes in human systems.

Many aspects are uncertain in the global warming policy model. The array of uncertain components can be summarized statistically by a standard error in the joint system of climate changes, ecosystem changes, and anthropogenic changes (Nordhaus 2008). The simulations of the above-described global climate policy model incorporating the uncertainty estimates would yield a range of estimates of carbon tax. Using the standard error in the carbon tax simulations, the upper bound and the lower bound of the carbon tax trajectory can be overlaid to the mean trajectory, as shown in Figure 4.7. The range of the optimal carbon tax in each time period provides decision-makers with critical information on policy uncertainty.

What is the size of the uncertainty in global warming policy-making? The calculation of the standard error in an optimal carbon tax for a time period is not simple because many distinct but interactive processes are involved in the global warming policy framework (Nordhaus 2008). That is, there are uncertainties in the climate system, ecological

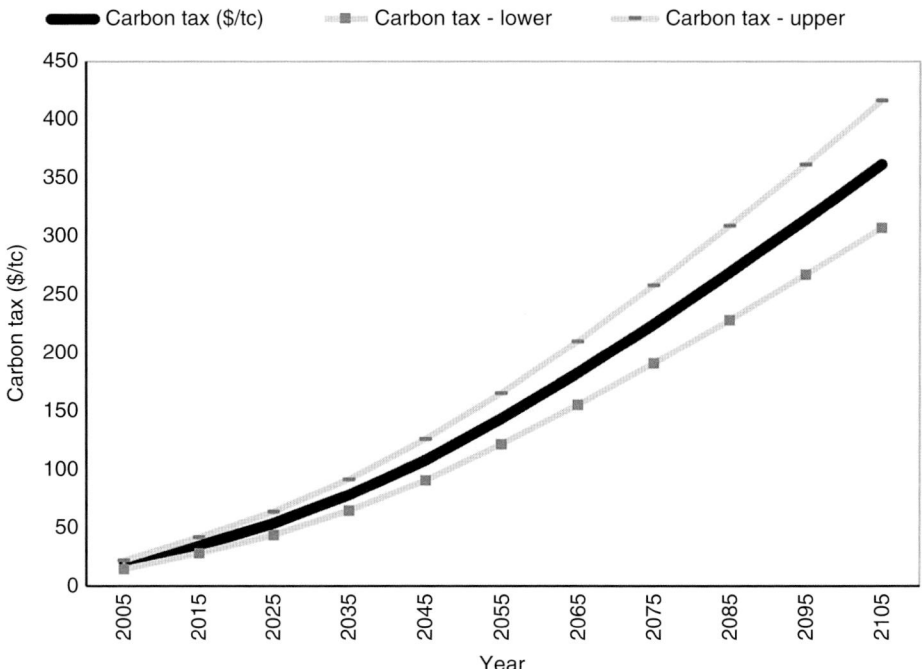

Figure 4.7 A trajectory of carbon tax with uncertainty.

systems, human systems, and economic systems. An aggregate uncertainty in the whole system is not easily calculated.

On the other hand, there are an array of behavioral as well as technological responses that interact with the effects that significantly reduce the uncertainty in global warming decisions (Seo 2017a). That individuals and communities, including the public sector, can and do adapt to changes in the climate system significantly reduces the vulnerability and uncertainties of the economic and human systems (Mendelsohn 2000; Seo 2010b, 2016a,d). Behavioral adaptation measures and strategies would be effective unless the present system of climate is shifting to another system abruptly (NRC 2013).

Many of the adaptation behaviors that are motivated by personal interests of individuals who take such actions are at the same time directed towards decreasing emissions of greenhouse gases (Seo 2015b, 2016c). An adoption of forest enterprises or grassland managements over slash-and-burn crop farming, for example, can increase absorptions of carbon dioxide in the atmosphere and prevent soil carbon released from cultivating crop lands (Seo 2010a, 2016a,b). An increase in hydroelectric energy generations and an increase in sea grasses, caused by global climate changes that increase intense rainfall and warmer oceans, will also lead to carbon-reducing adaptations (Seo 2016c, 2017a). This further reduces the size of uncertainty in climate change policy-making.

A third avenue of global warming policy responses which acts to reduce the magnitude of uncertainty is technological breakthroughs (Seo 2017a). A bundle of technological options is available to directly control the climate system: reflecting sunlight, albedo modification, carbon capture, and ocean iron fertilization (NRC 2015). Other technological options are low-carbon or zero-carbon technologies: electric vehicles, new lighting methods such as light-emitting diodes (LEDs), solar energy, nuclear fusion, nuclear fission with enhanced safety and storage, and hydroelectric energy production (MIT 2003, 2015; IPCC 2011). Many of these technologies are operational at present, which means that the high-end extreme climate change scenarios can be excluded from the range of climate policy uncertainty.

In summary, a Pareto optimal policy intervention, described in this section, is a policy instrument that is well equipped to address numerous problems of catastrophes that the market economy fails to address because of the publicness of a concerned catastrophe. Although the above-described policy framework is not explicitly developed to deal with a fat-tail or an extremely fat-tail catastrophe, it embeds a practical way to address it by providing the range and the upper limit in the keystone policy measure, that is, carbon tax, even where there is significant uncertainty (Nordhaus 2008, 2011). However, there is no principle in this approach with regard to whichever point in the range of carbon tax estimates could be picked, especially the upper bound estimate.

An extremely fat-tail catastrophe aspect of global warming, however, is addressed by behavioral adaptations with and without utilizations of technological breakthroughs (Seo 2017a). The large array of behavioral and technological options makes the distribution of a global warming catastrophe not to be fat, perhaps more precisely, not to be extremely fat. This again does not mean that the author is arguing that there is no possibility of an extremely long-tail event, and neither does the Pareto optimal policy framework.

This conclusion is at the heart of this book with regard to how the global society should deal with truly big catastrophes, say, humanity-ending, or even universe-ending catastrophes. In other words, this conclusion is applicable to the range of an extremely fat-tail catastrophic event surveyed in this book, including asteroid strikes, nuclear wars, high-risk physics and biological experiments, and AI.

4.8 Events of Variance Infinity or Undefined Moments

In the extreme of the extremely fat-tail distribution, there lies the catastrophe event with its variance being infinity or unbounded. Stated differently, the limit of the variance of a random variable with a fat-tail distribution is the variance infinity. The author explains the events of variance infinity in this section, which will certainly shed more light on the extremely fat-tail events.

The random variable with infinite unbounded variance has a solid foundation in the literature of statistics, for the explanation of which we need a short statistical introduction. In the probability and statistics literature, a random variable, X, is a variable that takes on a range of values with a probability of occurrence assigned to each value. The distribution of the random variable is expressed by the PDF or the CDF.

In a well-defined probability distribution such as the normal (Gaussian) distribution or the Poisson distribution, the PDF has a closed form. Most distributions of random variables do not have a closed form PDF. Only a number of random variables' distributions are known to have a closed form PDF and CDF (Hogg et al. 2012).

Of the distributions that have a well-defined PDF, the characteristics of the distribution are identified by the distribution's parameters. For the Gaussian distribution, for example, the mean and variance of the distribution, which are the parameters of the

Gaussian distribution, characterize the distribution completely. However, the mean and variance are not the only identifying parameters for a probability distribution of a random variable (Hogg et al. 2012).

In a statistical study of a physical, social, economic phenomenon, a researcher assumes that – or examines – the natural or social event of concern which is a random variable follows a well-defined probability distribution. The random variable, X, may be the prediction of the degree of global average temperature increase from a climate model, or the number of deaths by an earthquake event, or the magnitude of economic damages from a tropical cyclone. For example, the researcher may assume that the first event, i.e. the global average temperature increase, follows a Gaussian distribution, and the second random event, i.e. the number of deaths from an earthquake, follows a Poisson distribution (Nordhaus 2008; Seo 2015b).

The distribution of a random variable would typically have the feature that the realized values that are exceptional, i.e. extremely large or small, have a very low likelihood of realization. The farther apart the realized value from the mean of the distribution, the smaller the probability of realization. At certain threshold values, the probability of occurrence becomes approximately zero (Schuster 1984; Casella and Berger 2001).

An exceptional case is a Cauchy distribution. In the statistics literature, a Cauchy distribution is known as a pathological distribution in which the expected value and the variance of it are not defined or unbounded (Stigler 1999). The Cauchy distribution is defined as the PDF $f(x|x_0, \gamma)$ in which the two identifying parameters are x_0, a location parameter, and γ, a scale parameter.

The Cauchy distribution is also defined as the ratio of the two independently and identically distributed normal distributions (Casella and Berger 2001). Let U, V each be a random variable distributed independently with $N(0, 1)$. Then, the following ratio is distributed by the standard Cauchy distribution:

$$\frac{U}{V} \sim \text{Cauchy}(0, 1). \tag{4.17}$$

The PDF is well defined for the Cauchy distribution with the location parameter x_0 and the scale parameter $\gamma > 0$:

$$f(x|x_0, \gamma) = \frac{1}{\pi \gamma \left[1 + \left(\frac{x-x_0}{\gamma}\right)^2\right]} \tag{4.18}$$

From the PDF in Eq. (4.18), it is easy to write down the equations for the first moment, second moment, third moment, etc., of the Cauchy distribution. It can be proven handily in several steps that the moments, i.e. the mean, variance, etc. are not defined (see Casella and Berger 2001).

Traditionally, the Cauchy distribution has been interpreted as the distribution over x, given the scale parameter γ. A sensitive dependence of the Cauchy distribution on the scale parameter has not been analyzed. However, the scale parameter is an important variable for policy-making contexts, as will be shown presently.

Since the location parameter only shifts the location of the peak – i.e. the mode not the mean – of the distribution, for our purposes we can set it to zero or a fixed constant without loss of generality. Then, for a given value of $\widetilde{x}$, the probability density in Eq. (4.18) can be rewritten as the following:

$$f(\gamma|0,\widetilde{x}) = \frac{1}{\pi\gamma\left[1+\left(\frac{\widetilde{x}}{\gamma}\right)^2\right]}. \tag{4.19a}$$

$$\lim_{\gamma\to 0} f(\gamma|0,\widetilde{x}) = 0, \forall \widetilde{x}. \tag{4.19b}$$

Then, for any given value of $\widetilde{x}$, the probability approaches zero as the scale parameter γ becomes infinitely small, i.e. approaches zero, which is shown in Eq. (4.19b) above. Equation (4.19b) implies that the Cauchy distribution has a fat-tail only when the scale parameter is "significantly" larger than zero. In other words, the variance of the Cauchy distribution is finite for the values of γ which is close to zero.

A family of Cauchy distributions is shown in Figure 4.8. For all three, the location parameter is set to zero, while the scale parameter is varied from 2, 1, and 0.5. With the scale parameter of 2, the PDF shows a fat-tail distribution. With the scale parameter of 1, the PDF becomes significantly less fat. With the sale parameter of 0.5, the PDF shows a high peak at $x = 0$ and the density falls sharply as the x deviates from zero.

Note in the figure that in the range of $(-3 \leq x \leq 3)$ of the Cauchy random variable the probability of occurrence becomes 96.7%. This means that the probability of occurrence of an event outside this range is smaller than 3.3%. This is slightly larger than the tail probability of the Gaussian distribution, which is 1%. As the scale parameter becomes closer to zero, the PDF of the Cauchy distribution approximates the Gaussian distribution. When the scale parameter is set to $\gamma = 0.15$, the tail probability of the Cauchy distribution is equal to that of the Gaussian distribution, that is, the tail of $\{x||x| \geq 3\}$.

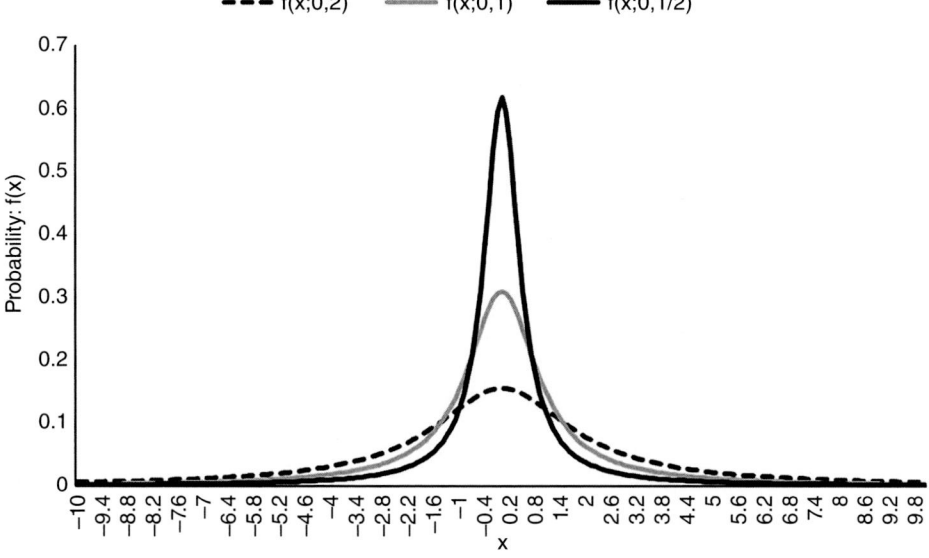

Figure 4.8 A family of Cauchy distributions with different scale parameters.

Before the author closes this section and moves on to the discussion of the dismal theorem of economics, it seems appropriate to mention here a terminology that has recently become popular with the public and is sometimes referred to as another way to state scientific concepts of infinite variance and undefined moments. The terminology is a "black swan," and you may recall it mentioned in connection with the fractal in Chapter 2 (Taleb 2007).

According to Taleb (2007), a black swan is an event with the following three attributes:

> First, it is an outlier, as it lies outside the realm of regular expectations, because nothing in the past can convincingly point to its possibility;
> Second, it carries an extreme 'impact';
> Third, in spite of its outlier status, human nature makes us concoct explanations for its occurrence after the fact, making it explainable and predictable.

Statistically, a black swan event can be best understood as a realized event from a random variable whose tail distribution is long-and-thin. So, when it occurs, it is recognized as if an outlier. It is a really long-and-thin-tail event, say, whose probability of occurrence is, say, 1 in 60 million years. For reference, the asteroid that resulted in the extinction of dinosaurs struck Earth 66 million years ago (Kaiho and Oshima 2017).

More critically, the definition of a black swan event does not insist, contrary to the author's claims, that a black swan event must be an event with infinite or undefined variance (Taleb 2005). What it defines is a long-and-thin tail event which lies outside regular expectations, whose occurrence is therefore unexpected and shocking. That is, it is a long-tail event, so an extreme event. Further, it is a thin-tail event, as it occurs so rarely. In other words, a black swan event can occur from a thin-tail distribution, not only from a fat-tail distribution (Schuster 1984).

From a fractal perspective, a black swan event can occur from a fractal geometry, but this does not exclude its occurrence from a nonfractal geometry (Mandelbrot 1997). Stated in the language of a power law distribution, a black swan event can be generated from a power law distribution, but can also be generated from a modified power law distribution with an additional parameter introduced in Chapter 2 that moderates the assumption of scale invariance embedded in the power law distribution (Gabaix 2009).

Should one be concerned about an event whose likelihood of happening is 1 in 60 million years? Unless it does cause an extremely large impact, one would not care much about it. But, a black swan event, as defined in the manner above, is an event with an extremely large economic impact. Like a large asteroid 10 km in diameter colliding with the planet, a black swan event is conceptualized to cause a massive damage (Chapman and Morrison 1994; NASA 2014).

From another perspective, the conceptualization of a black swan event fails to apprehend the cost side of the event. That "*nothing can convincingly point to the possibility of an event*" does not exclude that someone may describe the nature of the event in a probabilistic framework including the cost of prevention. A black swan event may be prevented without high cost, in which case a policy intervention may not be very cumbersome. In the case of a large asteroid collision, the cost of detection and prevention is not excessively high (NRC 2010).

What if the cost of preventing a black swan event is exceedingly high, say, at the level of loss of personal income by 50% annually for 20 years? In addition, let's further assume that the event arises from a fat-tail power law distribution and is, at the same time, expected to cause a humanity-scale catastrophe. How should a policy be designed to prevent the black swan event with these characteristics?

More formally, let's define the following catastrophe event with three characteristics, which is quite different from a black swan event:

> First, it arises from a fat-tail distribution, i.e. a power law or a fractal;
> Second, it causes a global-scale catastrophe;
> Third, there is no technological solution, so it is extremely costly to prevent the event.

What is the best way to address this type of event? This is at the heart of the questions raised in this book. The answer to the question is also provided throughout this book. A black swan event can be understood to be a subset of the event, so it is a dismal event, to be discussed presently.

4.9 Economics of Infinity: A Dismal Science

According to Weitzman, the phenomenon of global warming is a fat-tail event because there is an extremely large uncertainty in predicting the degree of global warming in a far future (Weitzman 2009). He argues that the probability of a 10° C increase in global average temperature or even a 20° C increase by 2200 cannot be made into an insignificant – a minute probability – event; as such, it cannot be excluded from a policy analysis as a nonconsequential forecast (Le Treut et al. 2007; IPCC 2014).

Weitzman argues that the exceptionally large uncertainty about the future climate cannot be even reduced because of the complexity in the climate system even if extraordinary resources and anthropogenic efforts were to be directed to the future science and economics of global warming. In other words, the expectation of the future average temperature increase has infinite variance and because of its power law distribution additional knowledge and resources are of no help at all.

Through the dismal theorem, Weitzman proposes that faced with such a catastrophic event whose distribution is fat-tailed, the global community as a decision-maker should devote limitless resources to prevent such a potential catastrophe. The dismal theorem is expressed via the following outcome equation:

$$\text{For any given } n \text{ and } k, \lim_{\lambda \to \infty} E[M|\lambda] = +\infty. \tag{4.20}$$

In Eq. (4.20), M is a "stochastic discounting factor," as defined by Weitzman, which is the amount of present consumption an economic agent would be willing to give up in the present period in order to obtain one sure unit of future consumption. This is a two-period model in which the future period is the world that is apart 200 years from the present period.

The parameters n, k are interpreted roughly to be the number of available empirical data/observations and the amount of prior knowledge, respectively. The two parameters are assumed to be fixed at a given point in time or at the time of decision-making.

The λ is called the VSL (Value of Statistical Life)-like parameter, approximately the value of the statistical civilization as we know it or the VSL on Earth as we know it.

Therefore, the dismal theorem in Eq. (4.20) states that the amount of sacrifice the present generation is willing to make in order to avert a future global warming which unfolds 200 years later in time is unbounded or infinite as long as the value of the human civilization on the planet is extremely large.

Despite many critiques it garnered with regard to its direct applicability to global warming policy-making (Nordhaus 2011; Yohe and Tol 2012; Seo 2017a), the dismal theorem provides an excellent exposition of the fat-tail distribution and its drastic policy implications at the extreme interpretation of its statistical attributes.

The dismal theorem assumes a standard utility function of an agent with a constant relative risk aversion (CRRA) (η):

$$U(C) = \frac{C^{1-\eta}}{1-\eta}. \tag{4.21}$$

In the above, C is, according to Weitzman's construction, a "reduced-form consumption" which is constructed to include the cost of global warming damages, the cost of abatement activities, and the cost of adaptation activities (Weitzman 2009). This is a critical assumption because these variables are unhinged in the proposed model from the realities of global warming and climate change. That is, there is no linkage, neither directly nor indirectly, of the reduced-form consumption variable (C) to global warming damages, abatements, and adaptations.

Let the present consumption be normalized to 1. Then, define Y as follows as the growth of consumption from the current period:

$$Y = \ln C. \tag{4.22}$$

From Eqs. (4.21) and (4.22), the stochastic discounting factor is defined, with the time preference parameter β since there are only two periods in the model, as follows:

$$M(C) = \beta \frac{U'(C)}{U'(1)} = \beta \exp(-\eta Y). \tag{4.23}$$

It is not difficult to understand the linkage of Eq. (4.23) to the dismal theorem in Eq. (4.20). With the PDF of Y being $f(y)$, the $E[M]$ is the Laplace transformation or the moment generating function of $f(y)$:

$$E[M] = \beta \int_{-\infty}^{\infty} e^{-\eta y} f(y) dy. \tag{4.24}$$

With the distribution of Y being the normal distribution parameterized by $N(\mu, s^2)$, the stochastic discounting factor becomes a familiar Ramsey form, with $\delta = -\ln \beta$:

$$E[M] = \exp\left(-\delta - \eta\mu + \frac{1}{2}\eta^2 s^2\right). \tag{4.25}$$

Let $Z \sim N(0, 1)$ and make an affine transformation to form $Y = sZ + \mu$. Then, the conditional PDF of y is

$$h(y|s) = \frac{1}{s}\phi\left(\frac{y-\mu}{s}\right). \tag{4.26}$$

The marginal or unconditional posterior-predictive PDF of y is

$$f(y) = \int_0^\infty h(y|s) p_n(s|y) ds. \tag{4.27}$$

In the above, the second term inside the integral is the posterior PDF of s, which can be written proportionally, following the Bayes theorem, as a prior PDF times the likelihood function $L(y|s)$:

$$p_n(s|y) \propto p_0(s) \prod_{j=1}^{n} h(y_j|s). \qquad (4.28)$$

The first term on the right-hand side is the prior PDF of s. For some number k, identifiable with the number of prior knowledge, the prior PDF can be written in the following general form which has scale invariance:

$$p_0(s) \propto s^{-k}. \qquad (4.29)$$

With $v_n = \sum_{j=1}^{n} (y_j - \mu)^2/n$, i.e. sample variance, the posterior-predictive PDF in Eq. (4.27) is the Student-t distribution, with $n+k$ degrees of freedom:

$$f(y) \propto \left(1 + \frac{(y-\mu)^2}{nv_n}\right)^{-(n+k)/2}. \qquad (4.30)$$

Then, the stochastic discounting factor in Eq. (4.25) becomes unbounded with the posterior-predictive distribution of Y given in Eq. (4.30), owing to the fat-tail in the Student-t distribution:

$$E[M] = +\infty. \qquad (4.31)$$

The model still needs to incorporate the value of all civilizations on Earth as we know them. Let's assume that the utility function takes the following CRRA form, with D being the starvation level of consumption below which an individual should die:

$$U(C|D) = \frac{C^{1-\eta} - D^{1-\eta}}{1-\eta}, \text{ for } C \geq D;$$

$$U(C|D) = 0, \text{ for } 0 < C < D. \qquad (4.32)$$

An individual in the model is a global decision-maker or a representative of Planet Earth. Let's normalize the current consumption as before to $C = 1$. Let $A(q)$ be the amount of extra consumption the individual requires within this period to exactly compensate for $P[C \leq D] = q$ within this time period. In the expected utility theory, $A(q)$ satisfies the following equation:

$$(1-q)U(1+A(q)|D) = U(1+D). \qquad (4.33)$$

Differentiating Eq. (4.33) with respect to q gives

$$-U(1|D) + U_1(1|D)\lambda = 0. \qquad (4.34)$$

In the above equation, $\lambda \equiv A'(0)$. The λ is the VSL-like parameter defined as the rate of substitution between consumption and mortality risk. Inverting Eq. (4.34) for the isoelastic utility function in Eq. (4.32) yields

$$D(\lambda) = [1 + (\eta - 1)\lambda]^{-1/(\eta-1)}. \qquad (4.35)$$

Putting together equations, the dismal theorem in Eq. (4.20) is derived. In simple words, given the number of prior knowledge and the number of empirical data on climate change fixed at a certain point in time, the amount of present consumption an individual is willing to sacrifice to obtain one sure unit of consumption in the future is infinite, as long as the VSL on Earth is infinite.

Rewriting the Eq. (4.20) explicitly in terms of the mean and the standard deviation of the prediction of consumption growth, y, yields the following alternative form of the dismal theorem which is at the heart of the proof of the theorem:

For any given n and k,

$$\lim_{\lambda \to \infty} E[M|\lambda] \propto \lim_{s \to \infty} \left\{ e^{-\eta z's} \frac{1}{s^{k+n}} \right\} = +\infty. \qquad (4.36)$$

Based on the conclusion in Eq. (4.20) and Eq. (4.36), Weitzman proposed a radical switch in global warming policy-making from a dynamic benefit-cost analysis to a generalized precautionary principle (Weitzman 2009). According to him, there is no need for a complicated painstaking benefit and cost calculation to be accounted for in the policy model of global warming. Uncertainty simply dominates all considerations. That is, unknowability and ignorance prevail in whatever circumstances.

A precautionary principle means that the world community should avert the catastrophic outcome at all costs. For example, the world community should end all emissions of greenhouse gases immediately. An immediate cessation of greenhouse gas-generating activities should apply to all the countries in the world regardless of economic realities in these countries.

4.10   Alternative Formulations of a Fat-tail Catastrophe

There are many other low-probability catastrophic events with horrific consequences that humanity is concerned about, e.g. asteroid strikes, strangelets, and AI. In other words, there are other long-tail events than global warming. However, experts agree that these events do not call for a draconian policy intervention. This is because these long-tail events are sufficiently thin-tailed or not sufficiently fat-tailed. In the dismal theorem, there is no discussion of how fat an event should be for the dismal theorem to be applied to a policy situation (Seo 2016c, 2017a).

This seems like a small inconvenience at first glance. However, this is a salient drawback which probably cannot be overcome in the contexts of policy practitioners. For a certain policy issue, e.g. global warming or asteroid strikes, should policy-makers rely on the dismal theorem? Or could they? The dismal theorem does not and cannot answer these questions. What it says is only that at the extreme end of uncertainty, the dismal theorem should apply.

The critical assumptions that underlie the dismal theorem were pointed out by global warming economists (Horowitz and Lange 2009; Karp 2009; Nordhaus 2011; Yohe and Tol 2012; Millner 2013; Seo 2016c, 2017a). Let me summarize them briefly here. This will be followed by a series of alternative mathematical formulations of the dismal theorem, from which the author derives an opposite policy conclusion.

First, the dismal theorem assumes that all future climate predictions, regardless of whether they are done by an expert or a casual observer or of whether they are based on less or more likely assumptions about the future, are all equally valid. Second, scientific efforts cannot narrow down the probability of the catastrophic event which is the end of all civilizations on the planet due to global warming, even with immense resources and time devoted. Third, the future extreme climate event will end all life on Earth. Fourth, behavioral responses by individuals, communities, governments, and international societies are of no significance in the context of the fat-tail event of global warming.

The dismal model does not have the cost of slowing down global warming as a variable in the model. This is because the catastrophic end of civilizations was assumed to be inescapable by the modeler. In reality, there are cheaper options and costlier options to deal with global warming (Seo 2017a). The key observation is that there are multiple rather costly options that can stop global warming catastrophe unfolding. A number of technological options are well documented, such as climate engineering through solar reflector, carbon scrubber capture from the atmosphere, and ocean fertilization (Martin et al. 1994; Lackner et al. 2012; NRC 2015).

Technological possibilities to contain the inexorable global warming trend include revolutions in the energy system. New lighting methods such as LEDs, electronic vehicles and battery revolutions, nuclear fusion energy technologies, and solar energy technologies are technologically possible now, but significantly more expensive than fossil fuel-based energy productions (NRC 2013; Akasaki et al. 2014; ITER 2015; MIT 2003, 2015).

The dismal model does not include behavioral aspects of individual decisions in response to global warming (Mendelsohn 2000; Seo 2015b, 2016c). Since global warming is assumed in the dismal model to unfold over more than two centuries, omission of behavioral adaptations leads to a major breakaway from the rest of the economic models on global warming (Seo 2010b, 2016a,b). As climate is changed in the coming decades in a gradual fashion, individuals will learn about how to adapt to or cope with changing climates (IPCC 1990, 2014; Yohe and Tol 2012).

In the world of the dismal theorem augmented by a gradually rising temperature over three centuries to come, it is certain that humanity will learn about how to adapt to and therefore reduce the impacts of even the high temperature discussed in the dismal theorem (Seo 2015a, 2017b). This has a critical implication with reference to the extreme case depicted in the dismal theorem. The possibilities of behavioral adaptations mean that the fat-tail distribution which embodies the infinity of variance shall not be materialized in reality. In other words, people and societies will learn to avoid the extreme case of the end of human civilizations in any case.

From this point on, the author will add alternative and more realistic assumptions to the framework of the dismal model, from which new conclusions will be shown to be possible. Each of the modifications suggested below will lead the dismal theorem in Eq. (4.20) to be invalid.

First, the utility function in the dismal model can take another form other than the CRRA class utility function, that is, a power law utility function. For example, let's say that it takes a generalized logistic functional form, defined on an appropriate range of consumption, in which an exponential process is embedded:

$$U(C) = \varrho_1 + (\varrho_2/(\varrho_3 + \varrho_4 e^{-Ct})^{1/\upsilon}), \tag{4.37}$$

where C is the level of consumption as before, t is time, and Greek letters are the parameters in the function to be set appropriately to fit the desired characteristics of the social utility function. An alternative form of the utility function can also take on a more established form as suggested by Millner (2013).

Second, let the number of prior knowledge and empirical data with regard to global warming be unlimited, which plays a vital role in forming a posterior function. This alternative assumption is more reasonable because new studies, findings, and observations are not bounded, since climate history is as long as we can think of, and so are experiences with climate. With this alternative assumption, the Student-t distribution in Eq. (4.15) converges to a normal distribution:

$$f(y) \xrightarrow{d} N(\mu, s). \tag{4.38}$$

Fixing the number of prior knowledge and empirical data to a certain value has the undesirable consequence of putting a cap on the possibility of an intellectual or technological solution for the global warming problem, based on the long history of experiences with the climate system by humanity.

Third, another way to formulate the prior PDF problem is whether the scale-invariant prior PDF, shown in Eq. (4.29), is in any way a desirable assumption for empirical modeling on global warming policy. The prior PDF in Eq. (4.29) is a power law distribution which is a generic form of the PDF when the uncertainty is so large that there is no way to reduce it (Mandelbrot 1997, 2001). To put this differently, the prior PDF is suitable when there is no knowledge at all on the distribution of a variable, say, an absolute dark age of knowledge (Mandelbrot and Hudson 2004; Taleb 2005). In the power law prior PDF, the variance of the variable is infinite, or, more accurately, undefined.

A more suitable prior PDF should capture the characteristic of scale variance. That is, the more knowledge we have, the smaller the scale becomes. Put differently, the more knowledge we obtain, we would expect a priori that the faster the tail probability would fall. With this more reasonable assumption, it may take the following form, with v, ω some constants:

$$p_0(s) \propto e^{-(vk+\omega s)} \tag{4.39}$$

Fourth, let the VSL on Earth be bounded by some numbers. Specifically, let's assume that it depends inversely on the magnitude of uncertainty in the climate system, i.e. instability of life, in addition to some fixed monetary value extraneous to the uncertainty (λ^o):

$$\lambda = f(s) = \lambda^o e^{-s}. \tag{4.40}$$

Fifth, let the size of uncertainty decrease with the length of time that the problem at hand unfolds and must be addressed, with a fixed magnitude, Ω:

$$s = f(T) = \Omega e^{-T}. \tag{4.41}$$

Sixth, let's assume that the consumption (or production) decreases due to the damage cost of global warming, denoted C^{gw}. Adaptation parameter, Å, decreases the damage cost of global warming. It can be interpreted as the fraction of the total damage that can be reduced by adapting sensibly and efficiently. Let's assume that the adaptation parameter depends on prior knowledge and empirical data in a functional relationship below:

$$\text{Å} = \frac{1 - e^{-(n+k)}}{1 + e^{-(n+k)}}, \tag{4.42}$$

$$Y = \ln\{C_1 - (1 - \text{Å})C^{gw}\}. \tag{4.43}$$

Seventh, a backstop technology is defined, by environmental and natural resource economists, as a technology that can eliminate all new emissions of greenhouse gases (Nordhaus 1973, 2008). These are, among other things, nuclear fusion, nuclear fission with safe storage, carbon capture and storage, climate engineering such as solar reflectors, ocean fertilization, and solar energy technologies. Given the availability of these backstops at the present time, we can bound the decrease of consumption by the least cost of backstop technologies. That is, if the decrease of consumption (or production) due to global warming becomes larger than the least cost of backstop technologies, it is Pareto efficient to adopt the least cost backstop.

Then, the growth of consumption can be expressed alternatively as follows, with C^{gw} the damage cost of global warming and C^b the least cost of the backstop technologies:

$$Y = \begin{cases} \ln(C_1 - C^{gw}), \text{if } C^{gw} < C^b; \\ \ln(C_1 - C^b), \text{if } C^{gw} \geq C^b. \end{cases} \tag{4.44}$$

Note in Eq. (4.44) that the cost of global warming cannot exceed the least cost of the backstop technologies, and hence it is bounded. This fundamentally undercuts the assumption in the dismal theorem that the consumption will approach zero due to a fat-tail global warming, so it is the conclusion of the theorem.

Eighth, let's assume that humanity's adaptation capacity is large enough to compensate all the damages from global warming in any form. This is a complete adaptation assumption. Or, we can assume that a portfolio of adaptation strategies compensates a certain percentage of global warming damages. In this compete adaptation scenario, changes in consumption and production caused by global warming are zero:

$$Y = \ln(C_1 - C^{gw,adapt}) \text{ where } C^{gw,adapt} = 0. \tag{4.45}$$

It is not difficult to draw an opposite conclusion from the dismal theorem by taking any one of these alternative formulations of the parameters and functions. In other words, under any of these alternative setups, Eq. (4.20) or Eq. (4.36) fails to hold the infinity or unboundedness of the stochastic discounting factor. In fact, it can be shown that the ultimate conclusion of the dismal theorem is indeed as follows:

$$\text{For a given } \lambda, \lim_{n+k \to \infty} E[M|n,k] = 0. \tag{4.46}$$

4.11 Conclusion

The economics of catastrophic events presented in this chapter provides a general framework of the economic aspects of catastrophes. It is not written for an application to a specific event but for a whole range of catastrophic events, both natural and man-made. The framework covers a long-tail event, an extremely long-tail event, a fat-tail event, and an extremely fat-tail event.

Broadly, the chapter is presented with three components. The first component is how to define catastrophic events or statistical expositions of salient characteristics of catastrophes. Concepts of a threshold value and a tail distribution are introduced.

The second component is how market and financial sectors respond to catastrophic risks. The author explains how insurances, options, futures, and CAT bonds are designed to help potential victims minimize the risk and losses in a variety of long-tail and/or fat-tail events. The third component is how a government should respond to catastrophic events. A Pareto optimal policy intervention is explained and illustrated through the problem of global warming policy-making. An alternative policy perspective that concentrates on the fatness of a tail distribution is elaborated and evaluated.

Let me conclude this chapter with a summary and some additional comments on the question of how a society should address a potential catastrophic event in an optimal way. First, when a threshold or a tipping point can be established with little disagreement for a particular catastrophic event, a strong-arm policy intervention is close to a Pareto-efficient policy which is also more politically feasible. But, when a tipping point is fuzzy and arguable, a strong-arm policy intervention tends to be neither efficient nor politically more feasible.

Second, even if we can expect a particularly catastrophic event, i.e. a long-tail event, it does not automatically call for a policy intervention. Market, financial, and behavioral responses will be sufficient in many situations for an individual and a community to cope with such a low-probability, high-damage event. This chapter explains in detail the workings of insurances, options, futures, and catastrophe bonds through which a potential victim can minimize and avoid a catastrophic loss.

Third, a governmental policy intervention or an international policy intervention is needed in a special case in which the distribution of a potential catastrophe event exhibits an extremely fat tail. In the fat-tail situations, that is, in the situations of large uncertainty, a Pareto optimal policy intervention should be designed by weighing the benefits and costs of a policy by accounting for long-term effects. A range of estimates of policy-target variables can be suggested by reflecting the range in the size of uncertainty.

Fourth, an extremely fat-tail situation may appear when the variance of the particular event appears infinite or undefined. That is, the uncertainty is too large to be made smaller through any means including extraordinary research and funding. In other words, the uncertainty takes on the shape of a scale invariant distribution. A situation with an extreme uncertainty may exist in real-world problems such as a strangelets-caused black hole or global warming.

This chapter emphasizes that this extreme uncertainty situation does not immediately call for a precautionary principle, i.e. protection against it at all cost. A Pareto optimal response to a potential extremely fat-tail event is suggested, which is a portfolio of efficient policy interventions, behavioral adaptations, and technological capabilities. The author explains how such an integrative approach can make the distribution of an extremely fat-tail event thinner using the example of global warming catastrophes, in other words, how it can make the scale invariance parameter scale variant.

In the next chapters, the author offers other examples of an extremely fat-tail event, e.g. a strangelets-caused black hole catastrophe, which has been made thinner through research and technological developments (Dar et al. 1999; Jaffe et al. 2000; Ellis et al. 2008).

This concludes the description of the economics of catastrophe events from the theoretical perspective. Chapter 5 presents the economics of catastrophic events from the empirical perspective, relying on empirical globally collected catastrophe data and empirical economic models of catastrophes.

References

Akasaki I, Amano H, Nakamura S (2014) Blue LEDs: Filling the World with New Light. The Nobel Prize in Physics 2014. http://www.nobelprize.org/nobel_prizes/physics/laureates/2014/popular-physicsprize2014.pdf.

Arrow, K.J., Cropper, M.L., Eads, G.C. et al. (1996a). Is there a role for benefit-cost analysis in environmental, health, and safety regulation? *Science* 272: 221–222.

Arrow, K.J., Cline, W.R., Maler, K.G. et al. (1996b). Intertemporal equity, discounting, and economic efficiency. In: *Climate Change 1995: Economic and Social Dimensions of Climate Change* (ed. J.P. Bruce, H. Lee and E.F. Haites). Cambridge, UK: Cambridge University Press.

Barrieu, P. and Albertini, L. (ed.) (2009). *The Handbook of Insurance-Linked Securities*. Chichester, UK: Wiley.

Barry, P.J., Elllinger, P.N., Baker, C.B., and Hopkin, J.A. (2000). *Financial Management in Agriculture*, 6e. Illinois: Interstate Publishers.

Black, F. and Scholes, M. (1973). The pricing of options and corporate liabilities. *Journal of Political Economy* 81 (3): 637–654.

Borch, K. (1962). Equilibrium in a reinsurance market. *Econometrica* 30: 424–444.

Broecker, W.S. (1997). Thermohaline circulation, the Achilles' heel of our climate system: will man-made CO_2 upset the current balance? *Science* 278: 1582–1588.

Buchanan, J.M. (1968). *The Demand and Supply of Public Goods*. Chicago: Rand McNally & Co.

Casella, G. and Berger, R.L. (2001). *Statistical Inference*, 2e. Independence, KY: Cengage.

Chapman, C.R. and Morrison, D. (1994). Impacts on the earth by asteroids and comets: assessing the hazard. *Nature* 367: 33–40.

Chicago Mercantile Exchange (CME) Group (2017) CME Group All Products – Codes and Slate. CME Group, Chicago. http://www.cmegroup.com/trading/products/#pagenumber=1&sortasc=false&sortfield=oi.

Cropper, M. (1976). Regulating activities with catastrophic environmental effects. *Journal of Environmental Economics and Management* 3: 1–15.

Cutler, D. and Zeckhauser, R. (2004). Extending the theory to meet the practice of insurance. In: *Brookings–Wharton Papers on Financial Services* (ed. R.E. Litan and R. Herring). Washington, DC: Brookings Institution Press.

Dar, A., Rújula, A.D., and Heinz, U. (1999). Will relativistic heavy ion colliders destroy our planet? *Physics Letters B* 470: 142–148.

Edesess, M. (2014). *Catastrophe Bonds: An Important New Financial Instrument*. Roubaix, France: EDHEC-Risk Institute, EDHEC Business School.

Ellis, J., Giudice, G., Mangano, M. et al. (2008). Review of the safety of LHC collisions. *Journal of Physics G: Nuclear and Particle Physics* 35 (11).

Emanuel, K. (2005). Increasing destructiveness of tropical cyclones over the past 30 years. *Nature* 436: 686–688.

Emanuel, K. (2008). The hurricane–climate connection. *Bulletin of the American Meteorological Society* 89: ES10–ES20.

Emanuel, K. (2013). Downscaling CMIP5 climate models shows increased tropical cyclone activity over the 21st century. *Proceedings of the National Academy of Sciences of the United States of America* (30): 12219–12224.

Evenson, R. and Gollin, D. (2003). Assessing the impact of the green revolution 1960–2000. *Science* 300: 758–762.

Fabozzi, F.J., Modigliani, F.G., and Jones, F.J. (2009). *Foundations of Financial Markets and Institutions*, 4e. New York: Prentice Hall.

Falk, I. and Mendelsohn, R. (1993). The economics of controlling stock pollution: an efficient strategy for greenhouse gases. *Journal of Environmental Economics and Management* 25: 76–88.

Fama, E. (1963). Mandelbrot and the stable Paretian hypothesis. *Journal of Business* 36: 420–429.

Froot, K.A. (ed.) (1999). *The Financing of Catastrophe Risk*. Chicago, IL: University of Chicago Press.

Fudenberg, D. and Tirole, J. (1991). Games in strategic form and Nash equilibrium. In: *Game Theory* (ed. D. Fudenberg and J. Tirole). Cambridge, MA: MIT Press.

Gabaix, X. (2009). Power laws in economics and finance. *Annual Review of Economics* 1: 255–293.

Halcrow, H.G. (1949). Actuarial structures for crop insurance. *Journal of Farm Economics* 31: 418–443.

Hawking S, Tegmark M, Russell S, Wilczek F (2014) Transcending complacency on superintelligent machines. *Huffington Post*. https://www.huffingtonpost.com/stephen-hawking/artificial-intelligence_b_5174265.html.

Hogg, R.V., Craig, A., and McKean, J.W. (2012). *Introduction to Mathematical Statistics*, 7e. New York: Pearson.

Horowitz J, Lange A (2009) What's Wrong with Infinity? A Note on Weitzman's Dismal Theorem. University of Maryland Working Paper. http://citeseerx.ist.psu.edu/viewdoc/download?doi=10.1.1.578.888&rep=rep1&type=pdf.

Intergovernmental Panel on Climate Change (IPCC) (1990) Climate Change: The IPCC Scientific Assessment. Cambridge University Press, Cambridge, UK.

Intergovernmental Panel on Climate Change (IPCC) (2011) Special Report on Renewable Energy Sources and Climate Change Mitigation. Cambridge University Press, Cambridge, UK.

Intergovernmental Panel on Climate Change (IPCC) (2014) Climate Change 2014: The Physical Science Basis, the Fifth Assessment Report of the IPCC. Cambridge University Press, Cambridge.

International Thermonuclear Experimental Reactor (ITER) (2015) The ITER Tokamak. https://www.iter.org/mach.

Jaffe, R.L., Buszaa, W., Sandweiss, J., and Wilczek, F. (2000). Review of speculative disaster scenarios at RHIC. *Review of Modern Physics* 72: 1125–1140.

Kaiho, K. and Oshima, N. (2017). Site of asteroid impact changed the history of life on earth: the low probability of mass extinction. *Scientific Reports* 7: 14855. doi: 10.1038/s41598-017-14199-x.

Kaplan T (2017) Senate Votes to Raise Debt Limit and Approves $15 Billion in Hurricane Relief. NYT 7 September.

Karp L (2009) *Sacrifice, Discounting, and Climate Policy: Five Questions*. Unpublished manuscript. http://are.berkeley.edu/~karp/fivequestionssubmit.pdf.

Kiger, M. and Russell, J. (1996). *This Dynamic Earth: The Story of Plate Tectonics*. Washington, DC: United States Geological Survey (USGS).

King RO (2013) The National Flood Insurance Program: status and remaining issues for Congress. CRS Report for Congress R42850. Congressional Research Service, Washington, DC.

Koopmans, T.C. (1965). On the concept of optimal economic growth. *Academiae Scientiarum Scripta Varia* 28 (1): 1–75.

Kossin, J.P., Olander, T.L., and Knapp, K.R. (2013). Trend analysis with a new global record of tropical cyclone intensity. *Journal of Climate* 26: 9960–9976.

Kurzweil, R. (2005). *The Singularity Is Near*. New York: Penguin.

Lackner, K.S., Brennana, S., Matter, J.M. et al. (2012). The urgency of the development of CO_2 capture from ambient air. *Proceedings of the National Academy of Sciences of the United States of America* 109 (33): 13156–13162.

Lakdawalla, D. and Zanjani, G. (2012). Catastrophe bonds, reinsurance, and the optimal collateralization of risk transfer. *Journal of Risk & Insurance* 79: 449–476.

Le Treut, H., Somerville, R., Cubasch, U. et al. (2007). Historical overview of climate change. In: *Climate Change 2007: The Physical Science Basis* (ed. S. Solomon, D. Qin, M. Manning, et al.). Cambridge, UK: Cambridge University Press.

Lenton, T.M., Held, H., Kriegler, E. et al. (2008). Tipping elements in the earth's climate system. *Proceedings of the National Academy of Sciences of the United States of America* 105: 1786–1793.

Lutz, W.K., Lutz, R.W., Gaylor, D.W., and Conolly, R.B. (2014). Dose–response relationship and extrapolation in toxicology: mechanistic and statistical considerations. In: *Regulatory Toxicology* (ed. X.-V. Reichl and M. Schwenk). Berlin: Springer.

Mahul, O. (1999). Optimum area yield crop insurance. *American Journal of Agricultural Economics* 81 (1): 75–82.

Mandelbrot, B. (1963). The variation of certain speculative prices. *Journal of Business* 36 (4): 394–419.

Mandelbrot, B. (1997). *Fractals and Scaling in Finance*. New York: Springer.

Mandelbrot, B. (2001). Scaling in financial prices: I. Tails and dependence. *Quantitative Finance* 1: 113–123.

Mandelbrot, B. and Hudson, R.L. (2004). *The (Mis)Behaviour of Markets: A Fractal View of Risk, Ruin, and Reward*. London: Profile Books.

Martin, J.H., Coale, K.H., Johnson, K.S. et al. (1994). Testing the iron hypothesis in ecosystems of the equatorial Pacific Ocean. *Nature* 371: 123–129.

Mas-Colell, A., Whinston, M.D., and Green, J.R. (1995). *Microeconomic Theory*. Oxford, UK: Oxford University Press.

Massachusetts Institute of Technology (MIT) (2003). *The Future of Nuclear Power: An Interdisciplinary MIT Study*. Cambridge, MA: MIT.

Massachusetts Institute of Technology (MIT) (2015). *The Future of Solar Energy: An Interdisciplinary MIT Study*. Cambridge, MA: MIT.

Meadows, D.H., Meadows, D.L., Randers, J., and Behrens, W.H. (1972). *The Limits to Growth*. New York: Universe Books.

Mendelshon, R. (2000). Efficient adaptation to climate change. *Climatic Change* 45: 583–600.

Mendelsohn, R., Dinar, A., and Williams, L. (2006). The distributional impact of climate change on rich and poor countries. *Environment and Development Economics* 11: 1–20.

Mendelsohn, R. and Williams, L. (2007). Dynamic forecasts of the sectoral impacts of climate change. In: *Human-Induced Climate Change: An Interdisciplinary Approach* (ed. M.E. Schlesinger, H.S. Kheshgi, J. Smith, et al.). Cambridge, UK: Cambridge University Press.

Millner, A. (2013). On welfare frameworks and catastrophic climate risks. *Journal of Environmental Economics and Management* 65: 310–325.

Mills, M.J., Toon, O.B., Turco, R.P. et al. (2008). Massive global ozone loss predicted following regional nuclear conflict. *Proceedings of the National Academy of Sciences of the United States of America* 105: 5307–5312.

Miranda, M.J. and Glauber, J.W. (1997). Systemic risk, reinsurance, and the failure of crop insurance markets. *American Journal of Agricultural Economics* 79: 206–215.

Molina, M.J. and Rowland, F.S. (1974). Stratospheric sink for chlorofluoromethanes: chlorine atom-catalysed destruction of ozone. *Nature* 249: 810–812.

National Aeronautics and Space Administration (NASA) (2014). NASA's efforts to identify near-earth objects and mitigate hazards. IG-14-030. In: . Washington, DC: NASA Office of Inspector General.

National Research Council (1997) Report of the Observer Panel for the U.S.–Japan Earthquake Policy Symposium. National Academies Press, Washington, DC.

National Research Council (2010) Defending Planet Earth: Near-Earth-Object Surveys and Hazard Mitigation Strategies. National Academies Press, Washington, DC.

National Research Council (NRC) (2013) Abrupt Impacts of Climate Change: Anticipating Surprises. National Academies Press, Washington, DC.

National Research Council (NRC) (2015). Climate intervention: reflecting sunlight to cool earth. In: . Washington, DC: National Academies Press.

Niehaus, G. (2002). The allocation of catastrophe risk. *Journal of Banking and Finance* 26: 585–596.

Nordhaus, W. (1973). The allocation of energy resources. *Brookings Papers on Economic Activities* 1973: 529–576.

Nordhaus, W. (1992). An optimal transition path for controlling greenhouse gases. *Science* 258: 1315–1319.

Nordhaus, W. (1994). *Managing the Global Commons*. Cambridge, MA: The MIT Press.

Nordhaus W (2002) The economic consequences of a war in Iraq. National Bureau of Economic Research Working Paper No. 9361. NBER, Boston.

Nordhaus, W.D. (2006). Paul Samuelson and global public goods. In: *Samuelsonian Economics and the Twenty-First Century* (ed. M. Szenberg, L. Ramrattan and A.A. Gottesman). Oxford, UK: Oxford University Press.

Nordhaus, W. (2007). A review of the "Stern review on the economics of climate change". *Journal of Economic Literature* 55: 686–702.

Nordhaus, W.D. (2008). *A Question of Balance: Weighing the Options on Global Warming Policies*. New Haven, CT: Yale University Press.

Nordhaus, W. (2011). The economics of tail events with an application to climate change. *Review of Environmental Economics and Policy* 5: 240–257.

Nordhaus, W. (2012). Economic policy in the face of severe tail events. *Journal of Public Economic Theory* 14: 197–219.

Nordhaus, W. (2013). *The Climate Casino: Risk, Uncertainty, and Economics for a Warming World*. New Haven, CT: Yale University Press.

Nordhaus, W. and Boyer, J. (2000). *Warming the World: Economic Models of Global Warming*. Cambridge, MA: The MIT Press.

Nuclear Energy Institute (NEI) (2016) Statistics. NEI, Washington, DC. https://www.nei.org/resources/statistics.

Pareto, V. (1896). *Cours d'Economie Politique*. Geneva: Droz.

Parson, E.A. (2007). The big one: a review of Richard Posner's catastrophe: risk and response. *Journal of Economic Literature* 45: 147–164.

Posner, R.A. (2004). *Catastrophe: Risk and Response*. New York: Oxford University Press.

Richter, C.F. (1958). *Elementary Seismology*. San Francisco, CA: W.H. Freeman and Company.

Samuelson, P. (1954). The pure theory of public expenditure. *Review of Economics and Statistics* 36: 387–389.

Samuelson, P. (1955). Diagrammatic exposition of a theory of public expenditure. *Review of Economics and Statistics* 37: 350–356.

Samuelson, P. and Nordhaus, W. (2009). *Economics*, 19e. New York: McGraw-Hill Education.

Schuster, E.F. (1984). Classification of probability laws by tail behavior. *Journal of the American Statistical Association* 79 (388): 936–939.

Seo, S.N. (2010a). Managing forests, livestock, and crops under global warming: a micro-econometric analysis of land use in Africa. *Australian Journal of Agricultural and Resource Economics* 54: 239–258.

Seo, S.N. (2010b). A microeconometric analysis of adapting portfolios to climate change: adoption of agricultural systems in Latin America. *Applied Economic Perspectives and Policy* 32: 489–514.

Seo, S.N. (2012). Decision making under climate risks: an analysis of Sub-Saharan farmers' adaptation behaviors. *Weather, Climate, and Society* 4: 285–299.

Seo, S.N. (2015a). Fatalities of neglect: adapt to more intense hurricanes? *International Journal of Climatology* 35: 3505–3514.

Seo, S.N. (2015b). Adaptation to global warming as an optimal transition process to a greenhouse world. *Economic Affairs* 35: 272–284.

Seo, S.N. (2016a). Modeling farmer adaptations to climate change in South America: a micro-behavioral economic perspective. *Environmental and Ecological Statistics* 23: 1–21.

Seo, S.N. (2016b). The micro-behavioral framework for estimating total damage of global warming on natural resource enterprises with full adaptations. *Journal of Agricultural, Biological, and Environmental Statistics* 21: 328–347.

Seo, S.N. (2016c). A theory of global public goods and their provisions. *Journal of Public Affairs* 16: 394–405.

Seo, S.N. (2016d). *Microbehavioral Econometric Methods: Theories, Models, and Applications for the Study of Environmental and Natural Resources*. Amsterdam, The Netherlands: Academic Press (Elsevier).

Seo, S.N. (2017a). *The Behavioral Economics of Climate Change: Adaptation Behaviors, Global Public Goods, Breakthrough Technologies, and Policy-Making*. London: Academic Press.

Seo, S.N. (2017b). Measuring policy benefits of the cyclone shelter program in the North Indian Ocean: protection from intense winds or high storm surges? *Climate Change Economics* 8 (4): 1–18. doi: 10.1142/S2010007817500117.

Seo, S.N. and Mendelsohn, R. (2008). Measuring impacts and adaptations to climate change: a structural Ricardian model of African livestock management. *Agricultural Economics* 38: 151–165.

Seo, S.N. and Bakkensen, L.A. (2017). Is tropical cyclone surge, not intensity, what kills so many people in South Asia? *Weather, Climate, and Society* 9: 71–81.

Shields DA (2013) Federal crop insurance: background. Congressional Research Service. R40532. US Congress: Washington, DC.

Shiller, R.J. (2004). *The New Financial Order: Risk in the 21st Century*. Princeton, NJ: Princeton University Press.

Shiller, R.J. (2005). *Irrational Exuberance*, 2e. Princeton, NJ: Princeton University Press.

Shiller, R.J. (2009). *The Subprime Solution: How Today's Global Financial Crisis Happened, and What to Do about it*. Princeton, NJ: Princeton University Press.

Skees, J.R., Black, J.R., and Barnett, B.J. (1997). Designing and rating an area yield crop insurance contract. *American Journal of Agricultural Economics* 79: 430–438.

Stigler, S.M. (1999). *Statistics on the Table: The History of Statistical Concepts and Methods*. Cambridge, MA: Harvard University Press.

Sumner, D.A. and Zulauf, C. (2012). *Economic & Environmental Effects of Agricultural Insurance Programs*. Washington, DC: Council on Food, Agricultural & Resource Economics (C-FARE).

Swiss Re Capital Markets (2017). *Insurance Linked Securities Market Update*. Zurich, Switzerland: Swiss Re.

Swiss Re Institute (2017). *Natural Catastrophes and Man-made Disasters in 2016: A Year of Widespread Damages*. Zurich, Switzerland: Swiss Re.

Taleb NN (2005) Fat tails, asymmetric knowledge, and decision making: Nassim Nicholas Taleb's essay in honor of Benoit Mandelbrot's 80th birthday. Wilmott Magazine (2005): 51–59.

Taleb, N.N. (2007). *The Black Swan: The Impact of the Highly Improbable*. London: Penguin.

Thom, R. (1975). *Structural Stability and Morphogenesis*. New York: Benjamin-Addison-Wesley.

Tol, R. (2009). The economic effects of climate change. *Journal of Economic Perspectives* 23: 29–51.

Turco, R.P., Toon, O.B., Ackerman, T.P. et al. (1983). Nuclear winter: global consequences of multiple nuclear explosions. *Science* 222: 1283–1292.

Udry, C. (1995). Risk and saving in northern Nigeria. *American Economic Review* 85 (5): 1287–1300.

United Nations Environmental Programme (UNEP) (1987) The Montreal Protocol on Substances that Deplete the Ozone Layer. UNEP, Nairobi, Kenya.

United Nations Environmental Programme (UNEP) (2016) The Montreal Protocol on Substances that Deplete the Ozone Layer. UNEP, Kigali, Rwanda.

United Nations Framework Convention on Climate Change (UNFCCC) (2010) Cancun Agreements. UNFCCC, New York.

United Nations Framework Convention on Climate Change (UNFCCC) (2015) The Paris Agreement. Conference of the Parties (COP) 21. UNFCCC, New York.

United States Congress (1978) Toxic Substances Control Act of 1978. US Congress, Washington, DC.

United States Department of Agriculture (USDA) (2017) Program Costs and Outlays. Risk Management Agency, USDA, Washington, DC. https://www.rma.usda.gov/aboutrma/budget/costsoutlays.html.

United States Environmental Protection Agency (US EPA) (1977) The Clean Air Act Amendments of 1977. US EPA, Washington, DC.

United States Environmental Protection Agency (US EPA) (1990) The Clean Air Act Amendments of 1990. US EPA, Washington, DC.

United States Environmental Protection Agency (2010) The 40th Anniversary of the Clean Air Act. US EPA, Washington, DC.

Utsu T (2013) Catalog of Damaging Earthquakes in the World (Through 2013). International Institute of Seismology and Earthquake Engineering, Tsukuba, Japan. http://iisee.kenken.go.jp/utsu/index_eng.html.

Vogel, S.A. and Roberts, J.A. (2011). Why the Toxic Substances Control Act needs an overhaul, and how to strengthen oversight of chemicals in the interim. *Health Affairs* 30: 898–905.

von Neumann, J. and Morgenstern, O. (1947). *Theory of Games and Economic Behavior*, 2e. Princeton, NJ: Princeton University Press.

Wall Street Journal (WSJ) (2016) The insurance industry has been turned upside down by catastrophe bonds. WSJ 8 August 2016. https://www.wsj.com/articles/the-insurance-industry-has-been-turned-upside-down-by-catastrophe-bonds-1470598470.

Warrick RA, Trainer PB, Baker EJ, Brinkman W (1975) Drought hazard in the United States: a research assessment. Program on Technology, Environment and Man Monograph #NSF-RA-E-75-004. Institute of Behavioral Science, University of Colorado, Boulder, CO.

Weitzman, M.L. (1998). The Noah's ark problem. *Econometrica* 66: 1279–1298.

Weitzman, M.L. (2009). On modeling and interpreting the economics of catastrophic climate change. *Review of Economics and Statistics* 91: 1–19.

Weitzman, M.L. (2011). Fat-tailed uncertainty in the economics of catastrophic climate change. *Review of Environmental Economics and Policy* 5: 275–292.

World Bank (2008) World Development Report 2008. Agriculture for Development. World Bank, Washington, DC.

World Bank (2014) World Bank issues its first ever catastrophe bond linked to natural hazard risks in sixteen Caribbean countries. World Bank, Washington, DC. http://www.worldbank.org/en/news/press-release/2014/06/30/world-bank-issues-its-first-ever-catastrophe-bond-linked-to-natural-hazard-risks-in-sixteen-caribbean-countries.

World Nuclear Association (WNA) (2017) Fukushima Accident. WNA, London. http://www.world-nuclear.org/information-library/safety-and-security/safety-of-plants/fukushima-accident.aspx.

Working, H. (1953). Futures trading and hedging. *American Economic Review* 43 (3): 314–343.

Working, H. (1962). New concepts concerning futures markets and prices. *American Economic Review* 52 (3): 413–459.

Wright, B. (2014). Multiple peril crop insurance. *Choices* 29 (3): 1–5.

Yohe, G.W. and Tol, R.S.J. (2012). Precaution and a dismal theorem: implications for climate policy and climate research. In: *Risk Management in Commodity Markets* (ed. G. Helyette). New York: Wiley.

Zipf, G.K. (1949). *Human Behavior and the Principle of Least Effort*. Cambridge, MA: Addison-Wesley.

5

Economics of Catastrophic Events: Empirical Data and Analyses of Behavioral Responses

5.1 Introduction

Chapter 4 described the economics of catastrophic events from the perspective of economic theoretical developments with their mathematical formulations and policy implications. This chapter continues the work of elaborating the economics of catastrophes but with an emphasis on empirical catastrophe data and economic analyses of them, as well as numerous adaptation strategies and measures that have occurred throughout the recorded historical data.

Of the catastrophic events, natural and man-made, dealt with in this book, the author relies on the two most catastrophic events that humanity frequently experiences: earthquakes and hurricanes (Kiger and Russell 1996; Emanuel 2008). The literature in these areas offers the most well-documented and sophisticated works on catastrophic events in terms of both science and economics (Knutson et al. 2010; Nordhaus 2010). Further, this literature has accumulated relatively reliable historical data valuable for the study of catastrophes, which is in contrast to many other areas of catastrophes that are deficient in such reliable data (Utsu 2013; NCEI 2016).

As highlighted throughout this chapter, the literature of empirical economic studies of these catastrophic events showcases an interdisciplinary characteristic in its approach. The science of hurricanes, also called tropical cyclones (TCs) or typhoons in different regions of the world, is well documented and in particular has made strides since the late 2000s in relation to global warming policy debates (Emanuel 2005; McAdie et al. 2009; Kossin et al. 2013). The economics of hurricanes couples scientific models of hurricanes with economic indicators such as monetary damages, insurance payments, and human fatalities (Nordhaus 2010; Mendelsohn et al. 2012). Studies of behavioral adaptations have emerged to explain the effectiveness of adaptation strategies through the effects of income growth and technological advances (Seo 2015, 2017; Bakkensen and Mendelsohn 2016; Seo and Bakkensen 2017).

One of the primary contributions of this chapter is the presentation of the empirical catastrophe data which have been recorded as early as the beginning of the twentieth century. The data presented are collected globally, capturing regional differences in physical as well as socioeconomic-policy characteristics, with a focus on TCs and earthquakes (Australian Bureau of Meteorology (ABOM 2011; IMD 2015b; JMA 2017). The catastrophe data are compiled from both quantitative statistics, e.g. earthquake Richter scales and cyclone maximum wind speed (MWS), and qualitative documents such as annual cyclone reports (IMD 2015a; JTWC 2017).

Natural and Man-made Catastrophes – Theories, Economics, and Policy Designs, First Edition. S. Niggol Seo.
© 2019 John Wiley & Sons Ltd. Published 2019 by John Wiley & Sons Ltd.

The chapter starts with a description of the scientific aspects of hurricanes, TCs, typhoons, and earthquakes which is a key element of an integrated economic model of each of these catastrophes. Section 5.2 is devoted to describing the science of explaining a genesis of a hurricane. Section 5.3 explains the efforts to construct an index of the destructive potential of a hurricane, which includes the Saffir–Simpson cyclone severity scale, the accumulated cyclone energy (ACE), and the hurricane power dissipation index (PDI) (Saffir 1977; Emanuel 2005). Section 5.4 reviews the efforts to identify the primary factors of destruction and death during catastrophe events, such as MWS, minimum central pressure (MCP), and storm surge (Mendelsohn et al. 2012; Seo and Bakkensen 2017). Section 5.5 elaborates hurricane models that forecast future hurricanes by the end of the twenty-first and twenty-second centuries (Emanuel 2013). Section 5.6 reviews the scientific literature of quantifying and predicting the magnitude of an earthquake (Richter 1958; USGS 2017).

An economic model of hurricanes couples the scientific models and outcomes with socioeconomic-policy variables and decisions. Economists have devised or relied upon various measures of economic impacts of hurricanes: total economic cost, insurance payments, the number of human lives lost, and destruction of capital assets (Pielke et al. 2008; Nordhaus 2010; Seo 2014, 2015). By associating economic indicators with scientific data, a hurricane intensity–damage function is estimated. Further, a hurricane intensity–fatality function is estimated, with applications of count data models. Estimated functional relationships would reveal what aspects of a hurricane lead to economic and human losses and to what extent. The literature of economic modeling is discussed in Section 5.7.

How much economic damages a hurricane causes and how many human lives a hurricane claims depend on how well a region is prepared and adapts. The economic literature shows that the higher the income, the higher the capacity of adaptation: with higher income and adaptation capacity, the number of lives lost or the magnitude of economic damages from the same-intensity hurricane is significantly reduced in all major cyclone ocean basins (Seo 2015; Seo and Bakkensen 2016; Bakkensen and Mendelsohn 2016). The income elasticity of hurricane damages will therefore capture the evidence and effectiveness of adaptation to catastrophic cyclones. The topics of adaptation are discussed in depth in Section 5.8.

The significance of income elasticity of hurricane damages means that a higher income enables affected individuals and communities to take certain adaptation measures and follow certain strategies. These measures and strategies of adaptation are discussed in Section 5.9, along with the economic models of adoptions of these strategies. Some measures are information-enhancing adaptations while others are structure-enhancing adaptations (Seo 2015; Seo and Bakkensen 2016). The former include an early warning system, a TC trajectory projection, and a storm surge advisory. The latter include a storm-resistant building, a cellar, and a cyclone shelter (Emanuel 2011; Seo 2017). A more active governmental intervention strategy is found in an evacuation order, under which a certain community is ordered to evacuate before a cyclone landfall (Seo 2015; Seo and Bakkensen 2017).

The author also showcases successful governmental and international policy interventions in response to catastrophic cyclones. As part of the cyclone preparedness program, the government of Bangladesh started the cyclone shelter program in the early 1990s and the World Bank started funding the restorations and constructions of the shelter program in the mid-2000s (Khan and Rahman 2007; World Bank 2007). By 2009, cyclone researchers reported that there were a sufficient number of cyclone shelters established across the coastal zones of the country, which had become effective in reducing the number of human fatalities significantly by the late 2000s (Paul 2009). The effectiveness of the cyclone shelter program was reported to be remarkable: a 75% reduction in the number of fatalities, given the same-intensity cyclone (Seo 2017). These topics are discussed in Sections 5.9 and 5.10.

5.2 Modeling the Genesis of a Hurricane

One of the most frequently asked policy questions with regard to a catastrophe event such as a hurricane strike is whether it will become more frequent because of changes in a particular variable (Webster et al. 2005; Landsea 2007; Landsea et al. 2009). A more drastic way to put this is to ask whether a more intense hurricane event will become more frequent (Landsea et al. 2006; Elsner et al. 2008; Kossin et al. 2013). An answer to this question cannot be given without a complete understanding of the genesis of a hurricane.

However, the genesis and physics of a hurricane may not be fully understood by hurricane scientists (Emanuel 2013). In the genesis and development of a hurricane, a large number of variables whose interactions are complex are involved. This is in fact one of the markers of a catastrophic event. Other catastrophes such as an earthquake or global warming are characterized by a similarly complex system.

Hurricanes, TCs, and typhoons refer to the same physical phenomenon. The term hurricane is used in the North Atlantic and the Northeast Pacific Ocean. The term tropical cyclone is used in the North Indian Ocean and the southern hemisphere. The term typhoon is used in the Northwest Pacific Ocean (McAdie et al. 2009).

One can rely on historical data to see if there is a trend in annual frequency of a hurricane event. However, the historical cyclone data before the 1970s are not reliable. Aircraft surveys of TCs began in the North Atlantic and the Northeast Pacific Ocean only after World War II. Although satellite recording of TCs began during the 1960s, all TCs were mapped by Earth-orbiting satellites without omission only by 1970 (Vecchi and Knutson 2008; Emanuel 2008).

Before the aircraft reconnaissance era, TCs were recorded through personal visual contacts from ships, islands, and coastal locations. Therefore, TC counts during this era are not precise and, as such, are controversial, although TC counts in the North Atlantic Ocean could be more accurate than those in other oceans owing to more frequent and dense shipping travels between the US and Europe (Holland and Webster 2007; Landsea 2007).

Before a hurricane genesis index is introduced, we need to understand several key concepts and general features of hurricanes. First, the basic mechanics of a TC is as follows. A TC is a giant rotating heat engine that is fueled by the flux of heat from the ocean and moves "erratically" with high speeds. In a positive feedback loop, the flux of heat generates surface winds in the ocean, surface winds then generate lower pressure, lower pressure generates an increase in heat flux, and an increased heat flux generates even stronger winds. A minimum pressure is reached at the eye of a storm. A cyclone rotates counterclockwise in the northern hemisphere and clockwise in the southern hemisphere.

According to the Saffir–Simpson hurricane intensity scale, to be explained in detail shortly, if the sustained wind speeds of a storm reach 33 m s^{-1} (equivalent to 74 m h^{-1} or 119 km h^{-1}), it is called a TC in the Pacific Ocean and a hurricane in the North Atlantic Ocean (Saffir 1977). When the MWS of a storm exceeds 17 m s^{-1} (equivalent to 39 m h^{-1} or 64 knots (nautical miles per hour)) but is below 33 m s^{-1}, it is called a tropical storm in the US (McAdie et al. 2009).

The interactions of a storm with the ocean and the atmospheric environment are governed by numerous factors and processes, of which sea surface temperature (SST) plays a critical role in the genesis and intensification of a cyclone. For a hurricane to be generated, the SST should be at least 26.5° C. Therefore, high-latitude ocean basins cannot be a place for hurricane genesis. In addition, SST places a thermodynamic upper limit on the TC intensity (Emanuel 2013).

A hurricane can be characterized by many variables, including maximum or average wind speeds, MCP, places of origin, cyclone tracks, wind shear (i.e. the difference in wind speed along a vertical straight line of a storm), hurricane size, vorticity (i.e. a measure of rotation), duration, accompanying rainfall, storm surge height, and landfall locations (NHC 2017b).

For example, Hurricane Sandy or Super-Storm Sandy that struck the US in 2012 was only a category 2 hurricane, but was the largest hurricane in history at more than 1000 miles in diameter (Blake et al. 2013). Super-Typhoon Haiyan that struck the Philippines in 2011 was forecasted to be a category 5 hurricane, with the strongest wind speeds ever recorded.

With this background knowledge, a hurricane genesis index can be presented. There is still no consensus theory that can explain the frequency of generations or rate of occurrences of TCs with precision of even an order of magnitude, despite the fact that there has been progress made on the theory of hurricane intensity (Emanuel 2008, 2013). Empirically, about 90 hurricanes develop annually around the globe, with a standard deviation of about 10.

The following index, the genesis potential index (GPI), is one of the theories and models of TC genesis, which gives readers a glimpse of the difficulty of modeling the genesis:

$$GPI = |10^5 \varphi|^{\frac{3}{2}} (M/50)^3 (1 + 0.1 \cdot V_{shear})^{-2}, \tag{5.1}$$

where φ is the absolute vorticity per second (s^{-1}), M is the relative humidity at 600 hPa (hectopascals or millibars), and V_{shear} is the magnitude of the vector shear from 850 to 250 hPa in meters per second (ms^{-1}) (Emanuel 2008).

Does the GPI have a predictive power? By and large, the index is not broadly validated. Nonetheless, the GPI was shown by theorists to explain the dependence of genesis rates, i.e. cyclone frequencies, on occurrences of the El Nino Southern Oscillation (ENSO) event (Camargo et al. 2007).

Figure 5.1 shows the annual number of hurricanes generated in the North Atlantic Ocean for the time period from 1880 to 2013. The data are from the Atlantic hurricane reanalysis project of the National Oceanic and Atmospheric Administration (NOAA) (Vecchi and Knutson 2008; NOAA 2016). The raw number shows that the annual hurricane frequency ranges from three to nine, with about six hurricanes per year on average. The solid line shows the number of hurricanes that made landfall in the US. About one to three hurricanes that are generated in the North Atlantic Ocean reached the country.

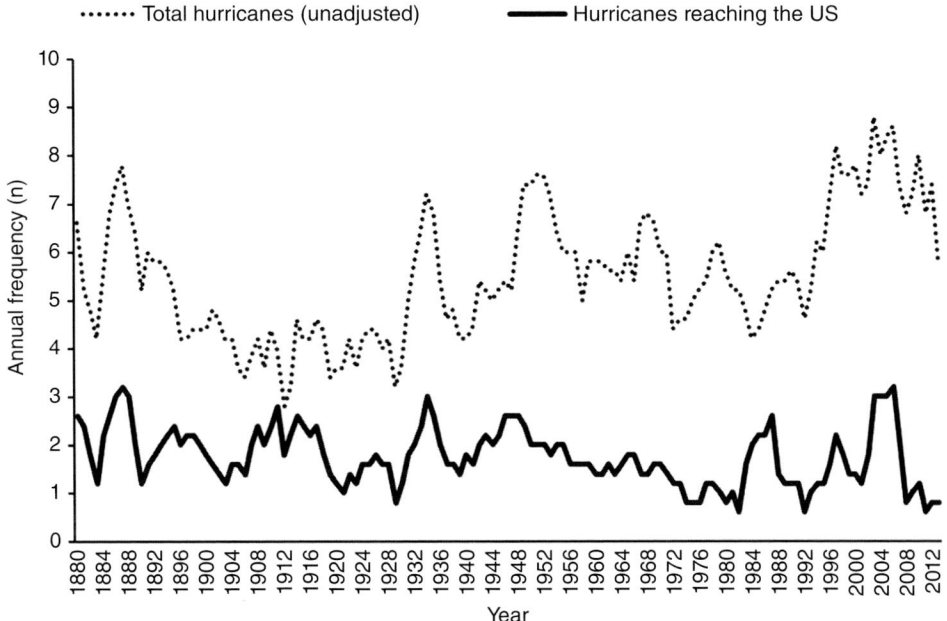

Figure 5.1 Hurricane frequency in the North Atlantic: 1880–2013. *Source*: National Oceanic and Atmospheric Administration's hurricane data re-analysis project (NOAA 2016).

5.3 Indices of the Destructive Potential of a Hurricane

What makes a hurricane destructive? The power, or intensity, or severity of a hurricane is measured in the science literature by various indicators and is a precursor of the destructiveness of a hurricane. The destructiveness is again identified by various measures including the magnitude of monetary loss, insurance payments, and the number of fatalities that result from a hurricane event.

For discussions of indices of destructive potential of a hurricane, let's start with the MWS, which has been relied upon by scientists to explain the severity or power of a hurricane. According to the Saffir–Simpson scale, hurricanes are classified into five categories according to the MWS, i.e. the highest 1-minute average surface wind speed, as follows (Saffir 1977):

- Category 1 TC if MWS $\geq$ 33 mps (74 m h^{-1} or 119 km h^{-1});
- Category 2 TC if MWS $\geq$ 96 m h^{-1};
- Category 3 TC if MWS $\geq$ 111 m h^{-1};
- Category 4 TC if MWS $\geq$ 130 m h^{-1};
- Category 5 TC if MWS $\geq$ 158 m h^{-1}.

Another widely applied index of the destructive potential of a hurricane is ACE. The ACE is an index used by the NOAA to describe the activity of an individual TC as well as an entire cyclone season, particularly in the Atlantic Ocean, by way of an approximation of the total energy exerted by a cyclone system over its lifetime. It is measured every 6 hours.

The ACE of a cyclone season (*i*) is calculated as the sum across individual cyclones (*j*) over the durations of individual cyclones (*k*) as follows:

$$ACE_i = 10^{-4} \sum_{j=1}^{J} \sum_{k=1}^{K} (V_{jk}^{max})^2, \qquad (5.2)$$

where V_{jk}^{max} is the MWS (in knots) and 10^{-4} is the scaling factor which makes the index more manageable.

In the calculation of the ACE, duration of a cyclone is a major factor. That is, a longer duration cyclone can have a higher ACE than a more powerful cyclone with a shorter duration. By contrast, the Saffir–Simpson scale changes from a lower category to a higher category, reaches a peak category, then changes from the peak category to a lower category during the lifetime of a hurricane.

The ACE index, as shown in Eq. (5.2), simply translates the kinetic energy, i.e. wind velocity, into the ACE measure. Therefore, it is not a direct calculation of the cyclone energy. An effort to measure the power of a cyclone more directly can be seen in the construction of a PDI (Emanuel 2008). The PDI is defined as a function of the MWS and the frequency of cyclones (Emanuel 2005):

$$PDI = 2\pi \int_0^\tau \int_0^{\gamma_0} C_D \rho |V|^3 \gamma \, d\gamma \, dt, \qquad (5.3)$$

where C_D is the surface drag coefficient, ρ is the surface air density, $|V|$ is the magnitude of the surface wind, and the integral is taken over a radius to an outer storm limit given by γ_0 and the lifetime of the storm τ. Approximating $C_D \rho$ as a constant, Emanuel defines a simplified PDI as a function of the MWS (V_{max}) measured at the conventional 10-m height and the duration of the storm:

$$PDI = \int_0^\tau V_{max}^3 \, dt \qquad (5.4)$$

One should note that there is a clear similarity as well as a major difference between the ACE in Eq. (5.2) and the PDI in Eq. (5.4). The similarity is that both measures of hurricane destructiveness are a function of the MWS. The clear difference lies in the order of magnitude with which each of these measures increases in response to the MWS. To put it concretely, the ACE increases with an elasticity of 2 in response to V_{max}, while the PDI increases with an elasticity of 3 in response to V_{max}. The correct size of the V_{max} elasticity will be discussed later in this chapter.

Changes in the PDI measured in the North Atlantic Ocean from 1949 to 2009 are plotted in Figure 5.2 using the data compiled by Emanuel (2017). It is overlaid by the SST during August–October in the Main Development Region (MDR), defined as the ocean region bounded by 6°N and 18°N, and 20°W and 60°W. For the whole period in the figure, the changes in PDI do not seem significant statistically, while yearly fluctuations of the PDI are salient.

Limiting our attention to the subset of the time period from 1980 to 2005, the PDI does reveal an increasing trend in the North Atlantic Ocean, perhaps in a statistically significant way, as reported by Emmanuel (2005, 2008). Also, an increasing trend of the SST seems even more evident during the period from 1980 to 2009.

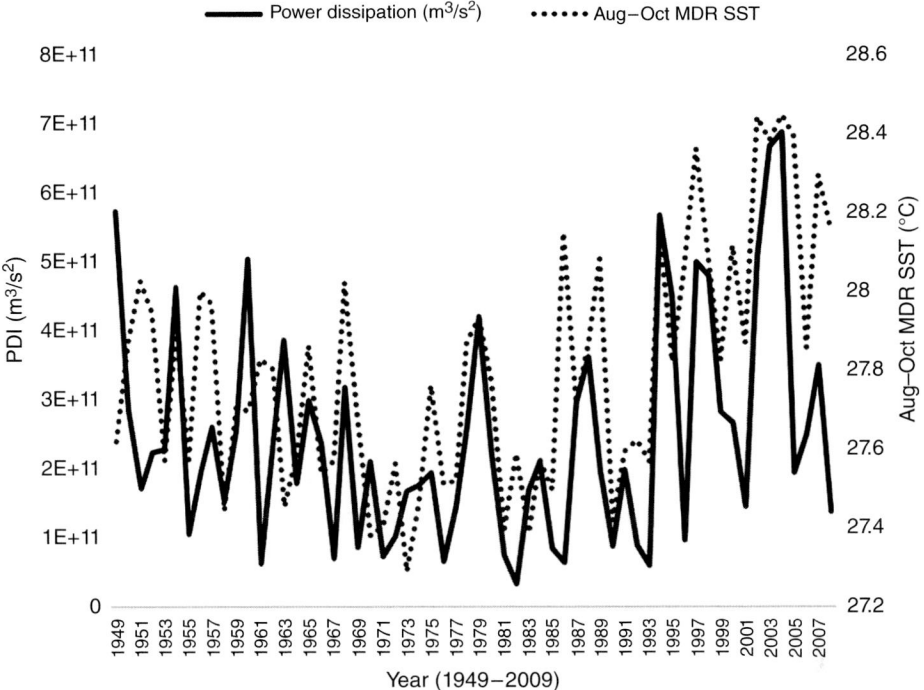

Figure 5.2 Changes in power dissipation index (PDI) and sea surface temperature (SST) from 1949 to 2009 in the North Atlantic Ocean. Main Development Region (MDR) is defined as the region bounded by 6°N and 18°N, and 20°W and 60°W. SST data are from the HADISST1 dataset from the UK Hadley Center averaged from August through October of each year. *Source*: Emanuel (2017).

5.4 Factors of Destruction: Wind Speeds, Central Pressure, and Storm Surge

The three indicators of TC destructive potential, i.e. the Saffir–Simpson scale, ACE, and PDI, are all based on the same variable: the MWS. The implicit assumption in the constructions of these indicators is that the MWS is the primary factor that causes destruction of built and natural environments as well as human fatalities.

With the same assumption, an early economic study by Nordhaus relies on the relationship between economic damages and MWS for estimating the destructiveness of hurricanes (Nordhaus 2010). However, it is more often not the MWS alone that makes a TC destructive or catastrophic.

Researchers noted that the MCP of a TC is a better indicator of how destructive it would turn out to be in terms of total economic damages (Mendelsohn et al. 2012). As explained in previous sections, a TC is generated and intensified by a pressure gradient which increases as the wind starts to rotate along the eye of a storm. Therefore, it is reasonable to think that the power and destructiveness of a TC depends on how low the central pressure at the eye of the storm becomes.

An empirical analysis of the relationship between economic damages and the hurricane intensity indicators reveals that the MCP of a TC is a better measure for explaining the economic damages than the MWS (Mendelsohn et al. 2012). Studies of TCs in the southern hemisphere as well as South Asia confirm that the relationship between economic damages or human fatalities and the MCP is highly significant (Seo 2014, 2015; Seo and Bakkensen 2016).

Lately, researchers reported that an exceptionally large number of hurricane fatalities from a TC is not primarily due to either MWS or MCP, two scientific indicators of hurricane intensity. In the US and Australia, the number of hurricane fatalities is in nearly all cases small, except for a few exceptional cases such as Hurricane Katrina in 2005 (Seo 2015; Bakkensen and Mendelsohn 2016; see also Figure 1.2 in Chapter 1). By contrast, the numbers of hurricane deaths are often exceptionally high in the North Indian Ocean that affects South Asia where Bangladesh and India are located (Seo and Bakkensen 2016).

It is found that the exceptionally large number of fatalities in this region is primarily caused by a high storm surge induced by a TC (Seo and Bakkensen 2017). As shown in Table 1.1 in Chapter 1, four of the five deadliest hurricanes in terms of the number of fatalities occurred in the North Indian Ocean. The deadliest, Cyclone Bhola in 1970, killed up to half-a-million people (IMD 2015a).

TC scientists in the South Asian region also have long been aware of catastrophic consequences of high sea surges induced by cyclones (Murty et al. 1986; Ali 1999). However, they by and large were unable to distinguish the effects of storm surges from those of high-speed winds (Paul 2009).

The scientific research to build a prediction model for storm surge heights during a cyclone event has been on-going, but has yielded only limited success (Hubbert et al. 1991; Jelesnianski et al. 1992; Dube et al. 2009; Surgedat 2015). In particular, there is no consensus prediction on the changes in the storm surges induced by future global warming (Knutson et al. 2010; IPCC 2014).

A hurricane damage–intensity function was first estimated by William Nordhaus using normalized hurricane damage and MWS as follows (Nordhaus 2010):

$$\ln(C_{it}/GDP_t) = \alpha + \beta \ln(MaxWind_{it}) + \delta Year_t + \varepsilon_{it} \tag{5.5}$$

In Eq. (5.5), GDP_t is gross domestic product at year t, C_{it} is damage from a hurricane, $MaxWind_{it}$ is the MWS of a hurricane, $Year_t$ is a year dummy, and ε_{it} is an error term.

The β captures how destructive an increase in the MWS of a hurricane is in terms of total damage. More specifically, it is an elasticity of the normalized damage in response to the MWS. That is, it captures the percentage increase in the normalized damage in response to the percentage increase in the MWS.

After testing a range of assumptions of the error term, Nordhaus proposed the "ninth power law of damages." That is, $\beta = 9$. This can be contrasted to the power of 2 in the ACE in Eq. (5.2) and the power of 3 in the PDI in Eqs. (5.3) and (5.4).

In place of the MWS, Mendelsohn and coauthors relied on MCP for estimating the hurricane damage–intensity relationship (Mendelsohn et al. 2012). They showed that the damage–intensity relationship is statistically more significant when the MCP is used for an indicator of hurricane intensity.

Applying Mendelsohn and coauthors' approach, the present author Seo estimated the hurricane damage–intensity relationship for hurricanes in the southern hemisphere using the MCP (Seo 2014):

$$\ln Damage_i = \alpha + \beta \ln MCP_i + \gamma_1 POP_i + \gamma_2 ALT_i + \gamma_3 LaNina_i + \gamma_4 STATE_i + \varepsilon_i, \quad (5.6)$$

where, in addition to the MCP, the population of a cyclone-hit region (POP), the altitude of a cyclone-hit region (ALT), a La Nina year dummy (LaNina), and the state of a cyclone-hit region (STATE) are entered as control variables.

After testing a set of assumptions on the error term, the author reports an estimate of 32 for the β term. That is, the elasticity of the total economic damage in response to the MCP is 32: a 1% increase in MCP results in a 32% increase in the total damage. The MCP elasticity of total economic damages is far larger in the southern hemisphere hurricanes than the MWS elasticity of total economic damages in the North Atlantic hurricanes.

The larger elasticity in the southern hemisphere hurricanes may be attributable to several factors. First, Nordhaus relied on total insurance payments as the measure of total economic damages, while the present author relied on actual reported losses by the ABOM (ABOM 2011). Second, the former relied on the MWS for the measure of intensity, while the latter relied on the MCP for the same (Nordhaus 2010; Seo 2014).

5.5 Predicting Future Hurricanes

Of major concern with regard to climate change is whether it will alter frequencies, intensities, and surge heights of the hurricanes that are generated under a warmer greenhouse world (Knutson et al. 2010). Predictions are made through to the end of the twenty-first century and the end of the twenty-second century, which vary across hurricane prediction models as well as across the climate scenarios based upon which TC prediction models are run (Nakicenovic et al. 2000; IPCC 2014). Further, future TC predictions vary across cyclone ocean basins to which these predictions are made.

According to the World Meteorological Organization (WMO)'s TC program, there are 9 TC ocean basins which are further divided into 13 TC ocean regions. There are 12 Regional Specialized Meteorology Centers (RSMC) and Tropical Cyclone Warning Centers (TCWC) (NHC 2017a). The 9 TC ocean basins are the Atlantic Ocean, Eastern Pacific Ocean, Central Pacific Ocean, Northwest Pacific Ocean, South Pacific Ocean, North Indian Ocean, Southwest Indian Ocean, Southwest Pacific Ocean, and Southeast Indian Ocean.

Table 5.1 summarizes the future TC predictions in the southern hemisphere ocean basins by a selected TC prediction model, under a range of climate change prediction models (Emanuel et al. 2008). Specifically, the ocean basins are the Southwest Pacific Ocean and the Southeast Indian Ocean where the TCs that affect Australia are generated. The future time period is the two preceding decades before year 2200.

Predictions are made under seven Climate Model Inter-comparison Project (CMIP) climate models, called the Atmospheric Oceanic General Circulation Models

Table 5.1 Projections of tropical cyclones in the southern hemisphere by 2200.

Climate model	Institution	Changes in TC intensity (~% changes)	Changes in TC frequency (~% changes)
CCSM3	National Center for Atmospheric Research, US	−1	−6
CNRM-CM3	Centre National de Recherches Météorologiques, Météo-France	−9	−21
CSIRO-Mk3.0	Australian Commonwealth Scientific and Research Organization, Australia	+9	−8
ECHAM5	Max Planck Institute, Germany	+5	−19
GFDL-CM2.0	NOAA Geophysical Fluid Dynamics Laboratory, US	−10	−22
MIROC3.2	CCSR/NIES/FRCGC, Japan	+13	−20
MRI_cgcm2.3.2a	Meteorological Research Institute, Japan	+11	−2

Source: Emanuel et al. (2008).

(AOGCMs), which are built by the following seven institutions (Taylor et al. 2012; Emanuel et al. 2008; Emanuel 2013): the CCSM3 model from the National Center for Atmospheric Research in the US, the CNRM-CM3 from the Centre National de Recherches Météorologiques, Météo-France, the CSIRO-Mk3.0 from the Australian Commonwealth Scientific and Research Organization in Australia, the ECHAM5 from the Max Planck Institute in Germany, the GFDL-CM2.0 from the NOAA Geophysical Fluid Dynamics Laboratory (GFDL) in the US, the MIROC3.2 from the CCSR/NIES/FRCGC Japan, and the MRI_cgcm2.3.2a from the Meteorological Research Institute (MRI) in Japan.

Across all seven climate models, a decrease in the frequency of hurricanes in the southern hemisphere is predicted, with a degree of change that varies widely from one climate model to another. More specifically, the MRI climate scenario results in a 2% decrease and the CCSM3 climate scenario results in a 6% decrease, whereas the GFDL climate scenario results in a 22% decrease and the CNRM climate scenario results in a 21% decrease in annual TC frequency.

Predictions of TC intensity changes are by and large inconclusive. Some climate models lead to a large increase in hurricane intensity, while other climate models lead to a large decrease in hurricane intensity. The MIROC climate scenario leads to a 13% increase in TC intensity and the MRI to a 11% increase. On the other hand, the GFDL climate scenario results in a 10% decrease in TC intensity and the CNRM climate scenario results in a 9% decrease in hurricane intensity by the end of the twenty-second century.

The projections of TC activities under a globally warmer world would be different in other TC ocean basins (Emanuel et al. 2008). In Table 5.2, projections of TC activities under the aforementioned seven climate models in South Asia by the end of the twenty-first century are presented.

The TCs that strike South Asian countries are generated in the North Indian Ocean basin which encompasses the Bay of Bengal and the Arabian Sea. Some of the most

5.5 Predicting Future Hurricanes

Table 5.2 Projections of tropical cyclones in South Asia by 2100.

Climate model	Institution	Changes in TC frequency (%)	Changes in TC intensity (%)
CCSM3	National Center for Atmospheric Research, US	+6	+13
CNRM-CM3	Centre National de Recherches Météorologiques, Météo-France	−21	−15
CSIRO-Mk3.0	Australian Commonwealth Scientific and Research Organization, Australia	−11	+8
ECHAM5	Max Planck Institute, Germany	−19	+2
GFDL-CM2.0	NOAA Geophysical Fluid Dynamics Laboratory, US	−13	−3
MIROC3.2	CCSR/NIES/FRCGC, Japan	−12	+20
MRI_cgcm2.3.2a	Meteorological Research Institute, Japan	+12	+2

vulnerable countries to TCs are located in South Asia, including Bangladesh, India, Myanmar, and Sri Lanka.

Predicted changes in annual TC frequency are not uniformly negative. Under some climate models such as the CCSM3 and the MRI, an increase in the TC frequency is forecast. Under the other climate models, a decrease in the TC frequency is forecast.

Predictions of changes in TC intensity are again split across the seven climate models, but predicted changes in TC intensity in the North Indian Ocean basin are much larger than those in the southern hemisphere ocean basins. The MIROC climate scenario results in a 20% increase in TC intensity, while the CNRM scenario leads to a 15% decrease by the end of the twenty-first century.

Future TC predictions by Emanuel and other scientists are based on simulation techniques (Emanuel et al. 2008). TC simulations crucially rely on two random or arbitrary components. One is downscaling of climate predictions. That is, to generate a TC, scientists need to know future climate conditions at a very fine spatial resolution (Nakicenovic et al. 2000; Taylor et al. 2012). Since the existing AOGCMs are capable of predicting future climate conditions at a gross resolution, e.g. a 1° latitude by 1° longitude grid cell, hurricane scientists must downscale the climate predictions to a much finer resolution cell. This process involves randomness.

Second, future TCs are then simulated by a "seeding" method. That is, a scientist sows a large number of seeds across the ocean basin of concern randomly and examines whether each seed meets a proper condition for it to develop eventually as a hurricane. Most of the seeds fail to develop into a TC, but some seeds succeed. Seeding is an arbitrary process in that it cannot replicate the process by which a TC is developed. To be more specific, scientists cannot know how and where seeds might fall in the actual world of TC generations.

Because of these random and arbitrary components, a single scientist's repeated simulations at different times can lead to different TC predictions. Notably, Emanuel's predictions of TC activities in 2013 are different from his 2008 TC predictions in a major way, due most likely to these uncontrolled components in the simulation method. Specifically, in contrast to the results in 2008, he predicted an increase in global TC frequency in the 2013 predictions (Emanuel 2013).

In addition to the discrepancies in the predictions by an individual researcher and a single TC prediction simulation method, there is variation in predictions across researchers and TC prediction models. Future TC predictions by Camargo (2013), Kossin et al. (2013), and Tory et al. (2013) yield different projections.

For example, examinations of future TC frequencies by Camargo across the CMIP5 climate models found no robust changes in TC frequency (Camargo 2013). Kossin and coauthors found a significant positive, albeit much reduced, trend in lifetime maximum intensity of the strongest storms during the period of 1982–2009 after homogenization of earlier period and later period storms (Kossin et al. 2013). Troy and coauthors, on the other hand, reported that TC frequency at the end of the twenty-first century will decrease across all the models examined by the authors by substantial percentages (Tory et al. 2013).

5.6 Measuring the Size and Destructiveness of an Earthquake

Before we move on to the analyses of economic, behavioral, policy outcomes of hurricane events, it would be worthwhile for us to take a look at the parallel literature of the science of earthquakes, another major catastrophic event which often kills thousands of people in each outbreak. Like the science of hurricanes, seismology science does not yet have the complete capacity to predict sufficiently early occurrence times and size of an earthquake.

In parallel with the cyclone literature, one of the fundamental research questions of seismology is how to define the size or destructive potential of an earthquake, based upon how to predict the times, locations, and sizes of earthquake occurrences. For the study of the literature of hurricane science as well as the broader literature of catastrophes, a review of the literature on earthquake science would turn out to be illuminating.

Earthquake scientists measure the size of an earthquake by examining seismic waves which are recorded on the instrument called the seismograph. The seismograph records a zig-zag trace in accordance with varying amplitudes of ground oscillations induced by an earthquake, from which time, location, and size of an earthquake are determined based on a range of assumptions and methods, to be explained presently (USGS 2017).

Three methods have been developed over time to quantify the magnitude, size, or destructiveness of an earthquake: a magnitude scale, a moment scale, and a radiated energy scale. Each of these scales has a unique emphasis. The first of these scales has a focus on maximum amplitude; the second on fault geometry; the third on energy release.

The Richter scale, the first commonly used empirical indicator of earthquake sizes, was developed in 1935 by Richter, a seismologist at the California Institute of Technology. Richter relied on the distance from an earthquake to a seismograph and the maximum signal amplitude recorded at the seismograph station to provide a quantitative index of the size or strength of an earthquake (Richter 1958).

More formally, the Richter magnitude scale (M_L) may be written as follows, with *Amp* being the maximum amplitude and δ being the correction factor which is a function of the distance (d) from the epicenter:

$$M_L = \log_{10} Amp - \log_{10} \delta(d). \tag{5.7}$$

The Richter magnitude scale is expressed as a logarithm of the maximum amplitude, with 10 as the base, which means that an increase by one in the magnitude scale corresponds to a 10-fold increase in the maximum amplitude. This holds true for other magnitude scales explained in the following.

His calibration held only for California earthquakes that would occur within 600 km of a seismograph. Further, the variations in earthquake focal depths were not corrected, since most California earthquakes occur within the top 16 km of the Earth's crust. Subsequently, researchers extended the Richter's original magnitude scale (M_L) to the earthquakes that would occur at any distance from a seismograph and at any focal depth as deep as 700 km below the ground (Engdahl et al. 1998).

Two magnitude scales were developed as an extension of the Richter scale: the body-wave magnitude scale and the surface-wave magnitude scale (USGS 2017). The former relies on the body-waves that are caused by earthquakes and travel into the Earth, while the latter relies on the surface-waves that are caused by earthquakes and travel through the uppermost layers of the Earth.

The body-wave magnitude scale (M_b) is defined by correcting for the distance, d (in degrees), between the epicenter and a seismograph station as well as the focal depth, h (in kilometers) of the epicenter, by a correction function, Ω:

$$M_b = \log_{10}\left(\frac{Amp}{T}\right) + \Omega(d, h), \tag{5.8}$$

where *Amp* is the maximum amplitude of ground motion (in microns, i.e. one in one million meters) and T is the corresponding period (in seconds), i.e. the time for one full cycle of a wave to complete.

The surface-wave magnitude scale is defined by correcting for the distance between the epicenter and a seismograph station:

$$M_S = \log_{10}\left(\frac{Amp}{T}\right) + 1.66 \cdot \log_{10} d + 3.30. \tag{5.9}$$

However, these extended magnitude scales of the Richter scale have turned out to seriously underestimate the size of an earthquake (USGS 2017). The primary reason is that larger earthquakes, having larger rupture surfaces, would systematically release more long-period energy. A more precise definition for the quantification of the physical impacts of an earthquake would therefore call for the fault geometry as well as the energy radiation during the event.

The moment scale is based on the concept of seismic moment (Hanks and Kanamori 1979). The seismic moment is defined in relation to the fundamental aspects of the fault geometry during an earthquake event. A fault is a fracture along which two blocks of the Earth's crust move. The seismic moment is defined as follows:

$$M_O = \mu \cdot s \cdot \tilde{d}, \tag{5.10}$$

where μ is the shear strength of the faulted rock, s is the area of the fault, and $\tilde{d}$ is the average displacement of the fault. Here parameters are obtained from the analysis of waveform in seismograms.

Because the calculation takes into consideration the fault geometry and the observer azimuth, the seismic moment is a more consistent indicator of the size of an earthquake than seismic amplitude. The moment scale is then a logarithm of the seismic moment:

$$M_W = \frac{2}{3}\log_{10}M_O - 10.7. \tag{5.11}$$

The third way to quantify the size of an earthquake focuses on total energy release. The total amount of energy radiated by an earthquake would provide a more accurate proxy for the potential damages of an earthquake to man-made structures.

The calculation of the total energy calls for the summation of energy flux across a broad suite of frequencies generated by an earthquake as it ruptures a fault. Due to instrumental and technical limitations, however, most estimates of energy releases have historically relied on the following relationship with the surface-wave magnitude scale:

$$\log_{10}E = 11.8 + 1.5 \cdot M_S. \tag{5.12}$$

A radiated energy scale (M_e), i.e. a seismic magnitude based on the total energy release, is then defined as a logarithm of energy radiation:

$$M_e = \frac{2}{3}\log_{10}E - 2.9. \tag{5.13}$$

From Eq. (5.13), for each one-unit increase in the energy release scale, M_e, the associated seismic energy (E) increases by about 32 times.

In Table 5.3, the annual number of earthquake events from 1990 to 2016 are presented by earthquake class as well as the annual number of earthquake deaths (Sipkin et al. 2000; USGS 2017). There are four classes: great earthquake (for moment scale 8+), major earthquake (for moment scale 7–7.9), strong earthquake (for moment scale 6–6.9), and moderate earthquake (for moment scale 5–5.9). In the US, the moment scale replaced the Richter scale in the 1970s, so the table uses the moment scale.

During the time period, a great earthquake occurred globally about once per year. A major earthquake occurred about 15 times per year, a strong earthquake about 139 times per year, and a moderate earthquake about 1510 times per year. As the earthquake category changes from a more severe to a less severe category, the number of events increases each time by about 10-fold.

The number of deaths from earthquakes varied greatly from one year to another. An examination of the table reveals that the number of deaths does not seem to depend on the number of occurrences of severe earthquakes, i.e. major or great earthquakes. For example, the year 2007 witnessed four great earthquakes but had one of the smallest numbers of casualties.

It rather depends on the locations where major earthquakes struck. The large number of deaths in 2004 was due to the South Asian earthquakes and tsunami in that year and the large number of deaths in 2010 was due to the Haiti earthquakes. In both years, the number of fatalities far exceeded 200 000.

This result indicates that, even for earthquake catastrophes, individual and social capacity is a key variable that determines the magnitude of destruction, or put differently, that reduces the number of human fatalities, besides the quantitative magnitude of an earthquake.

Table 5.3 Earthquake statistics, worldwide.

Year	Magnitude and class				Estimated deaths	Geographic locations of major events
	Great: 8+	Major: 7–7.9	Strong: 6–6.9	Moderate: 5–5.9		
1990	0	18	109	1617	52 056	
1991	0	16	96	1457	3 210	
1992	0	13	166	1498	3 920	
1993	0	12	137	1426	10 096	
1994	2	11	146	1542	1 634	
1995	2	18	183	1318	7 980	
1996	1	14	149	1222	589	
1997	0	16	120	1113	3 069	
1998	1	11	117	979	9 430	
1999	0	18	116	1104	22 662	
2000	1	14	146	1344	231	
2001	1	15	121	1224	21 357	
2002	0	13	127	1201	1 685	
2003	1	14	140	1203	33 819	
2004	2	14	141	1515	298 101	2004 Indian Ocean earthquake and tsunami
2005	1	10	140	1693	87 992	2005 Pakistan Kashmir earthquake
2006	2	9	142	1712	6 605	
2007	4	14	178	2074	708	
2008	0	12	168	1768	88 708	2008 China Sichuan earthquake
2009	1	16	144	1896	1 790	
2010	1	23	150	2209	226 050	2010 Haiti earthquake
2011	1	19	185	2276	21 942	
2012	2	12	108	1401	689	
2013	2	17	123	1453	1 572	
2014	1	11	143	1574	756	
2015	1	18	127	1419	9 624	
2016	0	16	130	1550		
Time period average	1.00	14.59	138.96	1510.67	35 241	

Source: USGS (2017).

5.7 What Causes Human Fatalities?

Of the numerous effects of TCs that incur economic damages, the most feared is the size of human fatalities that a TC would cause. A cyclone landfall around the Bay of Bengal, e.g. Bangladesh, India, Myanmar, and Sri Lanka, often kills tens of thousands of people

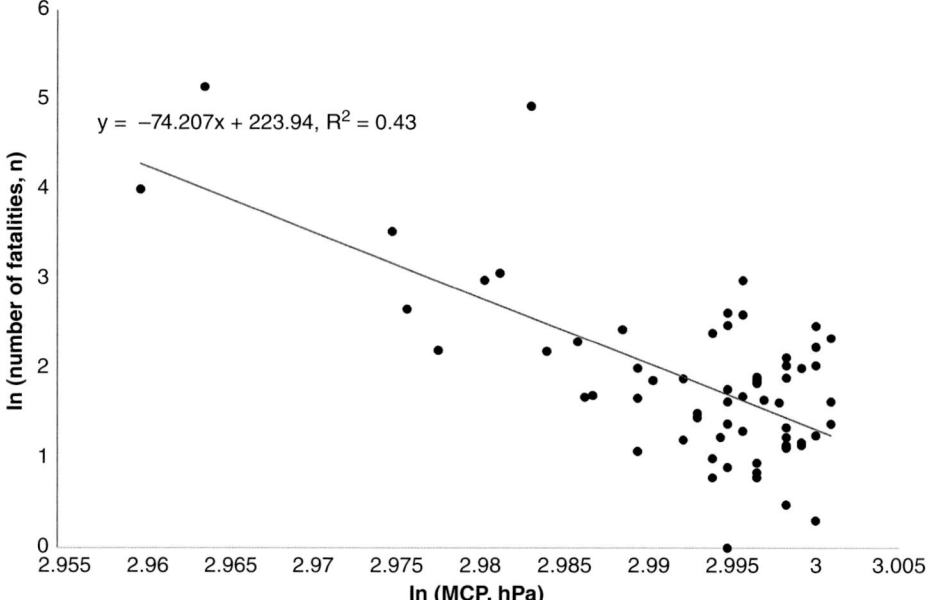

Figure 5.3 The fatality–intensity relationship of tropical cyclones in South Asia.

and displaces millions of people. Sometimes, a TC kills hundreds of thousands of people (Seo and Bakkensen 2016).

The catastrophic nature of TCs generated in the North Indian Ocean that affect South Asian countries is illustrated in Figure 5.3. The figure draws the ln (number of fatalities) against the ln (MCP) of each TC generated in the North Indian Ocean that made landfall and had at least one human casualty. It is drawn using the TCs generated during the 23-year period from 1990 to 2012 (IMD 2015a,b).

A linear trend line is overlaid on the two-dimensional log–log plot, which shows a linear increase in ln (fatalities) in response to a decrease in ln (MCP). The elasticity estimate is −74. That is, without noncyclone factors controlled, the elasticity is

$$\varepsilon = \frac{d \ln(fatalities)}{d \ln(MCP)} = -74. \tag{5.14}$$

This means that a 1% decrease in the MCP resulted in a 74% increase in the number of fatalities. Although this is an astonishing vulnerability of the South Asian countries to TCs, the elasticity estimate is a biased statistic since it removes the effects of other factors such as income and geography. Of the cyclones shown in the figure, the three most catastrophic ones resulted in the deaths of more than about 10 000 people: more specifically, 138 000, 84 000, and 9887 fatalities. Cyclone Bhola which hit Bangladesh in 1970, not included in the figure, caused up to 500 000 deaths (Ali 1999; IMD 2015a).

The large number of fatalities shown in Figure 5.3 is more exceptional than it is normal. Of the cyclones generated in the southern hemisphere ocean basins that have made landfall along the Australian coast since 1970, the most destructive TC was Cyclone Tracy in 1974, but it resulted in only 65 persons' death (ABOM 2011; Seo 2015). In the US, the number of hurricane fatalities from an individual cyclone very rarely exceeded

10 deaths. An exception was Hurricane Katrina in 2005 which killed 1836 people (Pielke et al. 2008; McAdie et al. 2009).

To explain the number of fatalities during a hurricane event with a set of explanatory variables, count data models are applied to cyclone fatality data from a TC ocean basin (Seo 2015, 2017; Seo and Bakkensen 2017). Count data statistical models are rooted on the Poisson distribution, the most basic count data distribution (Poisson 1837). From the Poisson distribution, other more sophisticated count data models are derived, e.g. a high-dispersion count data model and an excess zero count data model (Hilbe 2007).

Let C_i be the number of fatalities from a TC, μ_i be the mean of the random variable, and x_i be the vector of covariates. Then, the C_i is independently Poisson distributed when, for non-negative integer values of the random variable,

$$P[C_i = c_i | x_i] = \frac{e^{-\mu_i} \mu_i^{c_i}}{c_i!}. \tag{5.15}$$

The mean of the Poisson distribution is expressed by

$$E[c_i | x_i] = \mu_i = \exp(x_i' \beta). \tag{5.16}$$

In the Poisson count data model, the mean and the variance of the random variable are exactly the same. However, many count data variables exhibit a high dispersion, that is, the variance is larger than the mean. A negative binomial (NB) distribution (Hausman et al. 1984; Cameron and Trivedi 1986) as well as a Conway–Maxwell–Poisson distribution (Conway and Maxwell 1962; Shmueli et al. 2005) were developed to capture overdispersion in the count data.

For the NB distribution, the mean can be written as a generalized Poisson distribution as follows (Hausman et al. 1984):

$$E[c_i | x_i, \tau_i] = \mu_i \tau_i = \exp(x_i' \beta + \varepsilon_i). \tag{5.17}$$

Let's assume that τ_i follows a *gamma*(θ, θ) distribution with $E[\tau_i] = 1$, $V[\tau_i] = 1/\theta$. For $\theta > 0$, the PDF of the gamma distribution is then

$$g(\tau_i) = \frac{\theta^\theta}{\Gamma(\theta)} \tau_i^{\theta - 1} \exp(-\theta \tau_i). \tag{5.18}$$

The PDF of the NB distribution is written, with $\kappa = 1/\theta$:

$$f(c_i | x_i) = \frac{\Gamma(c_i + \kappa^{-1})}{c! \Gamma(\kappa^{-1})} \left(\frac{\kappa^{-1}}{\kappa^{-1} + \mu_i} \right)^{\kappa^{-1}} \left(\frac{\mu_i}{\kappa^{-1} + \mu_i} \right)^{c_i}, c_i = 0, 1, 2, \ldots \tag{5.19}$$

From Eq. (5.17), the NB distribution is a gamma mixture of a Poisson distribution, with its mean and variance defined as follows (Cameron and Trivedi 1986):

$$E[c_i | x_i] = \mu_i; V[c_i | x_i] = \mu_i [1 + \kappa \mu_i]. \tag{5.20}$$

Since the variance is expressed as a quadratic function of the Poisson mean, the NB model is referred to as the NEGBIN2 model. The κ is the dispersion parameter: a nonzero value of this parameter means overdispersion, while a zero value means the Poisson model.

The Poisson model, the Conway–Maxwell model, and the NB model can be further extended to account for excess zeros in the count data. These extended models are called the zero-inflated Poisson model and so on. The zero-inflated models are a two-stage model in which in the first stage the probability of being a zero count is modeled (Lambert 1992; Winkelmann 2000).

An NB model was applied to study hurricane fatality count data in the southern hemisphere ocean basins, North Indian Ocean, and Northwest Pacific Ocean. From Eq. (5.17), the log-link function of the hurricane fatality model is written as follows for the southern hemisphere ocean basins (Seo 2015):

$$\ln \mu_i = \alpha + \beta \mathrm{MCP}_i + \gamma \mathrm{INC}_i + \delta G_i + \varepsilon_i. \tag{5.21}$$

In the above, μ is the mean of the number of TC fatalities, MCP is minimum central pressure, INC is income per capita, G a set of vulnerability measures such as elevation and population, and ε the error term, and the subscript refers to an individual TC.

The estimate of β in Eq. (5.21) captures the sensitivity of TC fatality to TC intensity. More specifically, it is the growth rate of TC fatality in response to a marginal change in TC intensity:

$$\beta = \frac{d \ln \mu_i}{d \mathrm{MCP}_i} = \frac{d\mu_i/d \mathrm{MCP}_i}{\mu_i}. \tag{5.22}$$

The estimate of β in the southern hemisphere ocean which affects Australasia, based on the 41-year TC data during the period from 1970 to 2010, is -0.0575, which is robust across different error assumptions (Seo 2015). This is interpreted that a 1-millibar decrease in the MCP leads to a 5.75% increase in the number of fatalities.

In the North Indian Ocean which affects South Asian countries such as India, Bangladesh, Myanmar, and Sri Lanka, based on TC data during the period from 1990 to 2012, the estimate of β is, in the model that takes into account the level of storm surge, estimated to be -0.03 (Seo and Bakkensen 2017). This is interpreted that a 1-millibar decrease in the MCP results in a 3% increase in the number of fatalities, excluding the effects of the storm surge.

In the Northwest Pacific Ocean which affects East and South-East Asia, the β estimate was -0.018 (Seo 2018). That is, a 1-millibar decrease in the MCP results in a 1.8% increase in the number of fatalities. This lower sensitivity may be attributable to a larger income effect, γ, arising from the high-income country Japan which has established a resilient system to cope with frequent typhoons.

5.8 Evidence of Adaptation to Tropical Cyclones

Individuals as well as societies who are vulnerable to TCs are observed to have taken numerous measures and strategies in order to deal with the destructiveness of TCs. The evidence of adaptation is abundant, and this is studied by many scientists by way of income elasticity of hurricane fatality (Seo 2015; Bakkensen and Mendelsohn 2016). That is, the higher the income of an affected region, the lower the number of fatalities, from the same-intensity storm. Also for the same-intensity cyclone, the higher the income of an affected region, the lower the damage cost.

The income effect in the hurricane fatality model in Eq. (5.21) is captured by the parameter estimate of γ:

$$\gamma = \frac{d \ln \mu_i}{d \mathrm{INC}_i} = \frac{d\mu_i/d \mathrm{INC}_i}{\mu_i}. \tag{5.23}$$

5.8 Evidence of Adaptation to Tropical Cyclones

Table 5.4 Estimates of intensity, income, and surge effects.

	Southern hemisphere ocean basin	North Indian Ocean basin	North Indian Ocean basin	Northwest Pacific Ocean basin
	Australasia	South Asia	South Asia: a surge-intensity dependence model	East and South-East Asia
Intensity effect ($-\beta$)	0.0575	0.03	0.07	0.0182
Income effect (γ)	−0.0485 (per 1000 AUD)	−0.029 (per 1000 INR)	−0.027 (per 1000 INR)	−0.053 (per US$1000)
Storm surge effect (ρ)			0.099	

The estimate of γ suggests the growth rate of the number of fatalities in response to a one-unit increase in the personal income per capita. The larger the absolute value, the higher the effectiveness of employed adaptation measures and strategies. The smaller the value, the lower the effectiveness of adaptation.

As summarized in Table 5.4, the estimate of γ in the aforementioned southern hemisphere TC data that affect Australasia is −0.0485 (Seo 2015). This is interpreted that a one-unit increase, i.e. a 1000 Australian dollar (AUD) increase, in income per capita leads to a 4.85% decrease in the number of fatalities, given the same-intensity cyclone. This means that the income effect through increased adaptation capabilities by and large offsets the detrimental impact of the increase in hurricane intensity on the number of human fatalities.

Analogously, the estimate of γ in the aforementioned South Asian TC data from the North Indian Ocean is −0.029 (Seo and Bakkensen 2016). For each 1000 Indian Rupee (INR) increase in income per capita, the number of fatalities declines by 2.9%. This again offsets by and large the impact of a one-unit hurricane intensity increase, i.e. 1-millibar decrease in the MCP.

Similarly in Table 5.4, the income elasticity estimate of γ in the East and South-East Asian typhoon data coming from the Northwest Pacific Ocean basin is −0.054 (Seo 2018). For each US$1000 increase in personal income, the number of fatalities declines by 5.4%. Note that this income elasticity estimate is more than three times larger than the intensity elasticity estimate, $\beta = 0.018$.

In the Northwest Pacific Ocean basin, the income effect dominates the intensity effect on the number of fatalities. This is due to the super-wealth of Japan located in the Far East. Owing to the super-wealth, although the country faces super-typhoons quite frequently, the country suffers less than one-tenth of the fatalities suffered by the Philippines, a similarly typhoon-afflicted but poor nation in East Asia (Seo 2018). The per-capita income of Japan was US$47 600 in 2016, which is contrasted to the per-capita income of the Philippines, US$2700 in the same year.

As explained in previous sections, the height of storm surge is one of the primary killers of people under TCs, especially in low-lying South Asian countries such as Bangladesh and India. In such circumstances, the estimation of the hurricane intensity–fatality function shown in Eq. (5.21) can be biased because of unexplained or correlated storm surge effects (Seo and Bakkensen 2017).

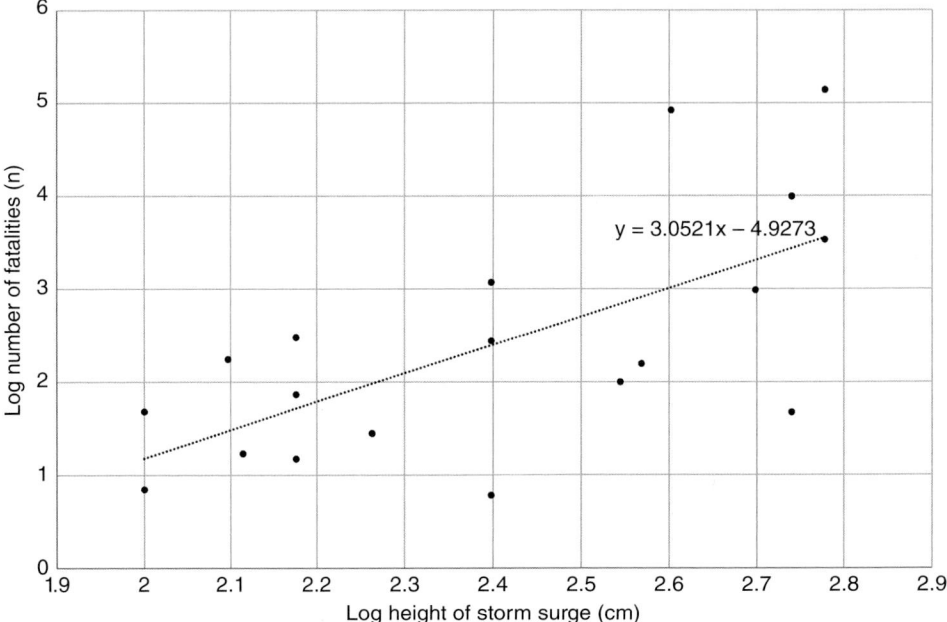

Figure 5.4 The surge–fatality relationship of tropical cyclones in the North Indian Ocean.

Figure 5.4 shows the relationship between the height of storm surge and the number of fatalities using the log-log functional form. From the aforementioned South Asian TC data from the North Indian Ocean (IMD 2015a), the figure shows the storms with reported height of storm surge and with at least one human fatality.

The trend line shows that the number of fatalities increases nonlinearly with the height of storm surge. That is, there is a linear relationship in the log-log plane. In this simplest functional relationship, the number of fatalities increases with an elasticity of +3.05 in response to the height of storm surge. For each 1% increase in the height of storm surge, the number of fatalities increases by 3.05%. This estimate of course does not account for other factors that affect cyclone fatality, and therefore it is biased.

Assuming that the height of storm surge is in part determined by the storm intensity, Seo and Bakkensen estimate the following surge–intensity dependence model (Seo and Bakkensen 2017):

$$\ln \mu_i = \alpha + \beta \text{MCP}_i + \gamma \text{INC}_i + \delta G_i + \rho \hat{\tau} + \varepsilon_i, \tag{5.24}$$

where $\text{SURGE}_i = \tilde{\alpha} + \tilde{\beta}\text{MCP}_i + \tau.$ (5.25)

In Eq. (5.24), $\hat{\tau}$ is the estimate of the error term in Eq. (5.25), which is calculated as the residual of the surge–MCP regression. The functional form in Eq. (5.25), i.e. the surge–MCP regression, can be either linear or nonlinear, depending on empirical evidence. $\hat{\tau}$ is interpreted as the level of surge unexplained by the MCP. In other words, it is the level of storm surge independent of the hurricane intensity.

The estimate of ρ from the aforementioned South Asian TC data is 0.099. That is, for each one-unit (10 cm in the study) increase in the independent storm surge, there is an increase in the number of fatalities by 9.9%. As shown in Table 5.4, the impact of storm surge on the number of fatalities dominates the impact of storm intensity thereof.

The high vulnerability of the South Asian population to the heights of storm surges is demonstrated in Table 5.4 in a stunning way. However, this does not mean that there is no way for an individual or a society to adapt to high storm surges, i.e. reduce the devastating impacts of high storm surges. A few studies have attempted to quantify the effectiveness of adaptation strategies to storm surges (Paul 2009; Seo 2017).

A prime example of adaptation to storm surges is the cyclone shelter program in Bangladesh. After Cyclone Bhola devastated Bangladesh in November 1970, killing as many as 500 000 people, the Bangladesh government initiated a cyclone shelter program as part of a cyclone preparedness program (Khan and Rahman 2007). By 2007, the cyclone shelter program was reported to have established a sufficient number of shelters across the coastal vulnerable zones of the country to be indeed quite effective against intense TCs.

A case for effectiveness of the program was made by way of TC Sidr which made landfall in Bangladesh in 2007 (Paul 2009). TC Sidr in 2007 was more intense than TC Bhola in 1970 in terms of both MCP and MWS. The former resulted in, however, the number of deaths two orders of magnitude smaller than that from TC Bhola, which was attributed by the author to the success of the cyclone shelter program that had lasted for more than three decades.

In addition to the governmental program, the World Bank, through the Emergency Cyclone Recovery and Restoration Project in 2007, provided additional funds to build 230 more shelters and repair 240 existing shelters in the country (World Bank 2007). The fund was used to ensure that a cyclone shelter is within an accessible distance in the event of a TC landfall for any affected individual, the key to the success of a cyclone shelter program.

Can we measure quantitatively the benefit of the cyclone shelter program through reduced number of fatalities in South Asia (Seo 2017)? It can be measured by applying the two-stage surge–pressure dependence model in Eqs. (5.24) and (5.25) to the cyclone shelter program. First, an indicator function of the cyclone shelter program, I_{CSP}, is created. The indicator function takes on the value of 1 when the cyclone shelter program is established to an effective level, and 0 if otherwise.

With this, in the first stage, Eq. (5.25) is run and the residual surge variable is created in the same manner as explained above. In the second stage, the effect of residual storm surge is calculated, using Eq. (5.24), with and without the cyclone shelter program by creating the interaction term with the cyclone shelter indicator:

$$\ln \mu_i = \alpha + \beta \mathrm{MCP}_i + \gamma \mathrm{INC}_i + \delta G_i + \rho_0 \hat{\tau} + \rho_1 \hat{\tau} \cdot I_{CSP} + \varepsilon_i. \tag{5.26}$$

The estimation of Eq. (5.26), presented in Table 5.5, shows the high effectiveness of the cyclone shelter program in South Asia. Parameter estimates of storm surge and the cyclone shelter program are shown in italicized numbers. Both the storm surge parameter estimate and the cyclone shelter program parameter estimate are significant at 5% level.

The estimate of ρ_0, the storm surge parameter, is 0.104, while the estimate of ρ_1, the cyclone shelter program parameter, is −0.076. When there is no cyclone shelter program, an increase in storm surge by one unit (i.e. 10 cm) leads to a 10.4% increase in the number of fatalities. With the effective cyclone shelter program in place, the

Table 5.5 An NB model for cyclone shelter program effectiveness (number of cyclone fatalities).

Explanatory variable	Estimate	P-value
Intercept	72.682	<0.0001
MCP (hPa)	−0.068	<0.0001
Income per capita (1000 INR in 2007)	−0.027	0.0002
Population density (km^2)	0.0002	0.665
Residual surge (10 cm)	*0.1037*	*<0.0001*
Residual surge × cyclone shelter program	*−0.076*	*0.019*
Model statistics		
Dispersion	7.45	

same increase in storm surge by one unit (10 cm) leads to only a 2.8%, i.e. 10.4% − 7.6%, increase in the number of fatalities.

The empirical results in Table 5.5 reveal that the cyclone shelter program reduces the number of fatalities by about 75%, given the same storm-surge-height cyclone. The table presents strong evidence of adaptation by humanity to even the most devastating aspect of TCs, i.e. the heights of storm surges.

5.9 Modeling Behavioral Adaptation Strategies

What underlies the strong and beneficial effects of income growth and the public program on the number of fatalities presented in Section 5.8? What specific actions does income growth enable communities vulnerable to hurricanes to take? What are the effective strategies and measures to varied aspects of hurricane occurrences? These are some of the questions raised with regard to behavioral aspects of TCs.

In the southern hemisphere, more specifically the Australian coast, an early warning system, a TC trajectory forecast, and advance evacuation orders are reported to have played important roles in reducing the number of fatalities attributable to TCs (Seo 2015). The probability of adoption of each of these adaptation strategies was shown to be sensitive to the personal income of an affected region at a specific time of landfall.

An early warning system was developed by the WMO's TC program through the aforementioned 13 RSMCs and TCWCs, before which national meteorological agencies were responsible for cyclone warnings. One of the RSMCs and TCWCs announces TC position and forecast every six or twelve hours depending upon the TC ocean basin, which also issues cyclone watches/warnings (NHC 2017a).

For the National Hurricane Center (NHC) which is one of the RSMCs for the North Atlantic Ocean basin, the TC warning system is composed of a TC advisory, gale warning, tropical storm warning, tropical storm watch, hurricane warning, hurricane watch, storm surge warning, and storm surge watch (NHC 2017b).

A TC advisory is issued before the issuance of TC watches and warnings. It details the TC locations, intensity, movement, and precautions that should be taken. A TC warning is an announcement that a TC with sustained wind speeds higher than 64 knots is expected within a specified area. A TC warning is issued 36 hours in advance of the expected onset of the TC.

A storm surge warning is issued when there is the danger of life-threatening inundation from rising water moving inland within a specified area in association with an ongoing or potential TC, a subtropical cyclone, or a post-TC. It is issued 36 hours in advance.

An advisory, watch, and early warning give residents within a specified area sufficient time and information needed for them to prepare for an upcoming TC landfall. They can take numerous precautionary actions such as boarding up exterior windows, closing all the doors and windows, remaining inside the house during a landfall, and evacuating before a landfall.

An early warning system can be more effective when the trajectory of an upcoming TC is projected with precision and announced sufficiently early. An accurate TC trajectory forecast reduces the uncertainty of an individual with regard to whether the TC will strike her/his community or not.

An early warning system and a TC trajectory forecast can be further augmented by an advance evacuation order in the case of an extremely severe hurricane, or expected inundation from high storm surges, or flooding from heavy accompanying rainfall. Under an intense hurricane, advance evacuation may be called for in communities without sturdy houses and buildings. Even in the case of a moderate hurricane, advance evacuation may be called for when it is accompanied by high sea surges or heavy rainfall, or where it strikes an easily flooded area due to geographic conditions.

With other factors kept the same, the welfare of the community that is struck by a specific TC can be denoted by π_{1j} for the community that adopts a specific adaptation measure (j), e.g. a TC trajectory projection technology, and by π_{0j} for the community that doesn't. Then, the community will prefer to adopt the adaptation measure if it improves the community's welfare:

$$\pi_{1j} > \pi_{0j}. \tag{5.27}$$

The two communities' welfare functions can be written as follows:

$$\pi_{1j} = \alpha_1 + \beta_1 x_1 + \mu_1;$$
$$\pi_{0j} = \alpha_0 + \beta_0 x_0 + \mu_0. \tag{5.28}$$

The probability of choosing the adaptation strategy j is then

$$\Pr(\pi_{1j} > \pi_{0j}) = \Pr(\alpha_1 + \beta_1 x_1 + \mu_1 > \alpha_0 + \beta_0 x_0 + \mu_0)$$
$$= \Pr(\mu_1 - \mu_0 < (\alpha_0 + \beta_0 x_0) - (\alpha_1 + \beta_1 x_1)). \tag{5.29}$$

The probability in the second line of Eq. (5.29) is the cumulative density function (CDF) of a random variable. Assuming a normal distribution or a Gumbel distribution for the random variable for μ_k which is identically and independently distributed, the probability is calculated as a Probit or a Logit (McFadden 1974; Train 2003). In the case of the latter, the probability is written succinctly as follows:

$$\Pr_j = \frac{\alpha_1 + \beta_1 x_1}{\sum_{k=0,1} \alpha_k + \beta_k x_k}. \tag{5.30}$$

A set of applications of the Probit model to the aforementioned TC data from southern hemisphere ocean basins is presented in Table 5.6 (Seo 2015). In the three TC adaptation

Table 5.6 Probit choice model of adopting a tropical cyclone adaptation strategy in southern hemisphere ocean basins. All figures are estimates.

Explanatory variable	Early warning issued	TC path forecast	Advance evacuation order + early warning + path forecast
Intercept	6.111	−6.308 2	8.575 8
MCP (hPa)	−0.008 7	0.003 1	−0.010 8[b)]
Population	−0.000 04	0.000 029	−5.40E-06
Income per capita (1000 INR in 2007)	0.029 7[a)]	0.037 7[a)]	0.022 3[b)]
Elevation	0.001 59	−0.000 41	−0.000 49
Summary statistics			
N	175	175	175
LR statistic (P-value)	0.13	0.01	0.04

a) Denotes significance at 5% level.
b) Denotes significance at 10% level.
Source: Adapted from Seo (2015).

measures examined, income per capita is a significant factor in an adoption decision of each of these measures. In other words, the three TC adaptation measures – early warning, a TC trajectory projection, a combined evacuation order – were adopted more frequently in the communities and times when the personal income was higher.

Notably, hurricane intensity, i.e. the MCP, is not a significant factor for explaining adoption of an early warning issue. Neither is it for adoption of a TC trajectory forecast. The result can be interpreted that regardless of the severity of an approaching cyclone, a TC trajectory forecast or an early warning was made when it approached a high-income locality, say, a rich town.

In the historical Australian context, a TC trajectory projection was more likely to be made when a TC approached higher income states of New South Wales and Queensland than when it approached lower income states of Northern Territory and Western Australia during the time period examined by the aforementioned study, i.e. 1970 to 2012.

The TC trajectory projection technology was not available during the early years of the time period but became available during the middle years. The technology has become more widely applied as the country's income has grown during the time period.

As explained in Section 5.8, TC-caused fatalities in the North Indian Ocean and South Asia are caused primarily by the heights of storm surges and subsequent inundation of villages. As such, effective adaptation strategies must be designed in ways to respond to a high storm surge as well as to high storm intensity. A strategy that is effective against the high intensity of a cyclone may turn out to be ineffective, or even amplify risks, as a response to a high surge. For example, an underground cellar is an effective shelter against high-intensity winds, but may not be a safe shelter against a high surge of seas because it can be easily flooded.

A list of effective adaptation strategies against high storm surges includes permanent relocation of vulnerable populations, temporary evacuation, zoning and building codes for high-vulnerability zones, physical barriers such as polders and levees, cyclone shelters, and post-disaster assistances. The most effective strategies will be different from one locality to another depending upon local geography, cultural history, socioeconomic status, and policy objectives.

A permanent relocation of people or a complete retreat from a low-lying coastal zone can ensure that vulnerable communities are moved out of TCs' harm's way. In many situations, however, it is not a feasible option. High population density, lack of governmental resources, and cultural factors are some of the obstacles in South Asian countries.

For instance, people in Bangladesh have populated TC hazard-prone low-lying coastal zones such as the Ganges–Brahmaputra delta for centuries because the river delta is fertile for rice and jute productions. Hence, it is nearly impossible to force them to relocate, even though several hundred thousand people were killed by the 2004 South Asian tsunami (Sanghi et al. 2011).

A temporary evacuation is a strategy that forces vulnerable population to safer locations temporarily through an evacuation order. An early warning and a TC trajectory projection normally precedes an evacuation order. So, the higher the precisions of an early warning issue and a TC trajectory prediction, the more effective the issue of an evacuation order is likely to be in reducing the number of cyclone fatalities.

An individual may purchase or build a high-wind-resilient house in the areas where cyclone strikes are not infrequent (Emanuel 2011). In low-lying areas that suffer not infrequently from inundations from sea surges caused by TCs, an individual may build or purchase a house in a piloti structure or a multistory apartment, which can serve as a cyclone shelter during a severe cyclone surge event (Paul 2009).

Another adaptation strategy is constructing a physical barrier. A coastal wall, a levee, or a polder can reduce the intensity by which seawater inundates coastal communities. The cost of constructing and maintaining a coastal wall can be significant, while that of constructing a polder is less costly. A green buffer, such as a mangrove forest, can provide partial protection from a cyclone storm surges. Unintended spillover effects, however, may ensue from construction of a coastal wall in some locations. For example, tourism may significantly suffer if coastal walls are built along the popular beaches or in high-value coastal lands (Ng and Mendelsohn 2006).

Using the aforementioned South Asian TC data from 1990 to 2015, a Probit choice model of adopting each of the following adaptation measures is estimated and shown in Table 5.7: a surge advisory, surge modeling, TC trajectory modeling by the Limited Area Model (LAM), more advanced TC trajectory modeling, and a cyclone shelter program (Seo and Bakkensen 2017).

The top panel in Table 5.7 shows the estimation results for adaptation strategies to mainly storm surges and the bottom panel shows the estimation results for adaptation strategies to mainly TC intensity. That is, a surge advisory, surge modeling, and a cyclone shelter program are responses to the risk of storm surges, while TC trajectory projection technologies are responses primarily to the risk of high-speed winds.

Commensurate with the southern hemisphere results in Table 5.6, adoption of a TC trajectory projection model with LAM or a more advanced TC trajectory projection model is more frequent with higher personal income of cyclone-affected regions. Also, TC intensity measured in MCP is not a significant factor in explaining the adoption of either of the TC trajectory projection models.

The top panel shows that adoptions of adaptation strategies to cyclone-caused storm surges were sensitive to the personal income of affected regions. The higher the income, the higher the probability of issuing a storm surge advisory. The same is true of TC

Table 5.7 Probit adoption model of adaptation strategies to cyclone-induced surges and cyclone intensity in South Asia.

(a) Adaptation strategies primarily to TC-induced surges

	Surge advisory		Cyclone shelter program		Surge modeling	
Explanatory variable	Estimate	P-value	Estimate	P-value	Estimate	P-value
Intercept	−1.53	<0.0001	−2.37	<0.0001	−0.53	0.03
Surge (cm)	0.027	0.78	0.122	0.29	−0.054	0.55
Income per capita (1000 INR in 2007)	0.034	<0.0001	−0.026	0.082	0.057	<0.0001
Population density (km^2)	−0.00008	0.82	0.0019	0.0004	−0.0009	0.02
Summary statistics						
Wald statistic (P-value)	<0.0001		<0.0001		<0.0001	

(b) Adaptation strategies primarily to TC intensity

	Cyclone trajectory modeling (LAM technology)		Advanced cyclone trajectory modeling (QLM, NMM technology)	
	Estimate	P-value	Estimate	P-value
Intercept	8.48	0.31	0.65	0.92
MCP (hPa)	−0.0086	0.31	−0.0017	0.79
Income per capita (1000 INR in 2007)	0.069	0.0004	0.04	<0.0001
Population density (km^2)	−0.0001	0.72	−0.0001	0.71
Summary statistics				
Wald statistic (P-value)	<0.0001		<0.0001	

QLM, Quasilinear model; NMM, nonhydrostatic mesoscale model.
Source: Adapted from Seo and Bakkensen (2016).

surge modeling. In high-income regions or times, these strategies were more likely to be implemented.

On the other hand, the cyclone shelter program implemented by the government of Bangladesh is more likely to be adopted in lower-income regions or times. The lower the income per capita of an affected region, the higher the probability of the program is to be adopted. Unlike the other storm surge adaptation measures, the cyclone shelter program is unique in that it is a government intervention program. The intervention is directed to poor and highly vulnerable communities (Khan and Rahman 2007).

Again, the three storm surge adaptation strategies are not adopted primarily because of higher storm surges. Instead, in the choice of surge adaptation strategies, personal income is a more dominant factor than the height of a storm surge. What this says is that the surge adaptation measures were adopted more often in higher-income regions or higher-income years than in poorer regions or times. Economy-wide technological changes, exogenous to historical hurricane events, have made these measures effective, available, and affordable to hurricane responders and potential victims.

5.10 Contributions of Empirical Studies to Catastrophe Literature

Following Chapter 4, this chapter presents the economics of catastrophic events with a focus on the empirical globally collected catastrophe data and the series of economic analyses of them. The literature of TCs is described at length through the datasets collected from the global hurricane records as well as TC ocean basin specific datasets. This is complemented by a brief presentation of the earthquake literature and data.

What are the critical messages or insights we can gain from the empirical studies of TC catastrophes and earthquakes – the two most catastrophic natural events that humanity has faced recurrently during the past century, as presented in this chapter – which hold broader implications for the catastrophe literature?

First, a complete and perfect understanding of the entirety of a catastrophic event seems almost impossible at the current stage of scientific development of the concerned event. For hurricanes, genesis, intensity changes, track changes, heights of storm surges, and global-warming-induced changes are only incompletely understood and predicted. The processes involved in a catastrophe event are Mandelbrot-complex and close to the chaos described by Edward Lorenz (Lorenz 1969a,b, Mandelbrot 1983). Century-long predictions of hurricane events are difficult, to say the least. Nonetheless, it should be noted that substantial progress has been made in the science of hurricanes (Emanuel 2013).

Second, economic damages and lives lost over the past century attributable to TCs and other natural catastrophes have decreased remarkably. In Figure 1.2 in Chapter 1, the annual number of US hurricane fatalities is shown from 1900 to 2016 (Blake et al. 2011; NOAA 2017). The figure shows a decreasing trend in the number of fatalities over the time period. It is also salient that the number of hurricane deaths during the post-1970 period is far smaller than that during the pre-1970 period. With the exception of Hurricane Katrina in 2005 which killed 1245 people, the annual number of fatalities since 1980 in the US did not exceed 62.

The reduction in the number of deaths has materialized despite the fact that individuals and governments were not fully aware of all aspects of TCs. Even without the perfect knowledge of numerous aspects of hurricane events, individuals and the public sector were able to reduce the deadly consequences of hurricane events, making use of available technologies, capacity building, and education.

The third message is that higher income enables individuals and communities to cope better with catastrophic events. In our analysis of the South Asian cyclone fatality data, in response to a 1000 INR increase in income per capita, the number of fatalities fell by about 3%. Similarly, there is strong income effect on the cyclone fatality of the southern hemisphere ocean basins as well as that of the Northwest Pacific Ocean basin (Seo 2015, 2018; Bakkensen and Mendelsohn 2016).

Higher income could mean many things to potential victims of a catastrophe. It may mean that an individual can locate her/his residence in a safer place; it may mean that an individual can choose a more sturdy and resilient house; it may mean that a potential victim has the means to escape in the event of a catastrophe, e.g. an automobile or a motorboat; it may mean that individuals have better and faster access to critical information.

The fourth message is that advances in science and technology play an important role in adaptation to catastrophic events. Advances in TC trajectory projection technology, TC storm surge modeling, and an early warning system have been largely successful in reducing the number of fatalities. These technologies would not have been possible without the success of science and a supercomputing technology. By 1980, satellite recording of cyclone events became near universal across the globe, making the hurricane research and policy discussions more precise and meaningful.

The fifth message is that the literature reviewed in this chapter indicates that governmental interventions can turn out to be a highly effective strategy in drastically reducing the devastating consequences of catastrophic events. The cyclone shelter program in Bangladesh which began during the 1970s and was aided by the World Bank has over time turned out to be successful in reducing the number of human deaths substantially (Paul 2009; Seo 2017). Although, the author cannot present an exact benefit-cost analysis of the cyclone shelter program by the Bangladesh government, it is evident that it has yielded far greater benefit in terms of reduced human deaths than the total cost of establishing the shelters which was in part provided by international organizations.

However, government interventions in the past in addressing catastrophes have resulted in more failures than successes. Since a detailed review of catastrophe policies is the subject of the next chapter, readers will have an opportunity to evaluate a whole range of policy measures for a variety of catastrophic events. In the policy area of TCs, for example, the National Flood Insurance Program which was designed to help the victims from hurricanes through a government-subsidized insurance resulted in perverse responses from individuals which led to a rapid increase in the government and taxpayer cost burden of the program (USGAO 2003; Knowles and Kunreuther 2014).

References

Ali, A. (1999). Climate change impacts and adaptation assessment in Bangladesh. *Climate Research* 12: 109–116.

Australian Bureau of Meteorology (ABOM) (2011) Tropical cyclones. http://www.bom.gov.au/cyclone/history.

Bakkensen, L.A. and Mendelsohn, R. (2016). Risk and adaptation: evidence from global hurricane damages and fatalities. *Journal of the Association of Environmental and Resource Economists* 3: 555–587.

Blake ES, Landsea CW, Gibney EJ (2011) The deadliest, costliest, and most intense United States tropical cyclones from 1851 to 2010 (and other frequently requested hurricane facts). NOAA Technical Memorandum NWS NHC-6. NOAA, Silver Spring.

Blake ES, Kimberlain TB, Berg RJ, Cangialosi JP, Beven II, JL (2013) Tropical Cyclone Report: Hurricane Sandy (AL182012). National Hurricane Center, Miami, FL.

Camargo, S.J. (2013). Global and regional aspects of tropical cyclone activity in the CMIP5 models. *Journal of Climate* 26: 9880–9902. doi: 10.1175/JCLI-D-12-00549.1.

Camargo, S.J., Emanuel, K., and Sobel, A.H. (2007). Use of a genesis potential index to diagnose ENSO effects on tropical cyclone genesis. *Journal of Climate* 20: 4819–4834.

Cameron, A.C. and Trivedi, P.K. (1986). Econometric models based on count data: comparisons and applications of some estimators and some tests. *Journal of Applied Econometrics* 1: 29–53.

Conway, R.W. and Maxwell, W.L. (1962). A queuing model with state dependent service rates. *Journal of Industrial Engineering* 12: 132–136.

Dube, S.K., Jain, I., Rao, A.D., and Murty, T.S. (2009). Storm surge modelling for the Bay of Bengal and Arabian Sea. *Natural Hazards* 51: 3–27.

Elsner, J.B., Kossin, J.P., and Jagger, T.H. (2008). The increasing intensity of the strongest tropical cyclones. *Nature* 455: 92–95.

Emanuel, K. (2005). Increasing destructiveness of tropical cyclones over the past 30 years. *Nature* 436: 686–688.

Emanuel, K. (2008). The hurricane–climate connection. *Bulletin of the American Meteorological Society* 89: ES10–ES20.

Emanuel, K. (2011). Global warming effects on U.S. hurricane damage. *Weather, Climate, and Society* 3: 261–268.

Emanuel K (2013) Downscaling CMIP5 climate models shows increased tropical cyclone activity over the 21st century. Proceedings of the National Academy of Sciences of the United States of America 110: 12219–12224.

Emanuel K (2017) Papers, Data, and Graphics Pertaining to Tropical Cyclone Trends and Variability. Personal website. MIT, Cambridge, MA. https://emanuel.mit.edu/papers-data-and-graphics-pertaining-tropical-cyclone-trends-and-variability.

Emanuel, K., Sundararajan, R., and Williams, J. (2008). Hurricanes and global warming: results from downscaling IPCC AR4 simulations. *Bulletin of the American Meteorological Society* 89: 347–367.

Engdahl, E.R., van der Hilst, R., and Buland, R. (1998). Global teleseismic earthquake relocation with improved travel times and procedures for depth determination. *Bulletin of the Seismological Society of America* 88: 722–743.

Hanks, T.C. and Kanamori, H. (1979). A moment magnitude scale. *Journal of Geophysical Research* 84 (B5): 2348–2350.

Hausman, J.A., Hall, B.H., and Griliches, Z. (1984). Econometric models for count data with an application to the patents–R&D relationship. *Econometrica* 52: 909–938.

Hilbe, J.M. (2007). *Negative Binomial Regression*, 251. New York, NY: Cambridge University Press.

Holland, G.J. and Webster, P.J. (2007). Heightened tropical cyclone activity in the North Atlantic: natural variability or climate trend? *Philosophical Transactions of the Royal Society* 365: 2695–2716.

Hubbert, G.D., Holland, G.J., Leslie, L.M., and Manton, M.J. (1991). A real-time system for forecasting tropical cyclone storm surges. *Weather Forecasting* 6: 86–97.

India Meteorological Department (IMD) (2015a). *Report on Cyclonic Disturbances in North Indian Ocean from 1990 to 2012*. New Delhi: Regional Specialised Meteorological Centre – Tropical Cyclones.

India Meteorological Department (IMD) (2015b) Best track data of tropical cyclonic disturbances over the North Indian Ocean Regional Specialised Meteorological Centre (RSMC) – Tropical Cyclones. IMD, New Delhi.

Intergovernmental Panel on Climate Change (IPCC) (2014) Climate Change 2013: The Physical Science Basis. Cambridge University Press, Cambridge, UK.

Japan Meteorological Agency (JMA) (2017) RSMC Best Track Data (1951–2017). JMA, Tokyo.

Jelesnianski, C.P., Chen, J., and Shaffer, W.A. (1992). SLOSH: sea, lake, and overland surges from hurricanes. *NOAA Technical Report NWS* 48.

Joint Typhoon Warning Center (JTWC) (2017). *Annual tropical cyclone reports*. Guam, Mariana Islands: JTWC http://www.usno.navy.mil/nooc/nmfc-ph/rss/jtwc/jtwc.html.

Khan, M.R. and Rahman, A. (2007). Partnership approach to disaster management in Bangladesh: a critical policy assessment. *Natural Hazards* 41 (1): 359–378.

Kiger, M. and Russell, J. (1996). *This Dynamic Earth: The Story of Plate Tectonics*. Washington, DC: USGS.

Knowles, S.G. and Kunreuther, H.C. (2014). Troubled waters: the National Flood Insurance Program in historical perspective. *Journal of Policy History* 26: 325–353.

Knutson, T.R., McBride, J.L., Chan, J. et al. (2010). Tropical cyclones and climate change. *Nature Geoscience* 3: 157–163.

Kossin, J.P., Olander, T.L., and Knapp, K.R. (2013). Trend analysis with a new global record of tropical cyclone intensity. *Journal of Climate* 26: 9960–9976. doi: 10.1175/JCLI-D-13-00262.1.

Lambert, D. (1992). Zero-inflated Poisson regression, with an application to defects in manufacturing. *Technometrics* 34 (1): 14.

Landsea, C.W. (2007). Counting Atlantic tropical cyclones back to 1900. *Eos, Transactions of the American Geophysical Union* 88: 197–200.

Landsea, C.W., Harper, B.A., Hoarau, K., and Knaff, J.A. (2006). Can we detect trends in extreme tropical cyclones? *Science* 313: 452–454.

Landsea, C., Vecchi, G.A., Bengtsson, L., and Knutson, T.R. (2009). Impact of duration thresholds on Atlantic tropical cyclone counts. *Journal of Climate* 23: 2508–2519.

Lorenz, E.N. (1969a). Atmospheric predictability as revealed by naturally occurring analogues. *Journal of the Atmospheric Sciences* 26: 636–646.

Lorenz, E.N. (1969b). Three approaches to atmospheric predictability. *Bulletin of the American Meteorological Society* 50: 345–349.

Mandelbrot, B. (1983). *The Fractal Geometry of Nature*. New York: Macmillan.

McAdie, C.J., Landsea, C.W., Neuman, C.J. et al. (2009). *Tropical Cyclones of the North Atlantic Ocean, 1851–2006*, Historical Climatology Series, vol. 6–2. Asheville, NC: NOAA.

McFadden, D.L. (1974). Conditional logit analysis of qualitative choice behavior. In: *Frontiers in Econometrics* (ed. P. Zarembka), 105–142. New York: Academic Press.

Mendelsohn, R., Emanuel, K., Chonabayashi, S., and Bakkenshen, L. (2012). The impact of climate change on global tropical cyclone damage. *Nature Climate Change* 2: 205–209.

Murty, T.S., Flather, R.A., and Henry, R.F. (1986). The storm surge problem in the Bay of Bengal. *Progress in Oceanography* 16: 195–233.

Nakicenovic N, Davidson O, Davis G, et al. (2000) Emissions Scenarios, a Special Report of Working Group III of the Intergovernmental Panel on Climate Change. IPCC, Geneva.

National Centers for Environmental Information (NCEI) (2016) International Best Track Archive for Climate Stewardship. NOAA, Silver Spring, MD. https://www.ncdc.noaa.gov/ibtracs.

National Hurricane Center (NHC) (2017a) Worldwide Tropical Cyclone Centers. NHC, Miami, FL. https://www.nhc.noaa.gov/aboutrsmc.shtml.

National Hurricane Center (NHC) (2017b) Tropical Cyclone Reports. NHC, Miami, FL. https://www.nhc.noaa.gov/data/tcr/.

National Oceanic Atmospheric Administration (NOAA) (2016). *Hurricane Research Division. Re-Analysis Project*. Silver Spring, MD: NOAA http://www.aoml.noaa.gov/hrd/hurdat/data_storm.html.

National Oceanic Atmospheric Administration (NOAA) (2017) Weather fatalities 2017. National Weather Service, NOAA, Silver Spring, MD. http://www.nws.noaa.gov/om/hazstats.shtml.

Ng, N.-S. and Mendelsohn, R. (2006). The economic impact of sea-level rise on nonmarket lands in Singapore. *Ambio* 35: 289–296.

Nordhaus, W. (2010). The economics of hurricanes and implications of global warming. *Climate Change Economics* 1: 1–24.

Paul, B.K. (2009). Why relatively fewer people died? The case of Bangladesh's Cyclone Sidr. *Natural Hazards* 50: 289–304.

Pielke, R.A., Gratz, J., Landsea, C.W. et al. (2008). Normalized hurricane damages in the United States: 1900–2005. *Nature Hazards Review* 9: 29–42.

Poisson, S.D. (1837). *Probabilité des Jugements en Matière Criminelle et en Matière Civile. Précédées Des Règles Générales du Calcul Des Probabilités*. Paris: Bachelier.

Richter, C.F. (1958). *Elementary Seismology*. San Francisco: W.H. Freeman and Company.

Saffir, H.S. (1977). *Design and Construction Requirements for Hurricane Resistant Construction*. New York: American Society of Civil Engineers.

Sanghi, A., Ramachandran, S., de la Fuente, A. et al. (2011). *Natural Hazards, Unnatural Disasters: The Economics of Effective Prevention*, 276. Washington, DC: World Bank Group.

Seo, S.N. (2014). Estimating tropical cyclone damages under climate change in the southern hemisphere using reported damages. *Environmental and Resource Economics* 58: 473–490.

Seo, S.N. (2015). Fatalities of neglect: adapt to more intense hurricanes? *International Journal of Climatology* 35: 3505–3514.

Seo, S.N. (2017). Measuring policy benefits of the cyclone shelter program in the North Indian Ocean: protection from intense winds or high storm surges? *Climate Change Economics* 8 (4): 1–18. doi: 10.1142/S2010007817500117.

Seo SN (2018) Will global-warming-fueled cyclone ravage East Asia? Nature Hazards Review (submitted).

Seo, S.N. and Bakkensen, L.A. (2016). Did adaptation strategies work? High fatalities from tropical cyclones in the North Indian Ocean and future vulnerability under global warming. *Natural Hazards* 82: 1341–1355.

Seo, S.N. and Bakkensen, L.A. (2017). Is tropical cyclone surge, not intensity, what kills so many people in South Asia? *Weather, Climate, and Society* 9: 71–81.

Shmueli, G., Minka, T.P., Kadane, J.B. et al. (2005). A useful distribution for fitting discrete data: revival of the Conway–Maxwell–Poisson distribution. *Journal of the Royal Statistical Society, Series C* 54: 127–142.

Sipkin, S.A., Person, W.J., and Presgrave, B.W. (2000). Earthquake bulletins and catalogs at the USGS National Earthquake Information Center. *IRIS Newsletter* 2000: 2–4.

SURGEDAT (2015) Global Peak Surge Map. SURGEDAT: The World's Surge Data Center, Louisiana State University. http://surge.srcc.lsu.edu/data.html#globalmap.

Taylor, K.E., Stouffer, R.J., and Meehl, G.A. (2012). An overview of CMIP5 and the experiment design. *Bulletin of the American Meteorological Society* 93: 485–498.

Tory, K.J., Chand, S.S., McBride, J.L. et al. (2013). Projected changes in late-twenty-first-century tropical cyclone frequency in 13 coupled climate models from phase 5 of the coupled model Intercomparison project. *Journal of Climate* 26: 9946–9959. doi: 10.1175/JCLI-D-13-00010.1.

Train, K. (2003). *Discrete Choice Methods with Simulation*. Cambridge, UK.: Cambridge University Press.

United States General Accounting Office (USGAO) (2003). *Flood Insurance: Challenges Facing the National Flood Insurance Program*. Washington, DC: USGAO.

United States Geological Survey (USGS) (2017). *Measuring the Size of an Earthquake*. Washington, DC: USGS https://earthquake.usgs.gov/learn/topics/measure.php.

Utsu T (2013) Catalog of Damaging Earthquakes in the World (Through 2013). International Institute of Seismology and Earthquake Engineering, Tsukuba, Japan. http://iisee.kenken.go.jp/utsu/index_eng.html.

Vecchi, G.A. and Knutson, T.R. (2008). On estimates of historical North Atlantic tropical cyclone activity. *Journal of Climate* 21: 3580–3600.

Webster, P.J., Holland, G.J., Curry, J.A., and Chang, H.R. (2005). Changes in tropical cyclone number, duration, and intensity in a warming environment. *Science* 309: 1844–1846.

Winkelmann, R. (2000). *Econometric Analysis of Count Data*. Berlin: Springer.

World Bank (2007). *Emergency 2007 Cyclone Recovery and Restoration Project*. Washington, DC: World Bank http://projects.worldbank.org/p111272/emergency-2007-cyclone-recovery-restoration-project?lang=en.

6

Catastrophe Policies: An Evaluation of Historical Developments and Outstanding Issues

6.1 Introduction

The previous two chapters offered a comprehensive elaboration of the economics of catastrophic events, from both the theoretical standpoints and the empirical applications to catastrophe events. A range of market/financial instruments developed in the area of events of catastrophic consequences was explained at length. Further, policy insights or recommendations from the economists' perspectives were introduced as a contentious point of debate among the economists and policy-makers. This was followed by the empirical discussion in Chapter 5 which presented empirical studies of economics of tropical cyclones and some aspects of earthquakes.

This chapter provides a review of national and international policies on an array of catastrophic events that have been of concern at different levels of policy negotiations and referred to frequently in the previous chapters of this book. The author will visit each of these events, to be discussed presently, and review in depth the policies and programs tried and implemented for the purpose of addressing each of these policy issues.

Of the whole array of policies reviewed in this chapter, a cohort of policies comes from the policy area of natural disasters and catastrophes. This includes earthquakes, hurricanes, asteroid and comet collisions, and severe droughts. Another cohort of policies addresses environmental catastrophes: criteria air pollutants, toxic and hazardous substances, and nuclear accidents. The third cohort of policies is concerned with the matters of global concern: nuclear and chemical wars, atmospheric ozone depletion, and global warming. The fourth family of policies is concerned with a very-low probability event or an event with very-low confidence, which includes the collapse of the entire universe from the Large Hadron Collider (LHC) experiments through a strangelets-caused black hole or the robots and artificial intelligence (AI) that can kill humans (Sandler 1997; Dar et al. 1999; Posner 2004; Sanghi et al. 2011; Hawking et al. 2014).

Across these families of catastrophes, there is a great deal of difference with regard to how well a public policy is defined and implemented, as well as how stringently and with how much legal force policy measures are designed. Among the most discussed legally binding policies that involve a catastrophic event are the US Clean Air Act (US EPA 1977, 1990), the Toxic Substances Control Act (TSCA) (US Congress 1978), the National Flood Insurance Program (NFIP) (FEMA 2012, 2014), the nuclear Non-Proliferation Treaty (NPT) (UNODA 2017a), the Montreal Protocol for ozone-depleting chemicals (UNEP 2016, 2017), and the Kyoto Protocol and the Paris Agreement for global warming gases (UNFCCC 1998, 2015).

Natural and Man-made Catastrophes – Theories, Economics, and Policy Designs, First Edition. S. Niggol Seo.
© 2019 John Wiley & Sons Ltd. Published 2019 by John Wiley & Sons Ltd.

That a regulation, a program, a law, or an international protocol is in place does not necessarily mean that the policy interventions with regard to a corresponding catastrophic event were successful. The author will highlight many of the policy interventions, past and present, which have yielded little effect or are seen as a failure that needs major revisions. It will also be highlighted that some policy measures have set lofty goals but lack proper tools and legal means to achieve them. On-going efforts to rectify or strengthen the existing regulations and programs, whenever available, will be also elaborated.

Sections 6.2–6.4 review policies of the catastrophes that result from natural disasters: asteroid and comet collisions, earthquakes, and hurricanes. The ensuing sections, Sections 6.5–6.7, are devoted to environmental catastrophes: nuclear accidents, criteria pollutants, and toxic chemicals. This is followed in Sections 6.8 and 6.9 by descriptions of policies on global-scale events: atmospheric ozone depletion and global warming. The final two sections, Sections 6.10 and 6.11 look at the emerging policy frameworks with regard to the event with a very-low probability of occurrence but a truly catastrophic humanity-ending consequence. This includes the LHC physics experiments that may give birth to a universe-ending black hole, and also AI and killer robots.

6.2 Protecting the Earth from Asteroids

Asteroids, meteoroids, and comets refer to different near-Earth objects (NEOs) which are of major concern to the global community because they can strike the Earth and severely harm people and natural systems thereof. In the worst-case scenario, a collision with a large asteroid could irreparably damage the Earth (Chapman and Morrison 1994). Before we discuss the programs and technologies that detect and divert NEOs, we need to begin with some basic knowledge about them.

An asteroid is a small-sized, naturally formed, solar system body that orbits the Sun, most of which reside in the so-called Asteroid Belt that is located in the region between Mars and Jupiter. A meteoroid is a fragment of an asteroid or comet that is broken off during collisions among them and, as such, is significantly smaller than an asteroid. More formally, it is an object less than 1 m-wide in diameter. A comet is, on the other hand, a small body of dusty material formed in the cold outer solar system which is embedded with icy volatiles such as water and carbon dioxide (NRC 2010).

When a meteoroid enters the Earth's atmosphere, it glows and produces a streak of light, the phenomenon of which is called a meteor or a shooting star. The vast majority of meteoroids that enter the Earth's atmosphere disintegrate before reaching the surface of Earth, but larger ones survive the entry and impact the Earth, and are then called a meteorite. Examinations of meteorites that have fallen to the Earth reveal that asteroids were formed about 4.5 billion years ago (NRC 2010).

The US asteroid protection program is anchored at the Planetary Defense Coordinating Office (PDCO) which was established recently in 2016 at the National Aeronautics and Space Administration (NASA). Planetary defense is the term used by the US government to refer to all the efforts and capacities for detecting, warning, and preventing potential asteroid or comet collisions with Earth, as well as for mitigating the effects of such collisions (NASA 2014).

For planetary defense activities, two parameters that identify asteroids, meteoroids, and comets are critical information for the determination of the level of policy actions. The first is how close these objects approach the planet. The second is how large these objects are in diameter.

The first parameter is an object's proximity to the planet. Of the asteroids, meteoroids, and comets that are existent in the solar system, a NEO is defined by the orbits of these objects around the Sun. When the orbit of an object brings it within 121 million miles (195 million kilometers) of the Sun, which is approximately 30 million miles (50 million kilometers) from the Earth's orbit, it is classified as an NEO.

As of 2016, more than 15 000 near-Earth asteroids have been discovered globally since 1998, the year when NASA began tracking them, with 95% of them discovered by NASA. Each week, about 30 new asteroids are discovered and added to the catalog. Asteroids are discovered mostly relying on ground-based telescopes. The number of NEOs discovered is about 50 000. It is predicted that about three-quarters of NEOs are still undiscovered (NASA 2014).

The second parameter is an NEO's size. Of the 50 000 discovered NEOs, roughly half are larger than 460 ft (140 m) in diameter. The 460-ft cut-off was determined by the NASA NEO survey team in 2003 to mean that the NEO whose size is smaller than 460 ft can only have regional effects. The NASA team suggested that a 984-ft (300-m)-wide NEO could have subglobal effects and a 0.6-mile (1-km)-wide NEO could have global effects. A re-evaluation of these cut-off points is under way (NASA 2014).

A collision with an asteroid whose size is larger than 1 km wide in diameter could have global impacts and possibly destroy the Earth irrecoverably. For reference, the asteroid that killed all dinosaurs on the planet 66 million years ago was 7.5 mile wide (about 10 km) in diameter (Kaiho and Oshima 2017).

An approach to the proximity of the planet of an NEO whose size is large than 1 km in diameter does not necessarily spell a doomsday for humanity. There are technologies that can be used for detection, diversion, or destruction of the NEO before it reaches the Earth, given that it may take many years or decades for the asteroid to reach the planet from the time of detection. What are these technologies?

Once an asteroid is identified as a potential threat as it is on a collision course with Earth, the most promising approach for preventing an eventual collision is deflecting the asteroid, rather than destroying it, by changing the velocity of the asteroid. If an asteroid can be detected a couple of decades earlier, altering the velocity as little as an inch per second several years in advance before it reaches the Earth's atmosphere would suffice to prevent an eventual collision (NASA 2014).

The two most promising techniques for deflecting a NEO, according to NASA, are the kinetic impactor and the gravity tractor. The kinetic impactor is a technique for hitting the asteroid with an object to slightly slow it down. The gravity tractor is a technique for gravitationally tugging on an asteroid by station-keeping a large mass near it (NASA 2014).

As stated above, the PDCO under NASA coordinates all activities related to planetary defenses against various NEOs, and hence it is the center of US governmental policy responses. The PDCO was established only a few years ago in 2016, but the NEO detection and tracking program by NASA began several decades ago in 1998. How much budget is allocated to the planetary defense activities?

Table 6.1 Historical budgets for US NEO observations and planetary defense.

Years	Budget	Notes
1998–2010	US$ a few million	The 1998 congressional directive: conduct a program to discover at least 90% of 1-km-diameter or larger NEOs within 10 years
2010	US$4 million	
2012	US$20.4 million	President Obama: a human mission to an asteroid
2014	US$40 million	
2016	US$50 million	

Source: NASA (2017).

Despite a potentially truly catastrophic consequence, say, the end of Planet Earth, the magnitude of public resources allocated to address this potentially catastrophic event does not amount to, even in the US, one-hundredth of 1% of the total federal budget. As summarized in Table 6.1, the US budget for the NEO observations and planetary defense program was only a few million dollars from 1998 to 2010. It subsequently increased to US$20.4 million in 2010 by the request of President Obama for NASA's new human mission to an asteroid, US$40 million in 2014, and US$50 million in 2016 (NASA 2017).

What the asteroid problem and policy responses tell us is that a low-probability high-damage event does not always call for a drastic policy action such as those espoused by the dismal theorists and the precautionary principle (Weitzman 2009). A planet-destroying asteroid collision is possible, and hence cannot be ruled out from policy considerations. However, the range of technological options explained above, although they may or may not turn out to be successful technologies, have made a radical policy intervention unnecessary.

To put it differently, a portfolio of available technologies has determined the nature of policy interventions as well as the appropriate level of policy responses when it comes to the asteroid problem.

Another aspect of the asteroid policy is that the protection of the Earth from a possible catastrophic asteroid collision is a global public good (Samuelson 1954; Nordhaus 2006). That is, the benefit of protection falls upon all nations on the globe. In the global-scale asteroid defined above, the benefit is neither excludable nor rivalrous. Unlike other global public goods, however, international societies are not cooperating extensively to deal with the global public good problem, for example, through the United Nations (UN) (Seo 2016c, 2017a).

This is due to the fact that the provision of the protection from asteroid collisions involves a best-shot technology (Hirshleifer 1983). That is, the success of providing the public good depends on the success of the best technology, while even the second-best technologies would mean nothing.

Other global public good problems, e.g. a global warming policy, involve in most cases a cumulative production technology. For a cumulative production technology, individual countries' commitments should add up to the sufficient global level of commitment (Nordhaus 1994). The characteristic of the production technology in the asteroid policy as the best-shot technology underlies the US' unilateral endeavors to detect, divert, and destroy the NEOs that pose a potential threat.

6.3 Earthquake Policies and Programs

Deadly earthquakes, along with severe tropical cyclones, are the most catastrophic natural events as measured by the number of human fatalities (Swiss Re Institute 2017). As explained in Chapter 5, precise predictions of earthquake outbreaks and magnitudes sufficiently ahead of the events are among the most difficult scientific challenges concerning natural catastrophes (Kiger and Russell 1996).

Earthquake policy responses are coordinated at a national or subnational level. The countries in which a significant fraction of the land area sits on the earthquake-prone zones and which often experience a large number of lost lives from an earthquake have sophisticated earthquake policies and programs, which include Japan, New Zealand, Chile, and California in the US.

The foundation of the US earthquake policy responses is the National Earthquake Hazards Reduction Program (NEHRP), authorized by the US Congress in 1977 and amended in 2004, Public Law 95-124 and Public Law 108-360, respectively (US Congress 2004). The NEHRP's stated purpose was to "reduce the risks to life and property from future earthquakes in the United States."

The NEHRP's role designated by the US Congress is to provide, for the public and private sectors, the scientific and engineering information, knowledge, and technologies necessary for earthquake responses, so that individuals and communities can prepare for an earthquake, but also reduce the costs of post-event damages and recovery efforts (US Congress 2004). As part of the NEHRP, grant programs are offered to universities and institutions through the National Science Foundation (NSF) for promoting basic research on earthquakes and responses to them.

The NEHRP is implemented through collaborations among the four US federal agencies: the National Institute of Standards and Technology (NIST), the Federal Emergency Management Agency (FEMA), the NSF, and the United States Geological Survey (USGS) (FEMA 2016). The NIST is the lead agency for the NEHRP programs.

Each of the four agencies takes action in the domain of its responsibilities which are designed to be distinct across the four agencies. As the lead agency, the NIST is responsible for developing earthquake-resistant designs and construction practices that can be implemented through building codes and engineering practices.

FEMA is responsible for developing and implementing effective earthquake risk reduction tools, as well as supporting the development of earthquake-resistant building codes and standards.

The NSF has responsibility for supporting basic research pertinent to improving the understanding of the causes, impacts, and responses to earthquakes in the fields of earth sciences, engineering, social sciences, and behavioral and economic sciences. The NSF's earthquake-related research is supported through research grants to individual universities and research institutions.

The USGS is responsible for, among other things, earthquake monitoring, data analysis, notifications, earthquake hazards assessments, and targeted research on earthquake causes and effects. Most significantly, the USGS provides the national seismic hazard maps and site-specific data (USGS 2017a).

In addition to the national programs, international collaborations have been an important aspect of the US earthquake policy responses, especially with Japan, a major earthquake-hit nation in the world (Greer 2012). Formal agreements between the two countries with regard to earthquake responses include the Japan–US Science and Technology Agreement (JUST) and the US–Japan Common Agenda for Cooperation for Global Perspective (Common Agenda) (NRC 1997).

Although it is possible that earthquake events are a globally connected phenomenon through the tectonic plates (Kiger and Russell 1996), there has been no global agreement or protocol for addressing earthquake disasters. Policy responses are made predominantly at the national level, and as such, response systems are mostly locally oriented, e.g. an earthquake-resistant building and a warning system.

Pertinent to the bilateral agreements between the US and Japan, in numerous earthquake-related areas, policy and research collaborations have been made between the two countries. For earthquake forecasting, warning, and hazard zonation, the two countries collaborate on real-time monitoring and seismic warning systems, earthquake predictions, and a probabilistic seismic hazard analysis. In the area of earthquake risk assessment and loss estimation, collaborations are made on loss estimation and disaster situation assessments. In the area of earthquake-resistant design construction, rehabilitation, and repair standards, collaborating topics include performance-based design and large-scale dynamic testing and simulation. Collaborations are also made in the research area of earthquake preparation, response, recovery, and mitigation (NRC 1997).

How do we evaluate the policy effectiveness of the NEHRP or various collaborations with Japan? Although it is beyond the scope of this book to provide a quantitative assessment, e.g. a benefit-cost ratio of the program, empirical evidence may be offered that these programs are indeed effective in mitigating earthquake damages. One is the historical earthquake magnitude and fatality statistics such as those presented in Table 5.3 in Chapter 5.

The table shows that, during the 27-year period from 1990 to 2016, the four most catastrophic years in terms of earthquake-caused fatalities were 2004, 2010, 2008, and 2005. Each of these years claimed 298 101, 226 050, 88 708, and 87 992 lives, respectively. These large numbers of human deaths occurred in low-income countries with poor earthquake-resilient infrastructure and warning systems: Bangladesh and India in 2004, Haiti in 2010, China in 2008, and Pakistan in 2005 (USGS 2017b).

By contrast, the most catastrophic earthquake that struck the US during the same time period was the Northridge, California earthquake in 1994, which killed 60 people. For Japan, the most catastrophic earthquake during this time period was the Great Hanshin earthquake in 1995 which killed 6434 people (Utsu 2013).

6.4 Hurricane, Cyclone, and Typhoon Policies and Programs

As elaborated at length in Chapter 5, a hurricane is a natural event which is as difficult to predict regarding many of its elements as an earthquake, including its origin, path, size, speed, surge, rainfall, and power (Emanuel 2008, 2013). In addition, a hurricane is a natural phenomenon as feared by people as an earthquake because of its potential for a catastrophic consequence through collapsed houses, flooded towns, floating debris, and human fatalities (Seo and Bakkensen 2017; Seo 2017b). These conditions make a government's policy interventions inevitable.

For tropical cyclone responses, an internationally coordinated system exists, which is managed at the level of tropical cyclone ocean basins and regions. There are globally 13 tropical cyclone ocean regions, each of which has one central agency that coordinates cyclone-related regional activities, either a Regional Specialized Meteorology Center (RSMC) or Tropical Cyclone Warning Center (TCWC) (NHC 2017). In addition, each hurricane-afflicted country has its own tropical cyclone policy and programs.

To begin, three different terms are used across the tropical cyclone ocean basins to refer to a hurricane. A hurricane is the term commonly used in North America and Europe. In the western North Pacific, it is called a typhoon or super-typhoon. In the Indian Ocean and the South Pacific Ocean, it is called a tropical cyclone (NHC 2017).

The name of an individual tropical cyclone is chosen by a corresponding RSMC from the predetermined list compiled from the lists submitted by member countries. The name is assigned once it is declared a tropical cyclone by the RSMC or TCWC according to the Saffir–Simpson hurricane scale (Saffir 1977).

As an international response to the deadly 1970 tropical cyclone in Bangladesh, Cyclone Bhola, the World Meteorological Organization (WMO)'s tropical cyclone program was created to establish nationally and regionally coordinated cyclone response systems to minimize the losses of lives and damages caused by tropical cyclones.

Under the WMO's tropical cyclone program, there are globally 12 RSMCs and TCWCs participating in the program, as summarized in Table 6.2 (NHC 2017). There are 13 tropical cyclone ocean regions and 9 tropical cyclone ocean basins. For each ocean region, there is one RSMC or TCWC. There are six RSMCs: Miami, Honolulu, Tokyo, New Delhi, La Réunion, and Nadi.

Besides the RSMCs and the TCWCs, the Joint Typhoon Warning Center (JTWC) located in Pearl Harbor, Hawaii is a joint command, established in 1959, of the United States Navy and the United States Air Force. For the US Department of Defense and other agencies of the US government, the JTWC has responsibility for issuing warnings for the tropical cyclones generated in the Pacific and Indian Oceans which include the Northwest Pacific Ocean basin, the South Pacific Ocean basin, and Indian Ocean basins (JTWC 2017).

The RSMCs, the TCWCs, and the JTWC play a vital role in hurricane preparedness and responses. They predict an upcoming tropical cyclone's trajectory and intensity, declare a tropical cyclone, assign it a name, issue advisories and early warnings, and recommend evacuation orders. In additions, they compile and maintain historical tropical cyclone data such as the best track data as well as detailed reports on individual cyclones (NCEI 2016). These records provide essential resources for research and policy discourses on tropical cyclones (Seo 2014, 2015a; Seo and Bakkensen 2017).

Each of the RSMCs and the TCWCs is composed of the member countries in a corresponding ocean basin that participate in the program. Each member country has its own tropical cyclone policy and programs. In the US, the national hurricane disaster response is anchored by the NFIP.

The US NFIP is an insurance program, overseen by FEMA, that provides residents of flood-prone zones with insurance protection subsidized by the government. It was

Table 6.2 Tropical cyclone RSMCs and TCWCs for ocean regions and basins.

Ocean basin	Ocean region	RSMC and TCWC
Atlantic Ocean; Eastern Pacific Ocean	1, 2	US National Hurricane Center: RSMC Miami
Central Pacific Ocean	3	US Central Pacific Hurricane Center: RSMC Honolulu
Northwest Pacific Ocean	4	Japan Meteorological Agency: RSMC Tokyo
North Indian Ocean	5	India Meteorological Department: RSMC New Delhi
Southwest Indian Ocean	6	Météo-France: RSMC La Réunion
Southwest Pacific Ocean; Southeast Indian Ocean	7	Australian Bureau of Meteorology: TCWC Perth
	8	Indonesian Agency for Meteorology: TCWC Jakarta
	9	Australian Bureau of Meteorology: TCWC Darwin
	10	Papua New Guinea: TCWC Port Moresby
	11	Australian Bureau of Meteorology: TCWC Brisbane
South Pacific Ocean	12	Fiji Meteorological Service: RSMC Nadi
	13	Meteorological Service of New Zealand Ltd: TCWC Wellington

Source: Modified from NHC (2017).

established by the National Flood Insurance Act of 1968 (Knowles and Kunreuther 2014).

The government-subsidized insurance for flood victims was conceived because private insurance companies had retreated from the flood insurance market due to heavy losses from offering private flood insurances following the Great 1927 Mississippi River Flood. After two catastrophic events hit the US in the mid-1960s, i.e. the Alaska earthquake in 1964 and Hurricane Betsy in 1965, the NFIP was established in 1968 by the aforementioned National Flood Insurance Act (Knowles and Kunreuther 2014).

However, a remarkable social change that has been occurring since the middle of the twentieth century in the US has put the government-subsidized NFIP program in trouble. Since the 1950s, the country has seen a striking population shift from inland counties to coastal counties and frequently hurricane-inflicted zones. As of 2014, the State of Florida's population has expanded from 1950 by 579% and that of Texas's population has grown by 226% during the same time period. As of 2010, 39% of all Americans lived in coastal shoreline counties, and population densities in coastal shoreline counties were six times greater than those in inland counties (Knowles and Kunreuther 2014).

The salient change in the distribution of the US population to shoreline counties has meant that more people and properties are placed in harm's way during hurricane landfalls, which in turn has meant a sharp increase in the government's burden for the NFIP subsidies. Empirical data show that 8 of the 10 costliest hurricanes in US history have hit the country since 2004, with total cost exceeding US$200 billion, and 17 of the twenty costliest flood events have occurred since 1995 (King 2013).

As shown in Table 6.3, as of 2010, the number of flood insurance policies sold amounted to 5.5 million policies in 20 000 communities. Since 2001, the number of claims made varied year by year from about 23 000 claims to 212 000 claims in 2005 when Hurricane Katrina hit the country (King 2013). In 2012, insured flood losses in the US reached US$58 billion (Swiss Re Institute 2017).

In addition to the aforementioned great population migration in the country, many other policy-relevant factors have contributed to the sharp increase in the cost of the NFIP, including repetitive flood loss properties, inaccurate floodplain maps, and lack of enforcement of a comprehensive building code (King 2013).

Owing to exceptionally destructive hurricane events, since Hurricane Katrina in 2005, the US Congress had to increase the NFIP's borrowing authority from US$1.5 billion to US$20.7 billion after Hurricane Katrina in 2005, from which US$17.7 billion was paid in total to policy-holders, as shown in Table 6.3. After the wide-ranging damages inflicted by Hurricane Sandy in 2012, the NFIP's borrowing authority had to be increased further to US$30 billion. As Table 6.3 shows, as of 2012, the NFIP had borrowed in total US$23.7 billion, repaid US$6 billion, and was in debt of $17.7 billion (King 2013).

In the summer of 2017 when severe Hurricanes Harvey, Irma, and Maria made landfalls in the US, President Trump signed the Hurricane Harvey Relief Bill with an aid package of US$15.3 billion, with additional aid to the victims of Hurricane Irma that hit Florida and Hurricane Maria that hit Puerto Rico (Kaplan 2017).

Addressing the need to contain a sharp increase in the government and taxpayers' cost burden for the NFIP, the US Congress passed in July 2012 the Biggert–Waters Flood Insurance Reform and Modernization Act (BW 2012), outlining a comprehensive reform of the NFIP (NRC 2015). This introduced a major reform of the NFIP. It phases out insurance subsidies and discounts by forcing the NFIP away from a government-subsidized flood insurance program to a system of insurance premiums that reflect the actual flood risks faced by potential victims (FEMA 2012).

In addition, the BW 2012 Reform Act required, among other changes, that the NFIP should draw updated floodplain maps, strengthen enforcements of building codes, and remove certain properties from the program that had incurred repeated losses from insurance subsidies.

A transformational change of the flood insurance program envisioned by the BW 2012 Reform Act was that it legislated to phase out pre-Flood Insurance Rate Map (FIRM) subsidized policies and grandfathered policies of the NFIP, but keep the NFIP risk-based policies. Before the BW 2012 Reform Act, the pre-FIRM subsidized policies had premiums less than those for the risk-based NFIP policies for the structures that were in place before a local FIRM was available. Grandfathered policies had allowed a property to continue to have a given rating class even if a new FIRM may indicate a higher level of flood risk (NRC 2015).

However, the BW 2012 Reform Act's radical transformation of the NFIP away from a mostly government-subsidized insurance program to a flood insurance system of risk-based premiums turned out to face many legal and political challenges. Some homeowners challenged the BW 2012 Reform Act by claiming that the new

Table 6.3 NFIP statistics on payments, borrowing, and cumulative debts.

Year (calendar or fiscal)	Number of policies in force (×10⁻³)	Total written premium (US$ million)	Total face value of coverage (US$ billion)	Total number of claims paid (×10⁻³)	Total payments made to policy-holders (US$ million)	Amount borrowed (US$ million)	Amount repaid (US$ million)	Cumulative debt (US$ million)
1978	1446	111	50	29	147			
1979	1843	142	74	70	483			
1980	2103	159	99	41	230	917a)	0	917a)
1981	1915	257	102	23	127	165	625	457
1982	1900	355	107	32	198	14	471	0
1983	1981	384	117	51	439	50	0	50
1984	1926	421	124	27	254	200	37	213
1985	2016	452	139	38	368	0	213	0
1986	2119	518	155	13	126	0	0	0
1987	2115	566	165	13	105	0	0	0
1988	2149	589	175	7	51	0	0	0
1989	2292	632	265	36	661	0	0	0
1990	2477	673	213	14	167	0	0	0
1991	2532	737	223	28	353	0	0	0
1992	2623	801	236	44	710	0	0	0
1993	2828	890	267	36	659	0	0	0
1994	3040	1004	295	21	411	100	100	0
1995	3476	1141	349	62	1295	265	0	265
1996	3693	1275	400	52	828	424	62	627
1997	4102	1510	462	30	519	530	240	917
1998	4235	1668	497	57	886	0	395	522
1999	4329	1720	534	47	754	400	381	541
2000	4369	1724	567	16	251	345	541	345
2001	4458	1740	611	43	1277	600	345	600
2002	4519	1802	653	25	433	50	640	10
2003	4565	1898	691	36	780	0	10b)	0
2004	4667	2041	765	55	2232	0	0	0
2005	4962	2241	876	212	17 713	300	75	225
2006	5514	2605	1054	24	640	16 660	0	16 885
2007	5655	2843	1141	23	612	650	0	17 535
2008	5684	3067	1197	74	3450	50	225	17 360
2009	5704	3202	1233	30	772	1988	348	19 000
2010	5559	3348	1227	27	708	0	500	18 500
2011	5585	3477	1264	65	1847	0	750	17 750

a) Before 1981.
b) Transactions in October 2002.
Source: Congressional Research Service by King (2013).

FEMA floodplain maps overestimated the actual flood risk faced by their properties. Homeowners in flood-prone zones argued that while the increased premiums are unjustified, they could not afford the planned phased increases in premium rates under the BW 2012 Reform Act (Knowles and Kunreuther 2014). Making things worse for the homeowners, the planned phase-out of the pre-FIRM subsidized policies caused property values in flood-prone zones to decline steeply.

In March 2014, the US Congress, unable to overcome the political challenges, passed the Menendez–Grimm Homeowner Flood Insurance Affordability Act (HFIAA), rolling back transformational provisions of the BW 2012 Reform Act, especially with regard to the pre-FIRM subsidized policies and grandfathered policies (FEMA 2014).

The Homeowner Affordability Act reinstates grandfathering by lowering premium rate increases on policies, preventing future rate increases, implementing a surcharge on all policy-holders, and repealing certain rate increases that have gone into effect and requiring FEMA to refund those policy-holders who overpaid premiums (FEMA 2014; US Senate 2014).

The Act also allows homebuyers to purchase pre-FIRM properties with pre-FIRM subsidy conditions and allows pre-FIRM homeowners to voluntarily purchase a new policy with pre-FIRM conditions (US Senate 2014).

6.5 Nuclear, Biological, and Chemical Weapons

Catastrophic consequences of nuclear explosions on the global community, through either a nuclear war or a series of accidents, have been proposed by many researchers (Turco et al. 1983; Mills et al. 2008). A nuclear war between the US and Russia, or between India and Pakistan, or between Iran and Israel – in some future – could result in an utter devastation unseen before by humanity.

Public perceptions of nuclear catastrophes are rooted in two types of nuclear events. One is a nuclear war among any of the nations that own nuclear weapons. The nuclear bombing of Hiroshima, Japan by the US which ended World War II, the Cuban missile crisis engendered by the cold-war arms race between the US and Russia, and North Korea's nuclear and hydrogen bomb missile tests are some of the past and on-going experiences by the public with regard to the threats of a nuclear war.

Another type of nuclear event is nuclear accidents. Representative historical events are the Chernobyl nuclear disaster in Ukraine (the former USSR) in 1986 and the Fukushima Daiichi nuclear disaster in Japan in 2011, caused by a magnitude 9.0 earthquake and the tsunami that followed. The Chernobyl disaster caused cancer in 4000–985 000 persons. In the Fukushima disaster, a 20-km evacuation zone was declared while 470 000 people were evacuated from the zone. These two nuclear accidents are the only nuclear events that have been classified as level 7 events in the International Nuclear and Radiological Event Scale (INES) (NCEI 2016).

There exist a number of scenarios of a truly global catastrophe caused by nuclear explosions put forth by concerned scientists. One of them is a nuclear winter hypothesis (Turco et al. 1983). The hypothesis states that the firestorms, smoke from cities, and forest fires caused by nuclear explosions would inject a massive amount of fine dust into the stratosphere which then attenuates solar radiation flux, resulting in subfreezing land temperatures.

Another is a massive ozone loss hypothesis (Mills et al. 2008). This states that a regional nuclear conflict could lead to a massive ozone loss in the stratosphere, which would pose catastrophic health threats such as skin cancer from the unblocked incoming ultraviolet radiation.

As of 2017, there are nine countries in the world that are recognized to have the capacity to build nuclear bombs or actually own them. There are five nuclear states that are formally recognized by the UN – the US, Russia, the UK, France, and China – all of whom are also permanent members of the UN Security Council. Additionally, four countries are known or believed to have nuclear weapons or the capacity to build them: India, Pakistan, North Korea, and Israel (UNODA 2017a).

International policy measures on nuclear weapons are encapsulated in the NPT, which came into force in 1970 and was extended indefinitely in 1995. As of 2018, the NPT is signed by 191 nations (UNODA 2017a). The treaty has established a safeguards system with the responsibility given to the International Atomic Energy Agency (IAEA). The IAEA verifies compliance of member nations with the treaty through nuclear inspections.

The objectives of the NPT are to prevent the spread of nuclear weapons and weapons technology, to promote cooperation in the peaceful uses of nuclear energy, and to further the goal of nuclear disarmament and general and complete disarmament (UNODA 2017a).

To achieve these goals, the NPT articles spell out specific obligations for the parties of the treaty. In Article 1 of the NPT, each nuclear-weapon State Party to the treaty is required to:

> undertake not to transfer to any recipient whatsoever nuclear weapons or other nuclear explosive devices or control over such weapons or explosive devices directly, or indirectly; and not in any way to assist, encourage, or induce any non-nuclear-weapon State to manufacture or otherwise acquire nuclear weapons or other nuclear explosive devices, or control over such weapons or explosive devices.

Article 2 of the NPT spells out the obligations of the non-nuclear-weapon States in parallel with Article 1 for the nuclear-weapon State Party:

> Each non-nuclear-weapon State Party to the Treaty undertakes not to receive the transfer from any transferor or whatsoever of nuclear weapons or other nuclear explosive devices or of control over such weapons or explosive devices directly, or indirectly; not to manufacture or otherwise acquire nuclear weapons or other nuclear explosive devices; and not to seek or receive any assistance in the manufacture of nuclear weapons or other nuclear explosive devices.

Article 3 of the NPT obligates non-nuclear-weapon States and nuclear-weapon States to follow the IAEA's safeguards system for the purpose of peaceful uses of nuclear energy:

1) Each non-nuclear-weapon State Party to the Treaty undertakes to accept safeguards, as set forth in an agreement to be negotiated and concluded with the International

Atomic Energy Agency in accordance with the Statute of the International Atomic Energy Agency and the Agency's safeguards system, for the exclusive purpose of verification of the fulfilment of its obligations assumed under this Treaty with a view to preventing diversion of nuclear energy from peaceful uses to nuclear weapons or other nuclear explosive devices.

2) Each State Party to the Treaty undertakes not to provide: (i) source or special fissionable material or (ii) equipment or material especially designed or prepared for the processing, use or production of special fissionable material, to any non-nuclear-weapon State for peaceful purposes, unless the source or special fissionable material shall be subject to the safeguards required by this Article.

Although nearly all members of the UN are parties of the NPT, four UN members never accepted the nuclear treaty: India, Pakistan, Israel, and the newly formed South Sudan. North Korea acceded but never complied, and withdrew from the treaty in 2003.

In addition to the NPT on nuclear weapons, the UN treaties on weapons of mass destruction include the Biological Weapons Convention (BWC) which entered into force in 1975 and the Chemical Weapons Convention (CWC) which entered into force in 1997. These conventions are structured in a similar framework as the NPT.

The objectives of the BWC, whose formal title is the Convention on the Prohibition of the Development, Production and Stockpiling of Bacteriological (Biological) and Toxin Weapons and on their Destruction, are stated in Article 1 of the treaty (UNODA 2017b):

Each State Party to this Convention undertakes never in any circumstances to develop, produce, stockpile or otherwise acquire or retain: (1) microbial or other biological agents, or toxins whatever their origin or method of production, of types and in quantities that have no justification for prophylactic, protective or other peaceful purposes; (2) weapons, equipment or means of delivery designed to use such agents or toxins for hostile purposes or in armed conflict.

With the entry into force of the CWC, the Organization for the Prohibition of Chemical Weapons (OPCWs) was established in 1997, whose secretariat is located in the Hague, the Netherlands. As of 2018, 189 nations have joined the CWC.

Article 1 of the CWC states the general objectives of the treaty, i.e. the Convention on the Prohibition, Production, Stockpiling and Use of Chemical Weapons and on Their Destruction (UNODA 2017c):

Each State Party to this Convention undertakes never under any circumstances: (a) To develop, produce, otherwise acquire, stockpile or retain chemical weapons, or transfer, directly or indirectly, chemical weapons to anyone; (b) To use chemical weapons; (c) To engage in any military preparations to use chemical weapons; (d) To assist, encourage or induce, in any way, anyone to engage in any activity prohibited to a State Party under this Convention.

The international treaty on nuclear weapons and other weapons of mass destruction – the NPT, the BWC, and the CWC – is among the strongest international regulations of all international policy agreements and treaties on multiple grounds, especially the NPT. First, participation rates are exceptionally high: almost all countries are members

Table 6.4 Treaties on nuclear, biological, and chemical weapons.

Treaties	Year of entry into force	Participation (number of parties)	Non-members	Notes
The Treaty on the Non-Proliferation of Nuclear Weapons (NPT)	1970	191	India, Pakistan, Israel, South Sudan, North Korea	The Comprehensive Nuclear-Test-Ban Treaty (CTBT), signed in 1996, has yet to enter into force
The Convention on the Prohibition of the Development, Production and Stockpiling of Bacteriological (Biological) and Toxin Weapons and on their Destruction	1975	179	Israel, Syria, Egypt, Chad, Central African Republic, and others	No formal verification regime to monitor compliance
The Convention on the Prohibition, Production, Stockpiling and Use of Chemical Weapons and on Their Destruction	1997	189	Israel, Egypt, South Sudan, North Korea	As of October 2016, about 93% of the world's declared stockpile of chemical weapons had been destroyed (according to the OPCW)

of these treaties (Table 6.4). More specifically, the NPT participation rate with 191 members in the treaty is around 99% among UN members.

Second, these treaties are signed as an international treaty, i.e. not as an agreement, which means that they are given legally binding force. A violator of the treaties will face international penalties and punishments. For the BWC, however, there is no formal verification regime to monitor compliance, at the time of writing.

Third, the terms of these treaties are strict prohibitions of regulated activities. That is, they require member nations to "never undertake under any circumstances" to develop, transfer, or stockpile concerned weapons of mass destruction.

The NPT is, however, not a one-sided imposition of international rules by the nuclear-weapon States Party on the non-nuclear-weapon States. In return for the non-nuclear-weapon States' participations in the treaty and forswearing of nuclear arms, nuclear-weapon States agree to share the benefits of peaceful nuclear technology and pursue nuclear disarmament aimed at the ultimate elimination of their nuclear arsenals (Graham 2014; UNODA 2017a):

> Parties to the Treaty in a position to do so shall also cooperate ... to the further development of the applications of nuclear energy for peaceful purposes, especially in the territories of non-nuclear-weapon States Party ... (Article 4, NPT); potential benefits from any peaceful applications of nuclear explosions will be made available to non-nuclear-weapon States Party ... [and] will exclude any charge for research and development (Article 5, NPT);

Each of the Parties to the Treaty undertakes to pursue negotiations in good faith on effective measures relating to cessation of the nuclear arms race at an early date and to nuclear disarmament, and on a treaty on general and complete disarmament under strict and effective international control (Article 6, NPT).

The success of the NPT in stopping nuclear arms proliferation across the globe has hinged on this central bargain between nuclear-weapon States and non-nuclear-weapon States (Graham 2014; Campbell et al. 2004). As of 2004, there were reported to be about 40 countries capable of building nuclear weapons, but most of them gave up developing them because of this bargain, more specifically, because of the protection, called a nuclear umbrella, as well as the commitments to general and complete disarmament of nuclear weapons provided by the nuclear-weapon States.

The NPT member countries that committed to nonpursuit of nuclear weapons, such as Egypt, Syria, Germany, Saudi Arabia, Turkey, Japan, South Korea, and Taiwan, have assured their neighbor countries that there is no urgent need to develop nuclear arsenals This has given the NPT every chance of success in achieving the nonproliferation goal (Campbell et al. 2004).

The success of the NPT in preventing proliferation of nuclear arms up to this point may not be sustainable if the nuclear-weapon States do not materialize their commitments to general and complete disarmament and do not give up future nuclear tests. In this respect, the Comprehensive Nuclear-Test-Ban Treaty (CTBT) which was signed in 1996 is a critical policy decision to be made by the global community. The CTBT has not yet entered into force, owing to the lack of commitment from nuclear-weapon States Parties (Table 6.4).

Another major challenge to the sustainability of the NPT lies in the zealous pursuit of nuclear arsenals and ballistic missiles that carry them in North Korea, and perhaps Iran at a future date, for the purpose of striking US territories. This may trigger neighboring countries to the two countries to pursue their own nuclear arms programs, for example, Japan, South Korea, Israel, Saudi Arabia, and other Middle Eastern countries (Graham 2004).

6.6 Criteria Pollutants: The Clean Air Act

Criteria pollutants are six of the most common air pollutants determined by the Clean Air Act, the cornerstone environmental law of the US, which are ground-level ozone, particulate matter, sulfur dioxide, nitrogen oxides (NOx), lead, and carbon monoxide (US EPA 1990, 2010).

Emissions of these pollutants from numerous sources including smoke-stacks and tail-pipes result in a variety of adverse human health effects through environmental disasters such as acid rain, smog, and fine particulate matter pollution. Further, releases of these pollutants result in many other harmful effects, including reduced visibility, damage to building, damage to crops and livestock, vegetation damages, and loss of species diversity (Mendelsohn and Olmstead 2009; Tietenberg and Lewis 2014).

In the catastrophe literature, environmental pollution is regarded as a prominent field on multiple grounds. The first is the large number of human deaths that result from environmental pollution, especially in developing countries such as India and China.

For example, pollution-related deaths amounted annually to 2.51 million people in India and 1.8 million people in China, according to a report by the Lancet Commission on pollution and health (Landrigan et al. 2018).

The second reason is existence of a dose-response relationship for a pollutant which describes the changes in the effect to an individual in response to differing levels of exposures or doses to a pollutant (Lutz et al. 2014). In a dose-response relationship of a criteria pollutant, a threshold or a tipping point is clearly established beyond which mortality results. As we extensively discuss in Chapters 2 and 3, the concept of a threshold or a tipping point is one of the keystone concepts of mathematical and philosophical catastrophe studies (Thom 1975).

For each of the criteria air pollutants, the US Environmental Protection Agency (EPA) has established the National Ambient Air Quality Standards (NAAQS), a set of thresholds defined based on the dose-response relationship of each pollutant. There are two NAAQS for each pollutant set by the EPA for achieving different purposes: primary standards and secondary standards.

The primary NAAQS are set to protect human health and the secondary NAAQS are set to protect public welfare from various harmful effects (US EPA 1990, 2010). The EPA is required to periodically conduct comprehensive reviews of the literature on human health and public welfare effects associated with criteria pollutants and revise the standards. The primary and secondary standards, updated to 2017, for the six criteria pollutants are summarized in Table 6.5 (US EPA 2017a).

Ground-level ozone (O_3) is a primary cause of smog. Ground-level ozone is formed through reaction of the pollutants emitted from a variety of human activities and changes in natural ecosystems, mainly volatile organic compounds (VOCs) and nitrogen oxides (NOx). NOx are released into the air when mobile vehicles and other stationary sources like power plants and industrial boilers burn fuels such as gasoline, coal, or oil (Mauzerall et al. 2005; US EPA 2017b).

VOCs are organic, i.e. containing carbon, chemicals that easily evaporate due to high vapor pressure. VOCs are numerous, of varied forms, and found everywhere. Examples of VOCs are gasoline, benzene, formaldehyde, solvents such as toluene and xylene, styrene, and perchloroethylene (or tetrachloroethylene). VOCs are released from automobiles and power plants burning fossil fuels such as gasoline, coal, or natural gas. Oil and gas fields and diesel exhausts are also sources of emissions of VOCs. Major sources of releases are solvents, paints, glues, and other common home and workplace products such as moth repellents, air fresheners, wood preservatives, aerosol sprays, degreasers, and dry-cleaning fluids (NIH 2017).

Ground-level ozone can pose various risks to human health, in addition to affecting forests and reducing yields of agricultural crops (Bell et al. 2004; Felzer et al. 2007; Wang et al. 2016). Repeated exposures to ozone can make people more susceptible to respiratory infections and lung inflammation. Long-term exposures can aggravate pre-existing respiratory diseases such as asthma (US EPA 2017b).

As of 2017 as shown in Table 6.5, the primary and secondary NAAQS for ground-level ozone which are not to be exceeded are 0.07 ppm in annual fourth-highest daily maximum 8-hour average concentration, averaged over three years. In 1971 when the NAAQS were determined for the first time, the NAAQS for ozone was 0.08 ppm which was not to be exceeded more than 1 hour per year (US EPA 2017a).

Table 6.5 NAAQS for criteria pollutants, as of 2017.

Criteria pollutants		Primary/ secondary standards	Averaging time	Critical levels	Regulations
Carbon monoxide (CO)		Primary	8 h	9 ppm	Not to be exceeded more than once per year
			1 h	35 ppm	
Lead (Pb)		Primary and secondary	Rolling 3-month average	0.15 µg m^{-3}	Not to be exceeded
Nitrogen dioxide (NO$_2$)		Primary	1 h	100 ppb	98th percentile of 1-h daily maximum concentrations, averaged over 3 years
		Primary and secondary	1 year	53 ppb	Annual mean
Ozone (O$_3$)		Primary and secondary	8 h	0.070 ppm	Annual fourth-highest daily maximum 8-h concentrations, averaged over 3 years
Particle pollution (PM)	PM2.5	Primary	1 year	12.0 µg m^{-3}	Annual mean, averaged over 3 years
		Secondary	1 year	15.0 µg m^{-3}	Annual mean, averaged over 3 years
		Primary/ secondary	24 h	35 µg m^{-3}	98th percentile, averaged over 3 years
	PM10	Primary and secondary	24 h	150 µg m^{-3}	Not to be exceeded more than once per year on average over 3 years
Sulfur dioxide (SO2)		Primary	1 h	75 ppb	99th percentile of 1-h daily maximum concentrations, averaged over 3 years
		Secondary	3 h	0.5 ppm	Not to be exceeded more than once per year

ppm, Parts per million; ppb, parts per billion; µg, microgram = one millionth (1×10^{-6}) of a gram.

Sulfur dioxide, along with NOx, is the primary pollutant that causes the environmental problem of acid rain (Likens and Borman 1974). Sulfur dioxide and NOx, once released into the atmosphere and transported by wind and air currents, react with water, oxygen, and other chemicals to form sulfuric and nitric acids. These acids eventually fall back on Earth in the form of rainfall or snowfall or by wind.

Further, sulfur dioxide is a precursor of fine particulates: by interacting with other air pollutants, it forms sulfate particles. Sulfate particles are major constituents of the fine particulate matter pollution, particulate matter 2.5 (PM2.5), which, as will be discussed shortly, are a pervasive threat to public health (Muller and Mendelsohn 2009; US EPA 2017b).

Primary polluters are electricity-generating power plants, which, by burning fossil fuels such as coal, natural gas, and oil, are responsible for about two-thirds of annual sulfur dioxide emissions in the US and about one-fourth of annual NOx emissions in the country (Mendelsohn 1980; Muller and Mendelsohn 2009). The rest of sulfur dioxide is emitted by vehicles and from other industrial processes.

Over half the annual NOx emissions in the US is accounted for by combustion of fuels through various modes of transportation – passenger cars, sport utility vehicles (SUVs), buses, trucks, and trains (US EPA 2014a).

Acid rain is the phenomenon of wet deposition of these acids accumulated in the air onto the ground and in water bodies. The wet deposition of acids can increase acidity in the affected lakes and other water bodies temporarily. The high acidity in the lakes and water bodies caused by the acid rain event can last for several days and as long as many weeks, harming fish and other aquatic life forms (Likens and Borman 1974).

In addition, sulfur dioxide in the atmosphere, in the form of very fine particulate matter, can harm public health. Repeated exposures to a high concentration level of sulfur dioxide in the air aggravate various lung problems of people with asthma and cause breathing difficulties in children and the elderly. An inhalation of a high concentration level of sulfur dioxide can even lead to damaged lung tissues and can be a cause of premature death (Smith and Huang 1995; Muller and Mendelsohn 2009; Burtraw and Szambelan 2009).

The primary NAAQS threshold for sulfur dioxide is, as shown in Table 6.5, set at 75 ppb (parts per billion), equivalent to 0.075 ppm, in 1-hour maximum concentration, calculated as the 99th percentile of 1-hour daily maximum concentrations, averaged over 3 years (US EPA 2017a).

The primary NAAQS threshold for nitrogen dioxide, one type of NOx, is set at 100 ppb, equivalent to 0.100 ppm, in 1-hour maximum concentration, calculated as the 98th percentile of 1-hour daily maximum concentrations, averaged over 3 years. Further, annual mean concentration must not exceed 53 ppb, equivalent to 0.053 ppm (US EPA 2010, 2017a).

Pollution of air particles or particle pollution (PM) is referred to as PM2.5 or PM10. PM includes very fine dust, soot, smoke, and liquid droplets. Sulfur dioxide and NOx released from the various pollution sources described above react with sunlight and water vapor to form liquid particles in the air. The particles are also released from fireplaces, wood stoves, unpaved roads, and crushing and grinding operations (US EPA 2017b).

The fine particulate matters are classified into PM2.5 or PM10 according to the size of the particle. PM10 is the particulate matter whose diameter is smaller than 10 μm and

PM2.5 is the particulate matter whose diameter is smaller than 2.5 μm. One micron, also called a micrometer, is one-millionth of 1 m or one-thousandth of 1 mm.

The finer the particles, the more harmful they tend to be and the more difficult to control. One of the reasons for this is that very fine particles can remain suspended in the air and get carried away long distances with prevailing winds from the sources of initial releases. Research in the US shows that over 20% of the particles that form haze in the Rocky Mountains National Park have been estimated to have originated from sources that are hundreds of miles away from the National Park (Muller and Mendelsohn 2009; US EPA 2017b).

Another reason for the greater concern regarding finer particles is that very fine particles such as PM2.5 can easily escape local plant-level control devices such as scrubbers and get carried farther from the sources by prevailing winds, which eventually get into human and animal lungs more easily (Mendelsohn 1980; Cropper and Oates 1992). They are more difficult to control and more damaging to public health.

These fine particles were reported to cause tens of thousands of human deaths in the US annually, by themselves or coupled with other pollutants, especially of the elderly and children, by damaging respiratory systems (Pope et al. 2002; Woodruff et al. 2006). Repeated exposures to a high concentration level in the air or indoors of fine particles may aggravate asthma, cause acute respiratory symptoms such as coughing, reduce lung function resulting in shortness of breath, and cause chronic bronchitis. Particularly susceptible are the elderly, children, and asthmatics, as well as individuals with pre-existing heart or lung diseases (Woodruff et al. 2006; Muller and Mendelsohn 2009).

Besides the human health effects, other effects of repeated exposures to high concentration levels of fine particles include visibility-reducing haze in cities and national parks, and soiled buildings and monuments. However, these are not considered catastrophic effects to the victims who experience such results (Mendelsohn and Olmstead 2009; Freeman et al. 2014).

As summarized in Table 6.5, the primary NAAQS threshold for PM2.5, as of 2018, is an annual mean concentration of 12.0 μg m^{-3}, averaged over 3 years. In addition, the primary and secondary standards require that the 98th percentile of 24-hour concentration should not exceed 35 μg m^{-3} averaged over 3 years (US EPA 2017a).

Lead, with the chemical symbol of Pb, one of the six criteria pollutants designated by the EPA, occurs naturally in small amounts in the rocks and soils of Earth's crust. Lead is widely used in products such as leaded gasoline, lead-based paint, ceramics, pipes, plumbing materials, solders, batteries, ammunition, cosmetics, roofing, scientific electronic equipment, and medical devices (US EPA 2010; NIH 2017).

In a human body, lead accumulates in bones, blood, and soft tissues, affecting almost every organ and system in the human body. According to the National Toxicology Program, exposure to lead has been associated with lung, stomach, and bladder cancer (NIH 2017). Lead exposure can affect adversely children's cognitive and physiological development. Among other health effects, lead exposure can exacerbate high blood pressure in adults (Nichols 1997).

Lead and lead compounds from house paint were banned in 1978 (NIH 2017). Lead in gasoline has been phased down by the EPA's rule, under the Clean Air Act Amendments of the 1970s, that required unleaded gasoline in new cars equipped with catalytic converters and by the EPA's individual facility performance standards in the 1980s under which refiners are forced to decrease the lead content in all gasoline (US EPA 1977; Nichols 1997).

As of 2018 (see Ta6.5), the primary and secondary NAAQS threshold for lead is 0.15 µg m^{-3} which must not be exceeded in rolling 3-month averages. Notably, the first NAAQS threshold for lead, which was set in 1978, was much higher at 1.5 µg m^{-3} (US EPA 2017a).

Finally, carbon monoxide (CO) is released when something is burned incompletely, e.g. natural gas, gasoline, liquefied petroleum gas, oil, diesel fuel, kerosene, coal, charcoal, wood, and wildfires (NIH 2017). In the outdoors, a primary source of carbon monoxide pollution is combustion of fossil fuels by motor vehicles and machineries. Indoors, major sources of carbon monoxide pollution are space heaters that rely on fossil fuels, damaged chimneys/furnaces, and gas stoves (US EPA 2010).

Carbon monoxide is a highly poisonous gas (NIH 2017). Inhalation of a high concentration may reduce the amount of oxygen in the bloodstream which is transported to body organs such as the heart and brain. Exposure to a very high concentration can occur indoors or in an enclosed environment such as a garage, vehicle, or tent. Such exposures can cause dizziness, unconsciousness, coma, and even death (US EPA 2010). Exposure by a pregnant woman to high concentrations may cause miscarriage or increased risk of damage to a developing fetus.

For carbon monoxide as one of the criteria air pollutants, the latest update in standards is that the primary NAAQS threshold is an 8-hour concentration of 9 ppm which should not be exceeded more than once per year. The secondary threshold is a 1-hour concentration of 35 ppm which again should not be exceeded more than once per year (US EPA 2017a).

The determinations of the NAAQS for the criteria pollutants are based on toxicology (Lutz et al. 2014). For each pollutant, repeated toxicological experiments are conducted on animals, e.g. mice and monkeys, whose results are extrapolated to humans, from which a (set of) thresholds is determined. Exposure to each of these pollutants beyond the determined threshold level, more specifically a primary standard, is deemed lethal to humans.

Implementing these thresholds, once determined, is another level of environmental policy challenges different from that of determining the thresholds (Baumol and Oates 1988). First of all, the number of sources of emissions of each of the criteria pollutants is enormous. For example, sources of sulfur dioxide releases which create the acid rain problem, among other things, are very large in number (Likens and Bormann 1974). Monitoring and regulating individual sources of emissions are virtually impossible or highly costly, to put it differently.

Further, there must be the least-costly way among available policy options to achieve the already-determined nationwide target for each pollutant. Policy-makers should be concerned about the total costs of various policy options and choose the least-cost policy (Tietenberg and Lewis 2014).

The least-cost policy instrument given the target level of abatements or emissions is called an emissions permit program because the total amount of emissions permits is determined and regulated (Montgomery 1972). It is also called a cap-and-trade program since it caps the total emissions of the concerned pollutant and allows polluting agents to trade permits among them. The US sulfur dioxide allowance program is one of the best known such programs.

For the implementation of the NAAQS for sulfur dioxide, the US Congress established the sulfur dioxide emissions allowance trading program for electricity-generating units through Title IV of the 1990 Clean Air Act Amendments, entitled "Acid Deposition Control" (US EPA 1990). The sulfur dioxide allowance trading program is referred to as a cap-and-trade program owing to its major features. It sets an annual cap on aggregate emissions of sulfur dioxide, allocates the total allowances across the polluters, and allows an individual polluter to trade allowances with another polluter in order to minimize its cost of complying with the regulation (Stavins 1998).

The sulfur dioxide allowance trading program in the US was implemented in two phases. In Phase I which began in 1995, 110 of the dirtiest coal-fired electricity-generating facilities, including about 374 electricity-generating units, were affected by the program. In Phase II which began in 2000, all other coal-fired facilities with a capacity greater than 25 MW were included – in total 1420 generation units (Burtraw and Szambelan 2009).

The aggregate emissions cap was set to reduce the annual emissions permanently by 10 million tons from the base-period annual emissions of 20 million tons. More specifically, in Phase I, the cap was set at 2.5 lb of sulfur dioxide per MMBtu (one million British Thermal Units) of energy used times an average fuel use during the base-period from 1985 to 1987. In Phase II, the cap was made more stringent, at 1.2 lb per MMBtu (US EPA 1990).

The sulfur dioxide allowance trading program by and large achieved the emissions reduction target. By the first year of Phase II, annual emissions fell by 40% from the 1980s' level (Ellerman 2003). According to the EPA, annual emissions fell from 17.3 million tons in 1980 to 10.2 million tons in 2005, to 5.2 million tons by 2010, and to a further 2.2 million tons by 2015 (US EPA 2017c). The amount of sulfur deposition has also shown a steady decrease since the 1970s (Burtraw and Szambelan 2009).

However, by 2012, the price of the sulfur dioxide allowance fell to virtually zero and the sulfur dioxide trading market crashed, after peaking at about US$1200 per ton in 2005 (Schmalensee and Stavins 2013). Figure 6.1 shows the history of the price of clearing bids (in 2016 US$) in the sulfur dioxide spot auction from 1993 to 2016, which was documented by the Clean Air Markets of the EPA (US EPA 2017c).

The crash in the sulfur dioxide trading market followed from legal challenges by the States and utilities. In 2008, the Circuit Court of Appeals for the District of Columbia vacated the Clean Air Interstate Rule (CAIR) under which the interstate sulfur dioxide allowance trading was made possible, primarily because the EPA does not precisely establish the relationship between pollution-source States and receptor States in achieving air quality standards, following a lawsuit by North Carolina and other States and utilities against the EPA. In 2010, the Obama administration proposed as an alternative state-specific emission caps for sulfur dioxide emissions, called the Cross-State Air Pollution Rule (CSAPR), which was invalidated by the US Court of Appeals for the DC Circuit in August 2012 (Schmalensee and Stavins 2013). Later, the US Supreme Court upheld the CSAPR rule by a 6:2 decision in 2014.

A cap-and-trade program is not necessarily a Pareto optimal (efficient) policy. A Pareto optimal policy is one that balances the social benefit and the social cost of abating a marginal unit of emissions at an optimal level of emissions, from which an optimal price of emissions and a socially optimal level of emissions are determined (Baumol and Oates 1988). The socially optimal level of emissions of a pollutant may

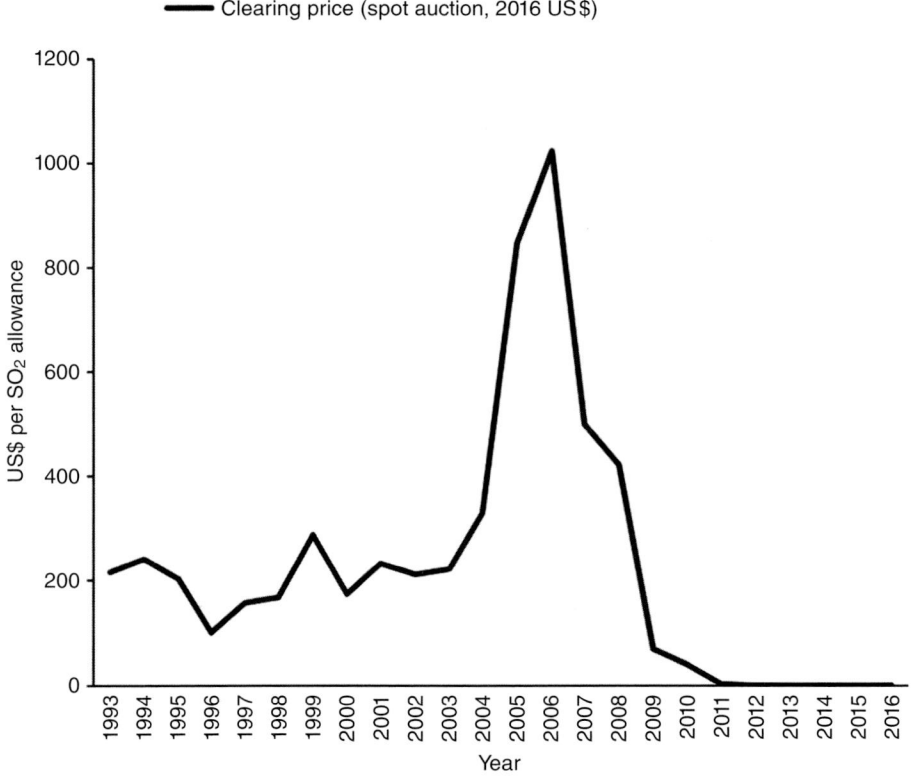

Figure 6.1 History of the sulfur dioxide allowance price of clearing bids from spot auction. *Source*: US EPA (2017c), accessed from https://www.epa.gov/airmarkets.

be different from the threshold value of a pollutant determined by the EPA through the toxicology experiments. It may be different also from the emissions cap set by, for example, the above sulfur dioxide allowance trading program.

6.7 Toxic Chemicals and Hazardous Substances: Toxic Substances Control Act

Another type of environmental catastrophes is caused by toxic chemicals and hazardous substances. Unlike the criteria pollutants described in Section 6.6, toxic chemicals can cause the death of an individual upon contact, act as a strong carcinogen inside a human body, or cause many other acute health problems of humans and animals.

The public fear of toxic chemicals was emphatically exposed by many catastrophe researchers including Rachel Carson, whose work was reviewed in detail in Chapter 3 of this book (Carson 1962). Historical catastrophic events caused by toxic chemicals include Itai-Itai disease in 1912 from cadmium poisoning, Minamata disease in 1956 from mercury poisoning, and the Bhopal disaster in 1984 from a pesticide plant in Bhopal, India which killed up to 10 000 people (Ministry of the Environment 2002).

The laws that regulate toxic chemicals and hazardous substances in the US include the TSCA of 1978, the Federal Insecticide, Fungicide, and Rodenticide Act (FIFRA) of 1947, and the Comprehensive Environmental Response, Compensation, and Liability Act (CERCLA) of 1980 (US Congress 1978).

The Superfund is the trust fund created by the CERCLA for the purpose of addressing the dangers of abandoned or uncontrolled hazardous waste dumps by developing a nationwide program for, among other things, emergency responses and site cleanups (US Congress 1980).

The FIFRA of 1947 is a federal regulation for pesticide distribution, sale, and use in the US, amended many times thenceforth, whose origin can be traced back to the 1910 Federal Insecticide Act (US Congress 2012). Under the FIFRA, all pesticides must be registered and licensed by the EPA before their sale and distribution. For registration, the applicant must demonstrate that the use of the pesticide will not generally cause unreasonable adverse effects on the environment.

The TSCA of 1978 gives the EPA the authority to regulate new and already existing "chemical substances and mixtures … (whose) manufacture, processing, distribution in commerce, use, or disposal may present an unreasonable risk of injury to health or the environment" (US Congress 1978). In the following, the author describes at length the US regulations on toxic chemicals and hazardous substances, with a focus on the TSCA.

For such harmful chemical substances and mixtures, the TSCA establishes the US government's policy position as follows (US Congress 1978):

1) Adequate data should be developed with respect to the effect of chemical substances and mixtures on health and the environment and that the development of such data should be the responsibility of those who manufacture and those who process such chemical substances and mixtures;
2) Adequate authority should exist to regulate chemical substances and mixtures which present an unreasonable risk of injury to health or the environment, and to take action with respect to chemical substances and mixtures which are imminent hazards.

The TSCA gives the EPA three major responsibilities: (i) to gather information on new and existing chemicals being manufactured in the US; (ii) to collect and produce data for use in assessing the risks of chemicals; and (iii) to properly control these chemicals deemed to present an unreasonable risk to health or the environment through rule-making that includes restrictions, labeling, and bans (Vogel and Roberts 2011).

The most powerful regulatory authority given to the EPA by the Act is specified in Section 6 of the TSCA. The Section on Regulation of Hazardous Chemical Substances and Mixtures gives the administrator of the EPA the authority to prohibit (or limit) the manufacturing, processing, or distribution in commerce of a certain chemical substance or mixture if s/he concludes that there is a reasonable basis to judge that the chemical substance or mixture, during any stage of production, distribution, use, or disposal of it, presents an unreasonable risk of injury to health or the environment.

In addition to the high burden of proof, Section 6 of the Act requires that the EPA administrator, in choosing from a pool of regulatory options to be imposed, must take the least burdensome ones if s/he decides to regulate or ban a certain chemical (US Congress 1978):

> The Administrator shall by rule apply one or more of the following requirements to such substance or mixture to the extent necessary to protect adequately against such risk using the least burdensome requirements.

The EPA, relying on the authority specified in Section 6, made numerous attempts to restrict toxic chemicals such as polychlorinated biphenyls (PCBs), chlorofluorocarbons (CFCs), dioxin, asbestos, hexavalent chromium, mercury, radon, and lead-based paint. However, the EPA was largely unsuccessful in regulating these toxic chemicals faced with a litany of lawsuits by chemical companies.

An illustrative case of the EPA's unsuccessful attempts is that of asbestos. Based on an extensive risk assessment for 10 years, the EPA issued a regulation in 1989 that would have banned almost all uses of asbestos. Asbestos producers immediately challenged the regulation in the court, which ruled against the regulation and the EPA. In 1991, the Fifth Circuit Court of Appeals vacated the EPA rule, on the grounds that the EPA had failed to meet the TSCA's high burden of proving with "substantial evidence" that an "unreasonable" risk existed and that the proposed regulation, which would ban most uses of asbestos, presented the "least burdensome" approach. The EPA chose not to appeal the court's decision, giving up its authority to place even limited restrictions on asbestos (Vogel and Roberts 2011).

The failure of the TSCA to regulate toxic chemicals is attributed to several factors. First, the burden of proof that the concerned chemical poses an unreasonable risk to health or the environment belongs to the EPA regulator, not chemical producers. The EPA's first task after passage of the TSCA in 1978 was, by Section 8 of the Act, to create an inventory of existing chemicals. The initial inventory listed 62 000 chemicals and the current inventory is greater than 83 000 chemicals. These existing chemicals were, by default, presumed to be safe and the burden of proving that a certain chemical poses an unreasonable risk fell on the environmental agency (Vogel and Robert 2011).

For new chemicals, the Act requires, under Section 5, producers to notify the EPA prior to manufacture by submitting a pre-manufacture notification including properties and toxicity of the chemical. However, it does not mandate any toxicity testing required of the chemical producer.

Second, under the TSCA, the EPA should show that the proposed rule to restrict the use of a chemical is the least burdensome approach. The requirement limits the EPA's authority to regulate toxic substances because chemical producers can challenge any regulation based on this requirement, as was evidently the case in the asbestos regulation (Vogel and Roberts 2011).

Notwithstanding the failures of the EPA to regulate toxic chemicals through the TSCA, a series of social, national, and international changes has provided renewed motivation for a stronger regulation of toxic chemicals. The first of these favorable backgrounds is changes in people's attitudes toward health problems posed by toxic substances. When the TSCA was passed in the 1970s, people's health concerns were limited to chemicals and substances that could be potential carcinogens (that cause cancer), mutagens (that cause genetic mutation), and teratogens (that cause defects in an embryo). Since then, people have become increasingly more concerned with everyday exposures to chemicals which may lead to chronic diseases (Vogel and Roberts 2011).

Second, despite the failures of the EPA in the US, State-level regulations have increased. During the period from 2003 to 2010, state legislatures passed 71 chemical safety laws aimed at restricting the use of specific chemicals, including lead in toys, polybrominated diphenylethers (PBDEs) in flame retardants, and bisphenol A (BPA) in baby bottles (Vogel and Roberts 2011). Several States have adopted their own chemical safety laws and given authority to develop a comprehensive regulatory program, while specific laws regulating toxic chemicals have been enacted in 32 States (NCSL 2017).

The third favorable backdrop is the European Union's adoption in 2006 of a comprehensive chemicals management program called the Registration, Evaluation, Authorisation, and Restrictions of Chemicals (REACH). Applying the "no data, no market" rule, the REACH program places responsibility on the industry to manage risks from chemicals and to submit safety information on chemical substances, which is deposited in a central database at the European Chemicals Agency (ECHA) in Helsinki (EC 2016c).

Amending the TSCA, the US Congress passed in 2016 the Frank R. Lautenberg Chemical Safety for the 21st Century Act with bipartisan support (Hamblin 2016). Among other things, the Frank Lautenberg Act increases pre-market testing requirements and funding to test priority chemicals. However, it also pre-empts States' ability to regulate toxic chemicals on the grounds that having different regulations in different States is difficult and costly. Under the Lautenberg Chemical Safety Act, a State cannot act on its own with regard to a chemical during the period when the EPA is testing the chemical.

6.8 Ozone Depletion: The Montreal Protocol

Thus far, we have reviewed two environmental catastrophes to which policy interventions have been made at a national or subnational level. Henceforth, two global-scale environmental catastrophe issues will be discussed: ozone depletion and global climate change. For both issues, a global-scale catastrophe that may result from either of the problems has been proposed by many researchers. Further, international treaties are in place for addressing each problem, binding member nations legally but with different force.

Ozone (O_3) on the ground is a pollutant that harms people and ecosystems, as explained in the Section 6.7 as one of the criteria pollutants. In the atmospheric zone of the stratosphere, however, the ozone layer plays a vital role in shielding people and ecosystems against the ultraviolet-B (UVB) radiation of the sun. In the 1970s, a group of scientists discovered that widely applied coolants for refrigerators and air-conditioners, i.e. CFCs and other gases, was depleting the ozone layer of the planet (Molina and Rowland 1974).

The depletion of the ozone layer can cause catastrophic damages by way of an increase in the UVB radiation that reaches the Earth's surface. Laboratory and epidemiological studies report that UVB radiation causes non-melanoma skin cancer and plays a major role in malignant melanoma development. In addition, UVB radiation has been linked to the development of cataracts, a clouding of the eye's lens (WMO 2014).

The ozone layer depletion poses a global challenge. Since the need for cooling for air-conditioning and refrigerating is ubiquitous, all countries are sources of significant releases of CFCs and other ozone-depleting substances (ODSs). A single country or a group of countries exclusively cannot prevent the depletion of the Earth's ozone layer, for which reason a policy intervention began at a global level even without country-level interventions.

International policy negotiations for addressing the ozone layer depletion are encapsulated in the Montreal Protocol on Substances that Deplete the Ozone Layer, a treaty signed in 1987. Through the Montreal Protocol, member countries agreed to phase out CFCs, the main chemical that depletes the ozone layer (UNEP 2016). As of 2018, 197 countries ratified the Protocol, including all member countries of the UN.

The Montreal Protocol is widely viewed as a successful international treaty for a global environmental issue because it was successful in almost banning the use of CFCs in developed countries. Under the Protocol, nations have phased out nearly all ODSs, and the total column ozone will recover to the benchmark 1980 levels by the middle of the twenty-first century over most of the globe with the full compliance of the Protocol (WMO 2014).

However, it has been overlooked that the success of the Montreal Protocol has worsened another global environmental problem, that is, a globally warming climate. In an effort to replace CFCs in response to the Protocol, chemical companies have gradually switched to hydrofluorocarbons (HFCs) and hydrochlorofluorocarbons (HCFCs). These replacement chemicals are, however, potent globe-heating gases with global warming potential (GWP) greater than 1000 times that of carbon dioxide (IPCC 2005, White House 2013).

However, owing to technological advances, there are multiple alternative coolants to the HFCs and HCFCs as well as the CFCs which do not contribute to global warming effects (EC 2016a). More specifically, hydrocarbons and other alternative coolants can be used as a replacement coolant and are in effect widely used in developed countries in Europe (White House 2013; Seo 2017a).

Being aware of the technical options that are also touted as being economical and safe, more than 170 parties to the Montreal Protocol signed an amendment to the Protocol to phase out Earth-warming HFCs in Kigali, Rwanda on October 2016, with only one month before the election of Donald Trump as President of the US (UNEP 2016). Unlike climate agreements, e.g. the Paris Agreement, the Montreal Protocol amendment is given the legal force of an international treaty.

The Kigali Amendment is expected to enter into force by January 2019 if at least 20 parties of the Protocol ratify the amendment. As of 1 October 2017, nine countries have ratified the amendment: Mali, the Federated States of Micronesia, the Marshall Islands, Rwanda, Palau, Norway, Chile, Tuvalu, and North Korea. Countries have the option to withdraw, after ratification, from the Protocol.

A major weakness of the Kigali Amendment may lie in the foundation of the Montreal Protocol. That is, the Kigali Amendment is not strongly tied to the fundamental mission of the initial Montreal Protocol which is to protect the ozone layer from depletion (UNEP 2016). Negotiators gathered at Kigali made the case that the Kigali Amendment can lower the degree of global warming by 0.5° C by the end of this century, which is, of course, not the fundamental objective of the initial Montreal Protocol. The point can be made that the phase-out of HFCs for the protection of the global climate system should be regulated by a global warming protocol, not by an ozone protocol.

In the Kigali Amendment, the planned global phase-out of HFCs is structured into three tracks defined by the economic status of participant countries. The first track is to be followed by developed countries including the US, the European Union (EU), and Japan. The second track is to be followed by most of the rest of the world, including China, Brazil, and all African countries. The third track is allowed to the world's hottest countries, including India, Pakistan, Iran, Saudi Arabia, and Kuwait.

For the first-track countries, production and consumption of HFCs should be frozen by 2018 and be reduced to about 15% of the 2012 level by 2036. For the second-track countries, the production of HFCs should be frozen by 2024 and be reduced to about 20% of the 2021 level by 2045. For the third-track countries, the use of HFCs should be frozen by 2028 and be reduced to around 15% of the 2025 level by 2047.

At the Kigali conference, particularly strong resistance to the amendment was expressed by India, where her people are just beginning to afford and enjoy the benefits of air-conditioning and refrigeration owing to recent economic successes in the country, which advanced economies have enjoyed and benefited from for so long (Barreca et al. 2016). With average temperature of the hottest month being higher than 40° C in many parts of the country, air-conditioning and refrigeration are a critical, often life-saving, technology for Indian people to overcome intolerable heat stress, monsoon rainfall, and aridity in dry seasons. As such, a higher cost of air-conditioners or refrigerators induced by the switch of cooling agents required by the Kigali Amendment is expected to have major implications on the country's economic well-being, which is the rationale for the 20-year moratorium for India and other third-track countries in the implementation of the Kigali Amendment.

From the catastrophe policy perspective, it can be argued that the Montreal Protocol was successful and has achieved its goal of preventing a global-scale catastrophe via UVB radiation. Further, it can be argued that its success is attributable to near-universal participation of the world's nations.

However, it should be noted that the universal participation was possible in the late 1980s because widespread uses of refrigerators and air-conditioners were possible only in rich countries of North America, Europe, and Japan. As such, the 1987 Montreal Protocol was expected to affect only those rich countries.

By 2016, when the Kigali Amendment was signed, air-conditioning and refrigerators have become essential amenities to developing countries' households across the globe, including China, India, South East Asia, Latin America, and many parts of Africa. It will be interesting to see whether the global community can turn the Kigali Amendment into a near-universal international treaty, like the initial Montreal Protocol.

As elucidated further in Section 6.9, a successful policy intervention for preventing a global-scale catastrophe can only be accomplished through global cooperation, that is, a near-universal participation rate, which is true in many, although not all, global-scale catastrophe issues (Barrett 2010; Nordhaus 2011a).

6.9 Global Warming: The Kyoto Protocol and Paris Agreement

By the 1980s, a truly global-scale environmental catastrophe was predicted by climate scientists through the phenomenon of global warming (Le Treut et al. 2007). The scale of destruction would be as severe as humanity-ending in unmitigated global warming. Further, the problem is truly pervasive at the humanity scale. That is, nearly every human activity is related to emissions of Earth-warming gases. This was recognized as a far worse environmental problem for humanity than ozone-layer depletion.

Like the above-described Montreal Protocol, policy negotiations for addressing the problem of global warming began at the international level, i.e. at the UN, before there were national policies in place. The United Nations Framework Convention on Climate Change (UNFCCC) is an international climate change treaty negotiated at the Earth Summit in Rio de Janeiro in 1992. In the treaty, the fundamental framework for addressing the climate change problem was expressed in terms of catastrophic consequences of projected changes in the climate system. More specifically, the objective of the treaty was declared as follows in Article 2 of the UNFCCC (UNFCCC 1992):

> The ultimate objective of this Convention … is to achieve … stabilization of greenhouse gas concentrations in the atmosphere at a level that would prevent dangerous anthropogenic interference with the climate system.

As of June 2018, the UNFCCC has near-universal membership with 197 parties to the Convention, i.e. the parties that ratified the UNFCCC. It has held 23 Conferences of the Parties (COPs) through the end of 2017, of which major outcomes are the Kyoto Protocol in 1997, the Copenhagen Accord in 2009, the Durban Platform in 2011, and the Paris Agreement (COP 21) in 2015.

A range of policy instruments advocated or adopted in the COPs, to be explained shortly, were often expressed in terms of thresholds in various indicators of global warming. Examples are a global atmospheric temperature ceiling, carbon emissions thresholds, and global atmospheric carbon budgets.

The first ever international climate treaty that specified limits of emissions of carbon dioxide was the Kyoto Protocol (UNFCCC 1998). It became an international treaty, i.e. a protocol, after required conditions were met. The Kyoto Protocol was the only legally binding international treaty on global climate change problems, but it became ineffective after failure in Copenhagen in 2009 to renew it and will be replaced by the Paris Agreement from 2020 (UNFCCC 2009).

The Kyoto Protocol's regulatory principle lied in setting and forcing the threshold (limit) of carbon dioxide emissions at the aggregate level. Specifically, nations agreed to a stabilization of carbon dioxide equivalent emissions at 5% below the 1990 level at the aggregate level in the first commitment period, from which individual countries' commitments are assigned (UNFCCC 1998):

> The Parties … shall, individually or jointly, ensure that their aggregate anthropogenic carbon dioxide equivalent emissions of the greenhouse gases … do not exceed their assigned amounts, … with a view to reducing their overall emissions of such gases by at least 5 per cent below 1990 levels in the commitment period 2008 to 2012. [Article 3 of the Kyoto Protocol]

In addition to carbon dioxide, the Protocol adds other greenhouse gases to the list of regulated chemicals under its implementation by way of converting other gases into carbon dioxide equivalents using the concept of GWP (IPCC 1990). Other greenhouse gases listed in Annex A of the Protocol are methane (CH_4), nitrous oxide (N_2O), HFCs, perfluorocarbons (PFCs), and sulfur hexafluoride (SF6).

Retrospectively, after two decades since the agreement in Kyoto, we can now make a verdict that the exclusion in the Kyoto Protocol of major developing countries that were then developing rapidly – including China, India, and Brazil – from the legal responsibility of cutting emissions of greenhouse gases has led to the Protocol's eventual demise (Nordhaus and Boyer 1999; MacCracken et al. 1999; Manne and Richels 1999; Norhaus 2001). The US Congress did not ratify the treaty, although the US initially agreed, citing the exclusion of China and India.

Excluding the US, China, India, and other developing countries, the first phase of the Kyoto Protocol was implemented during the 5-year period from 2008 to 2012 exclusively in the group of ratified countries that comprised mostly the EU, Japan, and Canada (EC 2016b). The negotiation for the second phase of the Protocol failed in the much-hyped Copenhagen Conference in 2009 owing primarily to disagreements between developed nations and developing nations on their respective responsibilities and financial aids from rich countries (Nordhaus 2010, 2011a; Seo 2012).

The Kyoto Protocol is going to be eventually replaced by the Paris Agreement achieved at the COP 21 held in Paris, France at the end of 2015 (UNFCCC 2015). The Paris Agreement accomplishes some of the goals set by the Durban Platform for Enhanced Action under which parties of the Convention agreed to launch an effort to have an agreement by the end of 2015 in which all countries take legal responsibilities for reducing emissions of greenhouse gases (UNFCCC 2011a). However, unlike the Kyoto Protocol, the Paris Agreement is not considered an international treaty which is legally binding to the ratified parties, for which reasoning the ratification of the Agreement did not require an approval from the US Senate.

For the Paris Agreement, countries of the Convention were asked to submit, before the meeting in Paris, voluntarily their intended commitments, called the Intended Nationally Determined Contributions (INDCs). Submitted and finalized INDCs vary greatly from one country to another. Essentially, there is no commitment by China and India in terms of carbon dioxide equivalent emissions reduction by the timeframe to 2030 or so (Seo 2017c).

In addition to the INDCs, the Paris Agreement is complemented by the financial instrument named the Green Climate Fund (GCF). The GCF plans to raise funds of the size of US$100 billion annually from 2020, which may increase to US$1 trillion per year soon afterwards, but the contributions to the GCF from the ratified countries are falling far short of the target expressed in the global climate talks (UNFCCC 2011b; GCF 2016; Seo 2017c).

From the policy perspective, the Paris Agreement is more specific than the Kyoto Protocol in its ultimate objective. Article 2 of the Paris Agreement declares the ultimate objective in concrete terms, keeping away from an equivocal expression such as "avoiding a dangerous anthropogenic interference" used in the UNFCCC (UNFCCC 1992, 2015):

> Holding the increase in the global average temperature to well below 2° C above pre-industrial levels and to pursue efforts to limit the temperature increase to 1.5° C above pre-industrial levels, recognizing that this would significantly reduce the risks and impacts of climate change.

Notwithstanding the acclaim by many politicians that the Paris Agreement is a landmark global climate policy or a turning point in global talks, the INDCs submitted by individual countries were never determined with consideration of the 2° C temperature ceiling, the ultimate objective of the Agreement. In determining the INDCs, countries adopted different baselines: 1990, 2005, or the Business-As-Usual (BAU) baseline. China, India, and South Africa have not committed to a reduction in the aggregate amount of emissions of carbon dioxide and other greenhouse gases. Roughly, Brazil has committed to a 43% reduction, the EU 40%, Russia 30%, the US 26%, Thailand 20%, and Vietnam 8%. Many countries' INDC commitments are expressed in contingence upon the delivery of financial aid through the GCF.

Why the 2°? It is the concrete threshold in the climate system determined by climate negotiators beyond which a dangerous anthropogenic interference is proposed to occur. In the language of catastrophe theories, the 2° ceiling is proposed by negotiators as a bifurcation point, a tipping point, a threshold, or a singularity (Thom 1975; Kurzweil 2005).

Let's have a closer look at the actual INDC proposals from several major countries: China, India, and the US. China's INDC proposal states its commitment in the language of emissions peaking:

> Based on its national circumstances, development stage, sustainable development strategy and international responsibility, China has nationally determined its actions by 2030 as follows: 1) To achieve the peaking of carbon dioxide emissions around 2030 and making best efforts to peak early; 2) To lower carbon dioxide emissions per unit of GDP by 60% to 65% from the 2005 level; 3) To increase the share of non-fossil fuels in primary energy consumption to around 20%; 4) To increase the forest stock volume by around 4.5 billion cubic meters on the 2005 level.

India's INDC proposal does not have a clause on the level of emissions reduction to which it commits. Instead, it proposes to increase energy production from nonfossil fuel resources, on the condition of technology transfer from developed nations as well as financial aid through the GCF:

1) To put forward and further propagate a healthy and sustainable way of living based on traditions and values of conservation and moderation;
2) To adopt a climate friendly and a cleaner path than the one followed hitherto by others at corresponding level of economic development;
3) To reduce the emissions intensity of its GDP by 33–35% by 2030 from the 2005 level;
4) To achieve about 40% cumulative electric power installed capacity from nonfossil-fuel-based energy resources by 2030 with the help of the transfer of technology and low-cost international finance including from the GCF;
5) To create an additional carbon sink of 2.5–3 billion tonnes of carbon dioxide equivalent through additional forest and tree cover by 2030.

The Paris Agreement entered into force on 4 November 2016 by satisfying the condition that more than 55 countries shall ratify the Agreement that account for more than 55% of total world emissions. As of June 2018, 168 parties have ratified the Agreement, including the US and China, out of 197 members of the UNFCCC.

However, the Trump Administration announced in a major speech at Rose Garden at the White House the decision to pull the US out of the Paris Agreement, saying that it is unfair to the US (White House 2017). A year earlier, the US Supreme Court blocked the Clean Power Plan from taking effect. In October 2017, the EPA began to take steps to repeal, on the grounds of the inconsistencies with the Clean Air Act, the Clean Power Plan, the main climate rule by the Obama Administration which is viewed as an inevitable policy instrument for the US to have any chance to accomplish its commitments to the Paris Agreement (US EPA 2014b, 2017d).

Without US participation and without substantial commitments by China and India until 2030 and beyond, the possibility for a major cut in global emissions of carbon dioxide and other greenhouse gases remains low (Seo 2017a). However, it does not necessarily mean that there will be no action at all in response to changing climates (Seo 2015b). People and the public sector will adapt to changing climates, while carbon-cutting practices will emerge from such adjusting behaviors (Seo 2016a,b,c).

In Chapter 4, the author describes the debate on how stringent a global climate policy should be in cutting greenhouse gases (Nordhaus 2008, 2011b; Weitzman 2009). Experts are still divided on the topic, which portends that future climate talks, internationally and nationally, will be as intense and controversial. A silver lining is that knowledge on adaptation to climate change has increased sharply since the late 1990s, which may cause the concerned or vulnerable parties to search for possible adaptation strategies to a gradual shift in the climate system. For those who look for adaptation options, technological breakthroughs as well as micro technologies that are already available will play vital roles (Seo 2017a).

6.10 Strangelets: High-Risk Physics Experiments

Besides global environmental catastrophes, the fear of a truly catastrophic event as a side-effect of remarkable technological innovations and advances has rapidly come to the forefront in the battle against catastrophes (Posner 2004; Hawking et al. 2014). In this section and the next, the author describes two such technology-induced catastrophes: high-risk physics experiments which may destabilize the planet and the universe, and AI which may threaten human existence.

Thanks to remarkable advances in technological and engineering capabilities, a large-scale physics experiment that could have a bearing on the existence of the planet and even the universe has become 'increasingly' possible. An example of such an experiment is the LHC experiments conducted by the European Organization of Nuclear Research (CERN) in which scientists from all major advanced countries are participating.

The LHC is a particle accelerator and collider through which scientists hope to test the state of the universe at the time of the origin of the universe, more specifically at the Big Bang, including the existence of a God particle called the Higgs Boson (Overbye 2013). It was built at the France–Switzerland border during the period from 1998 to 2008 and is argued to be the largest single machine in the world, 27 km in circumference and 175 m deep underground (CERN 2017a).

Some scientists have suggested that the scale of a potential catastrophe from the experiment could be truly catastrophic (for review see Dar et al. 1999; Jaffe et al. 2000). The catastrophe would unfold as follows. The particle accelerator may unintentionally produce strangelets – dark matter – which would then create a giant black hole. The black hole would then suck in the entire universe, ending the current universe with all its sentient inhabitants (Plaga 2009). The entire process would unfold swiftly, taking less than a split second.

Many argue that a potential LHC catastrophe is an example of a global catastrophe whose scale of destruction is as large as Earth-ending but whose small chance of occurrence cannot be eliminated (Posner 2004; Plaga 2009). Uncertainty on a potential catastrophe event is so large that the probability distribution of it is fat-tailed (Weitzman 2009).

Needless to say, there is no international treaty or agreement that purports to prevent such a potential catastrophe. Neither is there any specific policy for regulating the LHC-type experiments anywhere in the world. As such, there is not much to add by the author on regulatory measures. However, there is a fascinating scientific debate on the possibility or uncertainty of the potential LHC catastrophe.

According to the LHC Safety Assessment Group and the CERN that conducts the experiments, LHC particle collisions present "no" danger (Ellis et al. 2008). First, the group observes that such collisions occur naturally:

> We recall the rates for the collisions of cosmic rays with the Earth, Sun, neutron stars, white dwarfs and other astronomical bodies at energies higher than the LHC. The stability of astronomical bodies indicates that such collisions cannot be dangerous.

Second, the assessment group conducted the laboratory-controlled experiments from which creation of a strangelet is tested. The group concludes as follows that a number of adaptive mechanisms in the physical world would prevent the catastrophic event caused by a black hole even if it were to be created (Ellis et al. 2008):

> Specifically, we study the possible production at the LHC of hypothetical objects such as vacuum bubbles, magnetic monopoles, microscopic black holes and strangelets, and find no associated risks. Any microscopic black holes produced at the LHC are expected to decay by Hawking radiation before they reach the detector walls. If some microscopic black holes were stable, those produced by cosmic rays would be stopped inside the Earth or other astronomical bodies. The stability of astronomical bodies constrains strongly the possible rate of accretion by any such microscopic black holes, so that they present no conceivable danger. In the case of strangelets, the good agreement of measurements of particle production at RHIC with simple thermodynamic models constrains severely the production of strangelets in heavy-ion collisions at the LHC, which also present no danger.

These conclusions are widely shared by other scientists, the CERN itself, and the American Physical Society (Dar et al. 1999; Jaffe et al. 2000; Peskin 2008; CERN 2017b). According to one group, the possibility of an LHC universal catastrophe is absurdly small (Jaffe et al. 2000).

The experiments at the CERN are on-going and have not up until now yielded any disaster event. On the contrary, the experiments have successfully detected the Higgs

Boson, the long-sought-after particle by physicists, for which the Nobel Prize was awarded in 2013 to the originators of the concept of the particle (Nobel Prize 2013).

6.11 Artificial Intelligence

Another technology-induced fear of catastrophe has arisen from the advancement of AI technologies (Kurzweil 2005; Hawking et al. 2014). Applications of AI to everyday activities and workplaces are already wide-ranging. Among the most utilized are a medical check-up robot, an automated bank teller machine, a self-driving car, an AI financial portfolio manager, a robot killer employed in a battle, an automated doorman, an automated waitress in a restaurant, an automated automobile global positioning system (GPS) navigator, and a digital personal assistant Siri.

Owing to their ever-wider applications, which tend to replace human workers than to create human jobs, there is concern that robots will take over workplaces and replace human workers. Echoing the public concern, Bill Gates, former Microsoft founder and CEO, even proposed a robot tax as a way to slow down the pace of replacement of human laborers with AI workers (Quartz 2017).

A catastrophic scenario with regard to AI is that the AI robots become smarter than humans or they gain the capacity to attack humans. Indeed, some robots without any doubt outperform human competitors even today. For example, an AI robot handily defeats human champions of Chess or Go games. An AI robot also wins at Jeopardy, defeating human competitors.

In the literature of AI, singularity is accepted to be a landmark moment in human history. It is defined to be the point in time when the AI's brain surpasses the capacity of the human brain (Kurzweil 2005). Experts say that singularity will be reached in the next 10–30 years. Some says robots will be 100 times smarter than humans by about 2050 (Fox News 2017).

According to Hawking et al. (2014), the singularity point will be the biggest moment in human history, but can be the last moment as well. The authors forecast super-intelligent machines in the future outsmarting financial markets, out-inventing human researchers, out-manipulating human leaders, and developing weapons we cannot even understand (Hawking et al. 2014).

Nations have already begun competing to gain superiority in the field of AI as a way for military dominance in the world. Some experts, including Elon Musk at Tesla Motors, warned that the competitive race could lead to a third World War. The worst-case scenario is that not the leaders of countries but robots may be able to start a war on their own without human controls (CNBC 2017).

Unlike the LHC-caused catastrophe described in Section 6.10, there is an increasing call for an international response to AI threats. However, a specific policy dialogue has only begun at the UN with a narrow focus on killer robots employed in battles.

At the UN, killer robots are called the Lethal Autonomous Weapon Systems (LAWS) and are treated under the UN Convention on Certain Conventional Weapons (CCWs). The CCW meeting of experts on the LAWS was held in 2014 and 2015 for the first time (UNODA 2017d). There are numerous calls for international governance on the LAWS, including banning killer robots, for which international policy responses have not yet been made (Marchant et al. 2011; Sharkey 2012).

6.12 Conclusion

This concludes a wide-ranging review of the policies and regulations that were implemented to address the potential catastrophes that are covered throughout this book. This chapter shows that policy measures are better defined in some catastrophes than other catastrophic events. Some catastrophes require only local or national measures of intervention, while other catastrophic events call for a global cooperative protocol. A diverse basket of policy approaches to dealing with different types of catastrophe events can be verified throughout this review.

The review of the complex layers of catastrophe policy interventions reveals that successful policy interventions were rare rather than common. Policy interventions have failed because of varied reasons, as explained in this chapter. Many catastrophe policies are framed in terms of various thresholds of policy target variables, but when these thresholds cannot be met for various economic, political, and cultural reasons, policies have failed. In addition, unsuccessful policy interventions have resulted from perverse incentives created by a certain policy measure or program.

In the near future, major international policy interventions may soon emerge in novel areas of catastrophe research such as high-risk grand physics experiments, AI, and asteroids. Improved knowledge of these events and technological advances in dealing with these technological challenges may also eliminate the need for national or international entities to intervene.

References

Barreca, A., Clay, K., Deschenes, O. et al. (2016). Adapting to climate change: the remarkable decline in the US temperature–mortality relationship over the twentieth century. *Journal of Political Economy* 124: 105–159.

Barrett, S. (2010). *Why Cooperate?: The Incentive to Supply Global Public Goods*. Oxford: Oxford University Press.

Baumol, W.J. and Oates, O.A. (1988). *The Theory of Environmental Policy*, 2nde. Cambridge, UK: Cambridge University Press.

Burtraw, D. and Szambelan, S.J. (2009). *U.S. Emissions Trading Markets for SO2 and NOx. Resources for the Future Discussion Paper 09-40*. Washington, DC: Resources for the Future.

Bell, M.L., McDermott, A., Zeger, S.L. et al. (2004). Ozone and short-term mortality in 95 US urban communities, 1987–2000. *Journal of the American Medical Association* 292 (19): 2372–2378.

Campbell, K.M., Einhorn, R.J., and Reiss, M.B. (ed.) (2004). *The Nuclear Tipping Point: Why States Reconsider their Nuclear Choices*. Washington, DC: Brookings Institution Press.

Carson, R. (1962). *Silent Spring*. Boston, MA: Houghton Mifflin.

CERN (2017a). The Accelerator Complex. CERN, Geneva, Switzerland. https://home.cern/about/accelerators.

CERN (2017b). The Safety of the LHC. CERN, Geneva. https://press.cern/backgrounders/safety-lhc.

Chapman, C.R. and Morrison, D. (1994). Impacts on the earth by asteroids and comets: assessing the hazard. *Nature* 367: 33–40.

CNBC (2017). Elon Musk Says Global Race for A.I. Will Be the Most Likely Cause of World War III. CNBC, New York. https://www.cnbc.com/2017/09/04/elon-musk-says-global-race-for-ai-will-be-most-likely-cause-of-ww3.html.

Cropper, M.L. and Oates, W.E. (1992). Environmental economics: a survey. *Journal of Economic Literature* 30: 675–740.

Dar, A., Rujula, A.D., and Heinz, U. (1999). Will relativistic heavy-ion colliders destroy our planet? *Physics Letters B* 470: 142–148.

Ellerman AD (2003). Ex Post Evaluation of Tradable Permits: The U.S. SO2 Cap-and-Trade Program. Working Paper MIT/CEEPR 03-003. MIT, Cambridge, MA.

Ellis, J., Giudice, G., Mangano, M. et al. (2008). Review of the safety of LHC collisions. *Journal of Physics G: Nuclear and Particle Physics* 35 (11).

Emanuel, K. (2008). The hurricane–climate connection. *Bulletin of American Meteorological Society* 89: ES10–ES20. doi: 10.1175/BAMS-89-5.

Emanuel, K. (2013). Downscaling CMIP5 climate models shows increased tropical cyclone activity over the 21st century. *Proceedings of the National Academy of Sciences of the United States of America* 110 (30): 12219–12224. doi: 10.1073/pnas.1301293110.

European Commission (EC) (2016a). Climate-Friendly Alternatives to HFCs and HCFCs. EC, Brussels, Belgium. http://ec.europa.eu/clima/policies/f-gas/alternatives/index_en.htm.

European Commission (EC) (2016b). *EU ETS Handbook*. Brussels: European Commission.

European Commission (EC) (2016c). Registration, Evaluation, Authorisation & Restriction of Chemicals (REACH). EC, Brussels. http://www.hse.gov.uk/reach.

Felzer, B.S., Cronin, T., Reilly, J.M. et al. (2007). Impacts of ozone on trees and crops. *Comptes Rendus Geoscience* 339: 784–798.

Federal Emergency Management Agency (FEMA) (2012). *Biggert–Waters Flood Insurance Reform Act of 2012 (BW12) Timeline*. Washington, DC: FEMA.

Federal Emergency Management Agency (FEMA) (2014). *Homeowner Flood Insurance Affordability Act: Overview*. Washington, DC: FEMA.

Federal Emergency Management Agency (FEMA) (2016). *The FEMA National Earthquake Hazards Reduction Program: Accomplishments in Fiscal Year 2014*. Washington, DC: FEMA.

Fox News (2017). Robots will be 100 times smarter than humans in 30 years, tech exec says. http://www.foxnews.com/tech/2017/10/27/robots-will-be-100-times-smarter-than-humans-in-30-years-tech-exec-says.html.

Freeman, A.M. III, Herriges, J.A., and Cling, C.L. (2014). *The Measurements of Environmental and Resource Values: Theory and Practice*. New York: RFF Press.

Graham, T. Jr., (2004). *Avoiding the Tipping Point*. Washington, DC: Arms Control Association.

Green Climate Fund (GCF) (2016). *Status of Pledges and Contributions Made to the Green Climate Fund. Status Date: 23 July 2015*. Incheon, South Korea: GCF.

Greer, A. (2012). Earthquake preparedness and response: a comparison of the United States and Japan. *Leadership and Management in Engineering* 12 (3): 111–125.

Hamblin, J. (2016). Toxic substances will now be somewhat regulated: how the first ever update to the Toxic Substances Control Act of 1976 finally came to pass – and what it lacks. *The Atlantic*. May 26, 2016. Available at https://www.theatlantic.com/health/archive/2016/05/toxic-substances-control-act/484280/.

Hawking S, Tegmark M, Russell S, Wilczek F (2014). Transcending Complacency on Superintelligent Machines. *Huffington Post*. https://www.huffingtonpost.com/stephen-hawking/artificial-intelligence_b_5174265.html.

Hirshleifer, J. (1983). From weakest-link to best-shot: the voluntary provision of public goods. *Public Choice* 41: 371–386.

Intergovernmental Panel on Climate Change (IPCC) (1990). *Climate Change: The IPCC Scientific Assessment*. Cambridge, UK: Cambridge University Press.

Intergovernmental Panel on Climate Change (IPCC) (2005). *Special Report on Safeguarding the Ozone Layer and the Global Climate System: Issues Related to Hydrofluorocarbons and Perfluorocarbons*. Cambridge, UK: Cambridge University Press.

Jaffe, R.L., Buszaa, W., Sandweiss, J., and Wilczek, F. (2000). Review of speculative disaster scenarios at RHIC. *Review of Modern Physics* 72: 1125–1140.

Joint Typhoon Warning Center (JTWC) (2017). Annual Tropical Cyclone Reports. JTWC, Guam, Mariana Islands. http://www.usno.navy.mil/jtwc/annual-tropical-cyclone-reports.

Kaiho, K. and Oshima, N. (2017). Site of asteroid impact changed the history of life on earth: the low probability of mass extinction. *Scientific Reports* 7: 14855. doi: 10.1038/s41598-017-14199-x.

Kaplan T (2017). Senate Votes to Raise Debt Limit and Approves $15 Billion in Hurricane Relief. NYT 7 September.

Kiger, M. and Russell, J. (1996). *This Dynamic Earth: The Story of Plate Tectonics*. Washington, DC: USGS.

King RO (2013). The National Flood Insurance Program: Status and Remaining Issues for Congress. CRS Report for Congress R42850. Congressional Research Service, Washington, DC.

Knowles, S.G. and Kunreuther, H.C. (2014). Troubled waters: the National Flood Insurance Program in historical perspective. *Journal of Policy History* 26: 325–353.

Kurzweil, R. (2005). *The Singularity Is Near*. New York: Penguin.

Landrigan, P.J., Richard Fuller, R., Acosta, N.J.R. et al. (2018). The Lancet Commission on pollution and health. *The Lancet* 391: 462–512.

Le Treut, H., Somerville, R., Cubasch, U. et al. (2007). Historical overview of climate change. In: *Climate Change 2007: The Physical Science Basis. The Fourth Assessment Report of the Intergovernmental Panel on Climate Change* (ed. S. Solomon, D. Qin, M. Manning, et al.). Cambridge: Cambridge University Press.

Likens, G.E. and Bormann, F.H. (1974). Acid rain: a serious regional environmental problem. *Science* 184: 1176–1179.

Lutz, W.K., Lutz, R.W., Gaylor, D.W., and Conolly, R.B. (2014). Dose–response relationship and extrapolation in toxicology: mechanistic and statistical considerations. In: *Regulatory Toxicology* (ed. X.-V. Reichl and M. Schwenk). Berlin: Springer.

MacCracken, C.N., Edmonds, J.A., Kim, S.H., and Sands, R.D. (1999). The economics of the Kyoto Protocol. *The Energy Journal* 20 (Special Issue): 25–71.

Manne, A.S. and Richels, R.G. (1999). The Kyoto Protocol: a cost-effective strategy for meeting environmental objectives? *The Energy Journal* 20 (Special Issue): 1–23.

Marchant, G., Allenby, B., Arkin, R. et al. (2011). International governance of autonomous military robots. *Columbia Science and Technology Law Review* 12: 272–315.

Mauzerall, D., Sultan, B., Kim, N., and Bradford, D.F. (2005). NOx emissions from large point sources: variability in ozone production, resulting health damages and economic costs. *Atmospheric Environment* 39: 2851–2866.

Mendelsohn, R. (1980). An economic analysis of air pollution from coal-fired power plants. *Journal of Environmental Economics and Management* 7: 30–43.

Mendelsohn, R. and Olmstead, S. (2009). The economic valuation of environmental amenities and disamenities: methods and applications. *Annual Review of Resources* 34: 325–347.

Mills, M.J., Toon, O.B., Turco, R.P. et al. (2008). Massive global ozone loss predicted following regional nuclear conflict. *Proceedings of the National Academy of Sciences of the United States of America* 105: 5307–5312.

Ministry of the Environment (2002). Minamata Disease: The History and Measures. Government of Japan, Tokyo. http://www.env.go.jp/en/chemi/hs/minamata2002/index.html.

Molina, M.J. and Rowland, F.S. (1974). Stratospheric sink for chlorofluoromethanes: chlorine atom-catalysed destruction of ozone. *Nature* 249: 810–812.

Montgomery, D. (1972). Markets in licenses and efficient pollution control programs. *Journal of Economic Theory* 5: 395–418.

Muller, N.Z. and Mendelsohn, R. (2009). Efficient pollution regulation: getting the prices right. *American Economic Review* 99: 1714–1739.

National Aeronautics and Space Administration (NASA) (2014). *NASA's Efforts to Identify Near-Earth Objects and Mitigate Hazards. IG-14-030*. Washington, DC: NASA Office of Inspector General.

National Aeronautics and Space Administration (NASA) (2017). Planetary Defense Coordination Office. NASA, Washington, DC. https://www.nasa.gov/planetarydefense/overview.

National Centers for Environmental Information (NCEI) (2016). International Best Track Archive for Climate Stewardship. NOAA, Silver Spring, MD. https://www.ncdc.noaa.gov/ibtracs.

National Conference of State Legislatures (NCSL) (2017). *NCSL Policy Update: State Statutes on Chemical Safety*. Washington, DC: NCSL.

National Hurricane Center (NHC) (2017). Worldwide Tropical Cyclone Centers. NHC, Miami, FL. http://www.nhc.noaa.gov/aboutrsmc.shtml.

National Institutes of Health (NIH) (2017). Tox Town: Environmental Health Concerns and Toxic Chemicals Where You Live, Work, and Play. National Institutes of Health, Bethesda, MD. https://toxtown.nlm.nih.gov/text_version/chemicals.php.

National Research Council (NRC) (1997). *Report of the Observer Panel for the U.S.–Japan Earthquake Policy Symposium*. Washington, DC: National Academies Press.

National Research Council (2010). *Defending Planet Earth: Near-Earth-Object Surveys and Hazard Mitigation Strategies*. Washington, DC: National Academies Press.

National Research Council (2015). *Affordability of National Flood Insurance Program Premiums: Report 1*. Washington, DC: National Academies Press.

Nichols, A.L. (1997). Lead in gasoline. In: *Economic Analyses at EPA: Assessing Regulatory Impact* (ed. R.D. Morgenstern), 49–86. Washington, DC: Resources for the Future.

Nobel Prize (2013). The Nobel Prize in Physics 2013. François Englert and Peter W. Higgs. https://www.nobelprize.org/nobel_prizes/physics/laureates/2013.

Nordhaus, W. (1994). *Managing the Global Commons*. Cambridge, MA: MIT Press.

Nordhaus, W. (2001). Global warming economics. *Science* 294 (5545): 1283–1284.

Nordhaus, W.D. (2006). Paul Samuelson and global public goods. In: *Samuelsonian Economics and the Twenty-First Century* (ed. M. Szenberg, L. Ramrattan and A.A. Gottesman). Oxford, UK: Oxford Scholarship Online.

Nordhaus, W. (2008). *A Question of Balance: Weighing the Options on Climate Change*. New Haven, CT: Yale University Press.

Nordhaus, W. (2010). Economic aspects of global warming in a post-Copenhagen environment. *Proceedings of the National Academy of Sciences of the United States of America* 107 (26): 11721–11726.

Nordhaus, W. (2011a). The architecture of climate economics: designing a global agreement on global warming. *Bulletin of Atomic Scientists* 67 (1): 9–18.

Nordhaus, W. (2011b). The economics of tail events with an application to climate change. *Review of Environmental Economics and Policy* 5: 240–257.

Nordhaus, W. and Boyer, J.G. (1999). Requiem for Kyoto: an economic analysis of the Kyoto Protocol. *The Energy Journal* 20 (Special Issue): 93–130.

Overbye D (2013). Chasing the Higgs. NYT 4 March.

Peskin, M.E. (2008). The end of the world at the Large Hadron Collider? *Physics* 1 (14).

Plaga R (2009). On the potential catastrophic risk from metastable quantum-black holes produced at particle colliders. arXiv:0808.1415 [hep-ph].

Pope, C.A., Burnett, R.T., Thun, M.J. et al. (2002). Lung cancer, cardiopulmonary mortality, and long-term exposure to fine particulate air pollution. *Journal of the American Medical Association* 287 (9): 1132–1141.

Posner, R.A. (2004). *Catastrophe: Risk and Response*. New York: Oxford University Press.

Quartz (2017). The Robot that Takes your Job Should Pay Taxes, Says Bill Gates. Quartz, Washington, DC. https://qz.com/911968/bill-gates-the-robot-that-takes-your-job-should-pay-taxes.

Saffir, H.S. (1977). *Design and Construction Requirements for Hurricane Resistant Construction*. New York: American Society of Civil Engineers.

Samuelson, P. (1954). The pure theory of public expenditure. *Review of Economics and Statistics* 36: 387–389.

Sandler, T. (1997). *Global Challenges: An Approach to Environmental, Political, and Economic Problems*. Cambridge, UK: Cambridge University Press.

Sanghi, A., Ramachandran, S., de la Fuente, A. et al. (2011). *Natural Hazards, Unnatural Disasters: The Economics of Effective Prevention*. Washington, DC: World Bank.

Schmalensee, R. and Stavins, R.N. (2013). The SO_2 allowance trading system: the ironic history of a grand policy experiment. *Journal of Economic Perspectives* 27: 103–122.

Seo, S.N. (2012). What eludes international agreements on climate change? The economics of global public goods. *Economic Affairs* 32 (2): 74–80.

Seo, S.N. (2014). Estimating tropical cyclone damages under climate change in the southern hemisphere using reported damages. *Environmental and Resource Economics* 58: 473–490.

Seo, S.N. (2015a). Fatalities of neglect: adapt to more intense hurricanes? *International Journal of Climatology* 35: 3505–3514.

Seo, S.N. (2015b). Adaptation to global warming as an optimal transition process to a greenhouse world. *Economic Affairs* 35: 272–284.

Seo, S.N. (2016a). Modeling farmer adaptations to climate change in South America: a micro-behavioral economic perspective. *Environmental and Ecological Statistics* 23: 1–21.

Seo, S.N. (2016b). The micro-behavioral framework for estimating total damage of global warming on natural resource enterprises with full adaptations. *Journal of Agricultural, Biological, and Environmental Statistics* 21: 328–347.

Seo, S.N. (2016c). A theory of global public goods and their provisions. *Journal of Public Affairs* 16: 394–405.

Seo, S.N. (2017a). *The Behavioral Economics of Climate Change: Adaptation Behavior, Global Public Goods, Breakthrough Technologies, and Policy-Making*. London: Academic Press.

Seo, S.N. (2017b). Measuring policy benefits of the cyclone shelter program in the North Indian Ocean: protection from intense winds or high storm surges? *Climate Change Economics* 8 (4): 1–18. doi: 10.1142/S2010007817500117.

Seo, S.N. (2017c). Beyond the Paris Agreement: climate change policy negotiations and future directions. *Regional Science Policy and Practice* 9: 121–140.

Seo, S.N. and Bakkensen, L.A. (2017). Is tropical cyclone surge, not intensity, what kills so many people in South Asia? *Weather, Climate, and Society* 9: 71–81.

Sharkey, N.E. (2012). Evitability of autonomous robot warfare. *International Journal of the Red Cross* 94: 787–799.

Smith, V.K. and Huang, J.-C. (1995). Can markets value air quality? A meta-analysis of hedonic property value models. *Journal of Political Economy* 103: 209–227.

Stavins, R. (1998). What can we learn from the grand policy experiment? Lessons from SO_2 allowance trading. *Journal of Economic Perspectives* 12: 69–88.

Swiss Re Institute (2017). *Natural Catastrophes and Man-made Disasters in 2016: A Year of Widespread Damages*. Zurich, Switzerland: Swiss Re.

Thom, R. (1975). *Structural Stability and Morphogenesis*. New York: Benjamin-Addison-Wesley.

Tietenberg, T. and Lewis, L. (2014). *Environmental & Natural Resource Economics*. New York: Prentice Hall.

Turco, R.P., Toon, O.B., Ackerman, T.P. et al. (1983). Nuclear winter: global consequences of multiple nuclear explosions. *Science* 222: 1283–1292.

Wang, B., Shugart, H.H., Shuman, J.K., and Lerdau, M.T. (2016). Forests and ozone: productivity, carbon storage, and feedbacks. *Scientific Reports* 6 (22133): doi: 10.1038/srep22133.

United Nations Environmental Programme (UNEP) (2016). *The Montreal Protocol on Substances that Deplete the Ozone Layer*. Kigali, Rwanda: UNEP.

United Nations Environmental Programme (UNEP) (2017). *Montreal Protocol Marks a Milestone with First Ratification of Kigali Amendment*. Nairobi: UNEP.

United Nations Framework Convention on Climate Change (UNFCCC) (1992). *United Nations Framework Convention on Climate Change*. New York: UNFCCC.

United Nations Framework Convention on Climate Change (UNFCCC) (1998). *Kyoto Protocol to the United Nations Framework Convention on Climate Change*. New York: UNFCCC.

United Nations Framework Convention on Climate Change (UNFCCC) (2009). *Copenhagen Accord*. New York: UNFCCC.

United Nations Framework Convention on Climate Change (UNFCCC) (2011a). *The Durban Platform for Enhanced Action*. New York: UNFCCC.

United Nations Framework Convention on Climate Change (UNFCCC) (2011b). *Report of the Transitional Committee for the Design of Green Climate Fund*. New York: UNFCCC.

United Nations Framework Convention on Climate Change (UNFCCC) (2015). *The Paris Agreement. Conference of the Parties (COP) 21*. New York: UNFCCC.

United Nations Office for Disarmament Affairs (UNODA) (2017a). Treaty on the Non-Proliferation of Nuclear Weapons (NPT). http://www.un.org/disarmament/wmd/nuclear/npt.shtml.

United Nations Office for Disarmament Affairs (UNODA) (2017b). Convention on the Prohibition of the Development, Production and Stockpiling of Bacteriological (Biological) and Toxin Weapons and on their Destruction. http://disarmament.un.org/treaties/t/bwc.

United Nations Office for Disarmament Affairs (UNODA) (2017c). Convention on the Prohibition of the Development, Production, Stockpiling and Use of Chemical Weapons and on their Destruction. https://www.opcw.org/chemical-weapons-convention.

United Nations Office for Disarmament Affairs (UNODA) (2017d). Background on Lethal Autonomous Weapons Systems. https://www.unog.ch/80256ee600585943/(httppages)/8fa3c2562a60ff81c1257ce600393df6?opendocument.

United States Congress (1978). *Toxic Substances Control Act of 1978*. Washington, DC: US Congress.

United States Congress (1980). *Comprehensive Environmental Response, Compensation, and Liability Act of 1980*. Washington, DC: US Congress.

United States Congress (2004). *Earthquake Hazards Reduction Act of 1977 (as Amended in 2004)*. Washington, DC: US Congress.

United States Congress (2012). *Federal Insecticide, Fungicide, and Rodenticide Act of 1947*. Washington, DC: US Congress.

United States Environmental Protection Agency (US EPA) (1977). *The Clean Air Act Amendments of 1977*. Washington, DC: US EPA.

United States Environmental Protection Agency (US EPA) (1990). *The Clean Air Act Amendments of 1990*. Washington, DC: US EPA.

United States Environmental Protection Agency (2010). The 40th Anniversary of the Clean Air Act. US EPA, Washington, DC. http://www.epa.gov/airprogm/oar/caa/40th.html.

United States Environmental Protection Agency (US EPA) (2014a). National Emissions Inventory (NEI). US EPA, Washington, DC.

United States Environmental Protection Agency (US EPA) (2014b). *Carbon Pollution Emission Guidelines for Existing Stationary Sources: Electric Utility Generating Units*. Washington, DC: US EPA.

United States Environmental Protection Agency (US EPA) (2017a). Criteria Air Pollutants. US EPA, Washington, DC. https://www.epa.gov/criteria-air-pollutants.

United States Environmental Protection Agency (US EPA) (2017b). EPA's Report on the Environment. US EPA, Washington, DC. https://cfpub.epa.gov/roe.

United States Environmental Protection Agency (US EPA) (2017c). Clean Air Markets. US EPA, Washington, DC. https://www.epa.gov/airmarkets.

United States Environmental Protection Agency (US EPA) (2017d). *EPA Takes Another Step to Advance President Trump's America First Strategy, Proposes Repeal of "Clean Power Plan"*. Washington DC, USA: US EPA.

United States Geological Survey (USGS) (2017a). Seismic Hazard Maps and Site-Specific Data. https://earthquake.usgs.gov/hazards/hazmaps.

United States Geological Survey (USGS) (2017b). *Measuring the Size of an Earthquake*. Washington, DC: USGS.

United States (US) Senate (2014). Menendez-Led Flood Insurance Reforms Now Law. US Senate, Washington, DC. https://www.menendez.senate.gov/news-and-events/press/menendez-led-flood-insurance-reforms-now-law.

Utsu T (2013). Catalog of Damaging Earthquakes in the World. International Institute of Seismology and Earthquake Engineering, Tsukuba, Japan. http://iisee.kenken.go.jp/utsu/index_eng.html.

Vogel, S.A. and Roberts, J.A. (2011). Why the Toxic Substances Control Act needs an overhaul, and how to strengthen oversight of chemicals in the interim. *Health Affairs* 30: 898–905.

Weitzman, M.L. (2009). On modeling and interpreting the economics of catastrophic climate change. *Review of Economics and Statistics* 91: 1–19.

White House (2013). The President's Climate Action Plan. Executive Office of the President. The White House, Washington, DC.

White House (2017). Statement by President Trump on the Paris Climate Accord. The White House, Washington, DC.

Woodruff, T.J., Parker, J.D., and Schoendorf, K.C. (2006). Fine Particulate Matter (PM2.5) air pollution and selected causes of postneonatal infant mortality in California. *Environmental Health Perspectives* 114: 786–790.

World Meteorological Organization (WMO) (2014). *Scientific Assessment of Ozone Depletion 2014*. Geneva, Switzerland: Global Ozone Research and Monitoring Project – Report No. 55, WMO.

7

Insights for Practitioners: Making Rational Decisions on a Global or Even Universal Catastrophe

7.1 Introduction

We, as human beings, are faced with and forced to make decisions, small or large, at every moment, on numerous matters of unique characteristics. The possibility of a catastrophic event in a not-far-away future has given humanity big trouble in making appropriate decisions in the presence of such a possibility. This book provides a multidisciplinary review of the historical and frontier literature on catastrophes and chaos, whose purpose is to help individuals and societies make sensible decisions in such situations of a truly big forecasted catastrophe.

Human existence itself may be characterized as struggles against natural disasters and catastrophes as well as man-made catastrophes. We, as a species, have survived through natural disasters such as catastrophic earthquakes, volcano eruptions, deadly cyclones, tornadoes, great floods, extremely severe droughts, uncontrollable wild fires, and numerous infectious diseases (Kiger and Russell 1996; Emanuel 2005; Sanghi et al. 2011). We are also responsible for creating man-made catastrophes such as, inter alia, great wars, nuclear accidents and explosions, genocides, terrorism, and toxic chemicals (Turco et al. 1983; Swiss Re Institute 2017).

As humanity's technological capabilities are advancing markedly, potentially catastrophic events that have not been known before have come to enthrall people's imaginations and fear (Posner 2004; Hawking et al. 2014). Revolutions in physics, genetics, nanotechnology, and robotics may enable humans to rebuild the physical and biological world as we know it now and eventually the artificial intelligence (AI), i.e. the machines that have human-level or greater intelligence, which carries the risk of deadly viruses and out-of-control AI machines that have the capacity to kill humans and are deployed in battles (Kurzweil 2005; Marchant et al. 2011; Sharkey 2012; UNODA 2017).

The consequence of a technology-induced catastrophe may be as severe as the end of the entire universe or the end of humanity. An end of the universe may come by the "strangelets" or black matter which may arise from the Large Hadron Collider (LHC) experiments by the physicists who are attempting to recreate initial conditions of the universe at the time of the origin of the current universe, i.e. the Big Bang, that was predicted to create a large black hole which draws in the entire universe (Dar et al. 1999; Jaffe et al. 2000; Plaga 2009).

With advances in astronomy and space explorations, humanity has recently discovered many Earth-like planets on which humans may be able to habituate in the future. At the same time, astronomers have discovered the possibility of asteroids and comets

Natural and Man-made Catastrophes – Theories, Economics, and Policy Designs, First Edition. S. Niggol Seo.
© 2019 John Wiley & Sons Ltd. Published 2019 by John Wiley & Sons Ltd.

colliding with the Earth, significantly damaging or even destroying the planet (Chapman and Morrison 1994; NRC 2010).

Up to the present time, as reviewed in Chapter 6, these global or even universal catastrophic future events have received little attention as a global policy issue. On the other hand, there are other global catastrophic events that have received much attention from policy-makers, prime examples being global warming and nuclear proliferation. A catastrophic end to human civilizations caused by future extreme global warming has been forecast by many scientists through various causal mechanisms (Broecker 1997; Oppenheimer 1998; Mann et al. 1999; Lenton et al. 2008; Hansen et al. 2012). Global negotiations to mitigate Earth-heating gases and stabilize the global climate began in the late 1980s and have continued ever since (IPCC 1990, 2014; UNFCCC 1992, 2015). Nuclear treaties and remaining challenges and conflicts are explained at length in Chapter 6.

The central question that the author undertakes to address throughout this book is how individuals, societies, and the global community should make rational decisions, the concept of which is explained in this chapter, in the face of the wide range of potential catastrophic risks whose scale is truly global or even universal, which may or may not entail policy interventions.

As the first step toward searching for answers, Chapter 2 began with a critical review of the literature of scientific theories on catastrophes. The theory, elementary models of catastrophe, and applications to social sciences were described, followed by descriptions of chaos theories. This was accompanied by a review of literature on finding an orderly behavior in a system previously thought of as a chaotic system.

Chapter 3 followed up with a critical review of the literature of philosophical thoughts on catastrophic events, in which the author also included a selective review of some of the most influential environmental classics such as Rachel Carson's and Aldo Leopold's writings. One of the Indian classical philosophical thoughts and one of the western philosophical traditions were explained in the chapter, both of which have something to do with the creation and cessation of life.

Chapter 4 provided a comprehensive review of the literature of the economics of catastrophes. In the chapter, the market and financial instruments that were developed to deal with catastrophic economic outcomes were explained: insurance, options, futures, and catastrophe bonds. A major portion of Chapter 4 was devoted to elaborating a socially optimal policy intervention in the face of catastrophe risks. A detailed analysis of the generalized precautionary principle was given by way of the dismal theorem, and alternative formulations of the dismal model were provided.

Chapter 5 provided a series of empirical studies of the economics of catastrophic events, making use of the empirical data on tropical cyclones, plus earthquakes, collected from global ocean basins. It was emphasized that tropical cyclones have been the deadliest natural events in the past century in terms of the number of human fatalities. However, major causes of deaths differ from one ocean basin to another: high-speed winds, low minimum central pressure, storm surges in sea level, and accompanying rainfall and flooding. Empirical studies also reveal the effectiveness of numerous adaptation strategies taken either individually or by the public sector.

Chapter 6 provided a critical survey of national and international policies on the full array of catastrophic events and risks discussed throughout the book. Policies and programs on asteroids and comets, earthquakes, hurricanes, nuclear wars and chemical weapons, air and water pollutants, toxic and hazardous wastes, ozone depletion, global warming, high-risk physics experiments and strangelets, and AI and killer robots were reviewed. Major gaps in knowledge and policy frameworks in each of these concerned areas were pointed out by the author, applying the theories and insights gained from the previous chapters on science, philosophy, economics, and empirical studies of catastrophes. The author also evaluated successful as well as failed policy measures and interventions with regard to catastrophic events.

This final chapter of the book summarizes succinctly in Section 7.2 major gaps in knowledge, new findings, and important policy experiences which bear directly on the question of how to address a potential catastrophic event.

In Section 7.3, the author examines two recent opinion surveys of the most-feared catastrophic events for different groups of society: a group of "high minds" and a group of "low minds." The former is the opinion survey of Nobel Prize-winning scientists and the latter is the opinion survey of the American people (Chapman University 2017; THE/Lindau 2017).

Section 7.4 evaluates the entire range of catastrophic events that are often of much public concern and which are discussed throughout this book. Based on the evaluation, the section concludes which of these are most difficult to handle and which are most destructive when they befall.

This book has shown that the most difficult part of the question lies not in physical changes and technological capabilities but in making a rational decision in the face of one of these truly catastrophic events whose actual coming into being is judged to be of slight chance (von Neumann and Morgenstern 1947; Koopmans 1965; Nordhaus 2008).

Section 5 elaborates on the primary challenges in dealing with these truly global or even universal catastrophes and offers final insights on how to make a rational or optimal decision in the face of these challenges, adding to what has been elucidated in Chapters 4 and 5, the two economics chapters. As highlighted in previous chapters, a portfolio of behavioral adaptations that can be relied upon by individuals and societies as well as a bundle of technological capabilities that are advancing over time will be at the heart of the rational decision-making framework.

7.2 Lessons from the Multidisciplinary Literature of Catastrophes

It is not an exaggeration to say that philosophical musings about the nature of catastrophes, however primitive they were, existed from the very beginning of the recorded history of humanity, dating perhaps back to many millions of years ago. One of the earliest systematic philosophies of catastrophe is found in the Indian philosophical school which existed about 3000 years ago in ancient India that argued for random occurrences and dissolutions of things and living beings. But, other stories or relics of a catastrophic event are not hard to find in other cultures and regions as well.

A more rigorous scientific and mathematical conceptualization of catastrophes has waited a long time to emerge in the nineteenth century and more formally in the twentieth century (Poincaré 1880–1890). During this foundation period, numerous formal theories and models of a catastrophe have emerged, including the chaos theory by Lorenz, the catastrophe theory by Thom, elementary catastrophe models, the butterfly effect and the Lorenz attractor, the fractal theory, the Mandelbrot and Julia sets, the bifurcation theory, and Feigenbaum constants (Thom 1975; Lorenz 1963, 1969a, 1969b; Mandelbrot 1983, 2004).

The catastrophe and chaos literature kicked off by setting its sights on defining salient characteristics of a catastrophe and elucidating the emergence of a catastrophe by way of a simple set of mathematical equations. The literature has gradually turned its attention to the research endeavors of finding order in a chaotic system, but notable achievements in this line of work are sparse (Feigenbaum 1978).

This fascinating literature of mathematics on catastrophes and chaos also sends clear messages to applied scientists who attempt to apply the concepts and mathematical formulations to economic and social policy issues. It should be emphasized that moderation or regulation mechanisms in the catastrophe theory and models, which are commonly dominant aspects in economic and market interactions among agents, are by and large unheeded in catastrophe models.

As such, applications of the catastrophe theory and models may be justifiable in the study of some physical or perhaps biological systems, e.g. in the study of movements of celestial bodies (Poincaré 1880–1890). However, for the studies of behavioral and policy-driven systems where behavioral responses and regulation mechanisms play a decisive role in determining the behaviors of the system, any direct applications of the concepts and models of the catastrophe theory without modifications are inadequate (Seo 2015b, 2016c, 2017a).

Another distinction to make is that a chaotic system is not always a representation of a catastrophe. Rather, a chaotic system is more often a representation or misrepresentation of a complex system, or put differently, roughness. As an example, a fractal exists commonly in natural events. Clouds, trees, leaves, reefs, coasts, etc., are said to be a fractal (Mandelbrot 1983; Fractal Foundation 2009). These fractal events are considered to be complex, but also orderly natural phenomena rather than catastrophic events (Mandelbrot 2004; Frame et al. 2017).

Third, a fractal pattern may also be present in an anthropogenically driven system, which is captured by a power law function by a researcher who wants to describe the concerned system (Mandelbrot 1997; Taleb 2005; Gabaix 2009). It may be observed in financial markets and crashes (Mandelbrot 1997; Mandelbrot and Hudson 2004). However, researchers should be aware that there may lie an additional parameter in the fractal object whose role is to constrain a fractal expansion in a highly constrained system such as a human behavior or a social system.

Fourth, the author highlighted in Chapter 2 the endeavors to find the universal constant(s) in the chaotic system(s) by Feigenbaum and others (Feigenbaum 1978; Lanford 1982; Collet and Eckmann 1979, 1980; Boccaletti et al. 2000). What this line of research uncovers is that a chaotic system may appear to be chaotic or random-walks without any order but upon closer examinations may embed an orderly natural phenomenon underneath the guise of chaos.

Chapter 3 then reviewed selected but perhaps broadly encompassing philosophical and literary traditions that expressed consistent views on catastrophe events. Environmental thinkers such as Rachel Carson emphasize the deleterious and catastrophic effects of pollutants and toxic chemicals on biological systems and human health which could bring forth a doomsday for humanity (Carson 1962). However, interpreted from a slightly different angle, her writings remind humanity of the importance of managing these issues optimally in a rational manner.

As per Leopold's writings on the value of wildlife, it is an extreme view to argue that the value of a wild animal is equal to the value of a human being. It is equally extreme to argue that the wildlife should be managed in a nonhuman-centric perspective (Leopold 1949). To be more philosophically speaking, there are other ethical standpoints that are more persuasive, practical, and rational than Leopold's on the value of wildlife, which the author doesn't have to restate here.

In the western philosophical traditions, an interesting probability calculation was presented by Blasé Pascal in the medieval era, which has been appropriated with enthusiasm by many others since the much-acclaimed publication of *Pensées*. Pascal relies on the concept of infinity in order to force his argument, but the downside of this argument is also clear. In his argument, there is a certain event whose probability of occurrence is extremely low, but the reward is infinite (Pascal 1670). The other side of Pascal's wager is that an extremely unlikely catastrophic event may overwhelm all benefit-cost calculations in everyday matters because of the "infinite" losses to be expected if such an event should occur. Put differently, it would nullify all everyday decisions.

The flaws in Pascal's probability framework have also been made clear by many philosophers in the past. To point out a few of them, the concept of infinity has never been defined in a philosophical way, while exclusive reliance on the concept of infinity works to undermine the everyday reality faced by everyday decision-makers (Hájek 2012).

An Indian philosophy that was popular 3000 years ago in the country elucidates a catastrophic view of the world by formulating the inception and the termination of the universe by the workings of randomness or fortuitous occurrences. According to this tradition, things arise at random by pure luck and also dissipate at random by pure chance (Tipitaka 2010). The implication is that a sudden end of the world at some point would occur by pure chance, and, as such, no early prediction or preparation would be possible.

This philosophy may be the perfect conceptualization of a catastrophe among the whole array of concepts, models, and philosophies of catastrophes that are presented in this book. Seen through the mirror of this Indian philosophy, Thom's catastrophe theory and models are a modern approximation by mathematicians to this ancient Indian school (Thom 1975; Zeeman 1977).

As such, the Indian philosophical tradition could be subjected to the same critiques given to the catastrophe theory and models. That is, it lacks careful attention to regulation mechanisms in the system considered. To put it differently, it does not conform to the reality of causality, i.e. the cause-and-effect relationships so strong in all matters an individual human being considers and engages in every moment.

In Chapter 4, based on the foundations of mathematics and philosophy of catastrophes, the author elaborated the economics of catastrophes, in which major economic models are introduced and central economic questions are answered. The chapter explained the needs and complications of defining various thresholds, a tipping point, fat-tails, and variance infinity in a statistical distribution.

The author emphasized that there have long been market and financial instruments developed and utilized widely in preparation of catastrophic natural events. These range from insurances, options, futures, and catastrophe bonds (Shiller 2004; Fabozzi et al.

2009; Edesess 2014). Historical trends of these instruments indicate that purchases of these financial instruments have been increasing, which is to say that the percentage of insured or covered natural disasters and catastrophes has been increasing (Swiss Re Institute 2017). This means that individuals and societies have become more resilient against the potential risks of natural disasters.

The market/financial instruments are, however, an effective means for individuals or local communities only when a catastrophic event occurs at a local level or at an individual level. For the catastrophic events whose harms fall upon an entire nation or upon the entire global community, there arises the need for a policy intervention by the national government or a global political entity (Samuelson 1954; Nordhaus 1994, 2006). Alternatively, when the distribution of a potential catastrophe event is fat-tailed, market and financial instruments alone cannot be adequate for addressing the issue (Weitzman 2009).

Which is the best policy approach for dealing with a national-scale or a global-scale catastrophe with or without a fat-tail distribution has been a contentious policy question, especially since the late 2000s and in the context of global warming policy-making. A globally optimized dynamic social-welfare-maximizing policy intervention has long held the position of the cornerstone of all policy debates (von Neumann and Morgenstern 1947; Koopmans 1965).

In the policy framework thereof, an optimal level of the price of a ton of emissions of carbon dioxide is determined through empirical research and imposed across the globe in a harmonized way (Nordhaus 1992, 1994). With the impositions of carbon tax dynamically whose levels have been ramped up over the course of the twenty-first century, economic actors are forced to adjust their behaviors of emitting carbon dioxide and other greenhouse gases, which leads to an optimal trajectory of carbon emissions for the global community as well as an optimal trajectory of changes in the climate system.

An alternative policy intervention principle calls for a more drastic policy measure which is justified by the proponents of this approach from a totally destructive nature of the concerned catastrophic event even though the probability is very small (UNFCCC 2010). They argue that the observed trend of global warming today has a small probability of ending the human and nonhuman civilizations on the Earth permanently and such a small-probability event cannot be excluded from considerations. According to the proponents, this utterly destructive event cannot be assigned a very small probability of occurrence because of the high uncertainty in many chains of global warming debates, which manifests as a fat-tail in statistical distributions.

Assuming a specific form of a fat-tail distribution and selected forms of economic parameters and their relationships, these theorists propose that the global society should spend infinite resources and money to avoid a catastrophic event with a tiny probability of occurring. This principle is given the title of a generalized precautionary principle (Weitzman 2009).

The author provided a detailed critique of the dismal theorem which purports to provide the mathematical rationale for the generalized precautionary principle. In the cases where the tail distribution of a global catastrophic event is very fat, researchers should be aware of the conditions and situations that make the tail distribution thinner: behavioral, market, financial, governmental, and technological aspects (Seo 2015b, 2017a).

It is also suggested that a fat-tail distribution of a global catastrophic event does not necessarily necessitate a draconian change in policy approaches to abandoning the foundation of a benefit-cost analysis (Nordhaus 2011). The author provided an ensemble of alternative assumptions and formulations of parameters and functional relationships in the dismal model which would yield a different policy conclusion (Seo 2018).

In the presence of unusually high uncertainty on a large-scale policy issue, policy researchers should rather focus on the range of uncertainties. That is, they should put efforts into quantifying the range of estimates of policy-relevant variables which embody uncertainty. In addition, researchers can rely on the social welfare optimization framework, such as those in the aforementioned climate policy modeling, but with the full range of uncertainty estimates embedded into such a framework (Nordhaus 2008).

A large array of empirical economic studies on numerous catastrophic events provides ample evidence and insights on the formulations of the theories of economics of catastrophes and policy frameworks, which was the focus of Chapter 5. The author made use of a series of emerging empirical studies of hurricanes and tropical cyclones, along with those on earthquakes, for illustrating and highlighting the aspects of behavioral and policy dimensions of catastrophic events.

Tropical cyclones are the most catastrophic event that humanity has faced during the past century in terms of the number of lives lost, along with earthquakes (Swiss Re Institute 2017). Most catastrophic tropical cyclones have killed more than 100 000 persons, up to half a million fatalities in the most catastrophic cyclone, in South Asia, besides millions of people displaced and millions of animals killed during those severe cyclone events (Seo and Bakkensen 2017; Seo 2017c). Recent hurricanes such as Hurricane Harvey that hit Texas in the summer of 2017 and Hurricane Irma that hit Florida in the summer of 2017 were reported, at the time of landfalls, to be the most powerful hurricane ever recorded or the hurricane with the largest amount of rainfall ever accompanied (NHC 2017a).

Several clear messages pertinent to the economics of catastrophic events emerge from the empirical studies of tropical cyclones and behavioral responses to them. Frist, it seems almost impossible to have a complete understanding of a tropical cyclone event, as if it is a chaotic system envisioned by Lorenz. Genesis, wind speeds, central pressure, heights of storm surges, wind shear, storm size, vorticity, tracks, rainfall, frequency, and cyclone changes under a globally warmer world are only incompletely understood and predicted by scientists, despite the progresses made in the science of hurricanes since the late 2000s (Emanuel 2008, 2013; McAdie et al. 2009).

Second, despite the incomplete knowledge and predictions, the number of lives lost to tropical cyclones has declined saliently over the past century globally. As shown in Figure 1.2, the annual number of US hurricane fatalities from the beginning of the twentieth century to 2016 exhibits a decreasing trend (Blake et al. 2011; NOAA 2016). The decreasing trend is also observed in other ocean basins, e.g. the North Indian Ocean and the southern hemisphere ocean basins.

Third, the empirical studies show that the higher the income per capita, the better an individual and a community cope with tropical cyclone catastrophes. An analysis of the historical South Asian cyclone data, generated in the North Indian Ocean, shows that, per each 1000 INR (Indian Rupee) increase in income per capita, the number of fatalities falls by about 3% (Seo and Bakkensen 2017). The income effect is similarly strong in southern hemisphere cyclone fatality data as well as in the global cyclone fatality data (Seo 2015a; Bakkensen and Mendelsohn 2016).

Fourth, owing to more resources being available to higher-income individuals, richer individuals and communities are equipped with higher capabilities to cope with or be prepared for cyclone landfalls. With higher income, an individual can choose multiple locations of her/his residence in a safer place or choose a house built with structural safety measures against even the most severe cyclones expected. Further, a high-income individual has more options for evacuation in the event of a catastrophe, e.g. an automobile or a motor boat. A higher-income individual has better and faster access to critical information through, for example, internet connections or mobile phones. Owing to the higher capacities and more resources available, an individual can better adapt to cyclone events by making necessary behavioral changes (Seo 2015a; Seo and Bakkensen 2017).

Fifth, empirical examinations of historical tropical cyclone data and the associated fatalities and damages data reveal that technological advances have played a major role in reducing the number of cyclone fatalities, given the same intensity of cyclones. Airplane reconnaissance of cyclones, satellite monitoring, advances in tropical cyclone trajectory projection techniques, advances in tropical cyclone storm surge modeling, tropical cyclone advisory and early warning, and worldwide tropical cyclone centers seem most likely to have increased adaptation capacities of affected individuals and communities markedly (Emanuel 2008; Seo 2015a).

Sixth, the empirical literature points to a successful policy intervention in the face of a catastrophe. In poor-income countries where tropical cyclones had routinely killed tens of thousands of people who relied on immutable factors such as local geography and fertile lands for food productions, some policy interventions may have turned out to be highly effective in cutting the numbers of fatalities. The cyclone shelter program by the Bangladesh government aided by the World Bank's funding was presented by the author as an illustrative example of an effective governmental role in dealing with a catastrophe. Analysis shows that the cyclone shelter program has reduced the number of human fatalities by 75%, given the same level of storm surges caused by a tropical cyclone (Seo 2017c). It might be argued that the benefit of the program which is on-going far exceeds the cost.

Seventh, policy interventions must pay heed to what aspects of a catastrophic event, say, a tropical cyclone event or an earthquake event, are a primary killer of people or destroyer of assets and properties (Seo and Bakkensen 2017). For example, a heavy investment and subsidy by the government into a cyclone cellar program in Bangladesh might have failed dramatically while wasting immense resources. A cellar amplifies the risk in frequently inundated towns because of storm surges, not reduces it.

Similarly, a public policy intervention in response to an impending earthquake catastrophe may turn out to be misguided if it is based on a prediction of the earthquake using the magnitude scale of the earthquake. That is, another earthquake scale, i.e. the moment scale, may present a different but more realistic picture of the earthquake (USGS 2017).

Chapter 6 is devoted to the review and evaluation of the policies that prescribe governmental interventions for the purposes of preventing, minimizing, or eliminating economic damages and deaths from numerous catastrophic events. National policies and programs as well as international treaties and agreements on the full range of catastrophic events covered in this book are reviewed.

Unlike other reviews of the catastrophe policy literature which focus on a single catastrophe policy area, e.g. the National Flood Insurance Program in the US, the sulfur dioxide allowance trading scheme, and toxic and hazardous substances (Vogel and Roberts 2011; Schmalensee and Stavins 2013; Knowles and Kunreuther 2014), the author endeavored to provide a comprehensive review of all the major catastrophe policy areas that have received much attention or areas of concern by many scientists. The review includes policies on asteroids and comets, earthquakes, tropical cyclones, nuclear and biological weapons, criteria pollutants, toxic and hazardous chemicals, ozone depletion, global warming, high-risk physics experiments, and AI.

Of the array of policy measures implemented in these policy areas, some are credited to have achieved their respective goals. The nuclear Non-Proliferation Treaty (NPT) has been adopted by almost all members of the UN and is credited to have stopped proliferation of nuclear weapons in the world, although it remains to be seen whether major nuclear powers and non-nuclear-weapon states will be committed to the full implementation of the NPT (Graham 2004; Campbell et al. 2004).

The Montreal Protocol on ozone depletion is also judged to have achieved the cessation of the spread of ozone holes across the global atmosphere and the phasing out of ozone-depleting substances such as chlorofluorocarbons (CFCs), although it has to be seen whether the recent amendment in Kigali, Rwanda of the Protocol to include greenhouse gases as regulated gases will survive (WMO 2014).

Many policy interventions are evaluated to have failed to achieve their respective policy goals. The Toxic Substances Control Act (TSCA) is called by many a paper tiger owing to the inability to ban or limit toxic substances under the authority given by the Act to the Environmental Protection Agency (Vogel and Roberts 2011).

The National Flood Insurance Program may have encouraged people to relocate to coastal counties in the US which are vulnerable to severe hurricane attacks (Knowles and Kunreuther 2013). The consequence of the dramatic shifts of population to coastal communities has been the accumulation of federal debt, owing to the increase in government subsidy to hurricane victims (King 2013).

International conferences to design a global warming policy have faltered multiple times since the first meeting held in 1992. The Kyoto Protocol had a goal to limit carbon dioxide equivalent emissions at 5% below the 1990 level by the end of the first phase of the Protocol, which could not be achieved because of the lack of obligations imposed on China, India, the US, and many other countries (Nordhaus and Boyer 1999; Nordhaus 2010). Despite the achievement of the Paris Agreement, countries were allowed to participate on their own terms, without the legal responsibility of cutting emissions, and conditional on international financial transfers and aids (Seo 2017a, 2017b).

The failures to frame an international policy on global warming can be attributed to many factors, but most prominently to the disparity of the burden of any chosen policy framework across the nations (Seo 2012b, 2017a). This disparity can be ascribed in large part to the future impacts of climate change and global warming which are expected to vary greatly across nations (Mendelsohn et al. 2006; Tol 2009; Seo 2015c). The magnitude of the impacts again depends crucially on how an individual or a society will cope with the rise in temperature and changes in other climate factors, especially in agricultural and natural resource industries (Seo 2012a, 2016a, 2016b, 2016c).

Many areas of catastrophic concern attract very little governmental or international policy interventions. High-risk physics experiments that might engender strangelets, AI and killer robots, asteroids and comets are areas of great concern to many scientists (Dar et al. 1999; Jaffe et al. 2000; Kurzweil 2005; UNODA 2017; CNBC 2017), but no laws and regulations exist to prevent such events. These events are truly catastrophic events that could end human civilizations and the universe itself, but the small probability of such an event has dissuaded national governments and international organizations from intervening (Weitzman 2009; Nordhaus 2011; Seo 2018).

7.3 Fears of Low-Minds and High-Minds: Opinion Surveys

Throughout this book, a host of catastrophe events, both recurring ones and predicted ones, have been elaborated in detail. Some of these events are truly catastrophic, either humanity-ending or universe-ending. Are people really afraid of such doomsday events? Or are such events just a remote possibility which common people cannot and do not care about?

There are multiple opinion surveys available that offer a glimpse into people's minds, from which the author will summarize the results from two. One is the Chapman University Survey of American Fears and the other is the Times Higher Education (THE)/Lindau Survey of Nobel Laureates (Chapman University 2017; THE/Lindau 2017). These opinion surveys are of two different social groups with regard to the level of scientific knowledge each group commands.

Although these surveys may not have been scientifically/statistically sound, that is, taking care of all statistical issues, they nonetheless offer some relevant and interesting observations to readers of this book. The results are particularly fascinating when common people's perceptions which are starkly different from specialist scientists' perceptions are placed in comparisons with the types of catastrophic events presented throughout this book.

In Table 7.1, the top fears of average Americans in the US are presented for 2016 and 2017. The Chapman University Survey of American Fears asked in May 2017 a random sample of 1207 adults about their perceptions of 80 different fears. The types of fears surveyed cover a large variety of topics that include crime, government, environment, disasters, personal anxieties, and technology (Chapman University 2017). From the 80 fears, the top ten fears in 2016 as ranked by the percentage of respondents who stated that s/he is afraid or very afraid of an event are presented in the top panel of Table 7.1. To our surprise, the most feared event is by far corrupt government officials (60.6%). This is followed by terrorist attack, not having enough money for the future, terrorism, government restrictions on firearms and ammunition, people the respondent loves dying, economic/financial collapse, identify theft, people the respondent loves becoming seriously ill, the Affordable Care Act/Obamacare.

It is surprising that average Americans do not care about any of the catastrophic events that are explained throughout this book, e.g. AI, asteroids, cyclones, earthquakes, or strangelets and black holes. They are more feared about the government, losing people they love, money, and financial collapse.

The 2017 survey, presented in the bottom panel in Table 7.1, reveals quite a different story, reflecting the change in the US Presidency to Donald Trump at the beginning of the year. Notably, multiple environment-related fears ranked high in the table: pollution of rivers, oceans, and lakes (ranked third, with 53%), pollution of drinking water (ranked fourth, with 50%), global warming and climate change (ranked eighth, with 48%), and air pollution (ranked ninth, with 45%).

Table 7.1 The top fears of average Americans.

Rank	Top fears	Percentage afraid or very afraid
2016 Survey		
1	Corrupt government officials	60.6
2	Terrorist attack	41
3	Not having enough money for the future	39.9
4	Terrorism	38.5
5	Government restrictions on firearms and ammunition	38.5
6	People I love dying	38.1
7	Economic/financial collapse	37.5
8	Identity theft	37.1
9	People I love becoming seriously ill	35.9
10	The Affordable Care Act/Obamacare	35.5
2017 Survey		
1	Corrupt government officials	74.5
2	American Healthcare Act/Trumpcare	55.3
3	Pollution of oceans, rivers, and lakes	53.1
4	Pollution of drinking water	50.4
5	Not having enough money for the future	50.2
6	High medical bills	48.4
7	The US will be involved in another World War	48.4
8	Global warming and climate change	48
9	North Korea using weapons	47.5
10	Air pollution	44.9

Source: Chapman University (2017).

This is certainly owing to Trump's fast-paced and massive roll-backs of the Obama era environmental regulations such as the Clean Power Plan, the Keystone XL pipeline project, the methane rule, and the Waters of the United States rule, and the fuel economy standards, as well as Trump's decision to pull the US out of the Paris Agreement (US EPA 2017; White House 2017).

Nonetheless, economic and governmental fears still ranked high in 2017: corrupt government officials (75%), healthcare (55%), not having enough money (50%), and medical bills (48%). Other than the sentiments driven by political rage and divisiveness, unmoving fears of average Americans persist, that is, economic and financial distress that are experienced personally.

Another survey of opinions was conducted from the annual Lindau Nobel laureates meeting organization in Germany. Since 1901, fewer than 700 individuals have been awarded a Nobel Prize in science, medicine, or economics, of which 235 individuals are still alive. The survey asked the personal opinions of 50 of them regarding the biggest challenges faced by humanity (THE/Lindau 2017).

Table 7.2 Nobel laureates' ranking of the biggest challenges facing humanity (2017). Note: Some respondents gave more than one answer.

Rank	Biggest threats to humankind	Percentage of respondents
1	Population rise/environmental degradation	34
2	Nuclear war	23
3	Infectious disease/drug resistance	8
4	Selfishness/dishonesty/loss of humanity	8
5	Ignorance/distortion of truth	6
6	Fundamentalism/terrorism	6
7	Trump/ignorant leaders	6
8	AI	4
9	Inequality	4
10	Drugs	2
11	Facebook	2

Source: THE/Lindau (2017).

At first glance, the list of challenges or fears of Nobel laureates seems to be at odds with that of average Americans. The list has selfishness/dishonesty/loss of humanity ranked in fourth place, ignorance/distortion of truth ranked in the fifth place, fundamentalism/terrorism in sixth place, inequality in ninth place, and Facebook in eleventh place (see Table 7.2).

The list of biggest threats by the high-minds, say, the Nobel laureates who we believe to be highly intellectual and informed of scientific knowledge, is much closer to the taxonomy of catastrophic events described in this book. It has nuclear war ranked in second place, infectious diseases/drug resistance in third place, AI in eighth, and drugs in tenth.

The population rise/environmental degradation is ranked as the greatest challenge facing humanity, but it is not clear what this category actually means: Is it population rise and failure to feed people? Or toxic chemicals? Or air pollutants? Or climate change? It is also not clear whether the aggregated category of population rise/environmental degradation was put together by the survey organizers after the survey was conducted.

7.4 Planet-wide Catastrophes or Universal Catastrophes

The most critical policy question with regard to the literature of catastrophes is, with little doubt, how we should be prepared for the possibility of a truly global catastrophe, including a universal catastrophe, which tends to have a very small probability of actually coming into being. What are those truly global catastrophic events that some people are gravely concerned about but whose probabilities of occurrences are so small? Four such global-scale possibilities are described in the following: nuclear or chemical arms race and wars, a large asteroid collision, high-risk large-scale physics or biological experiments, and AI and killer robots (Posner 2004; Parson 2007; Hawking et al. 2014).

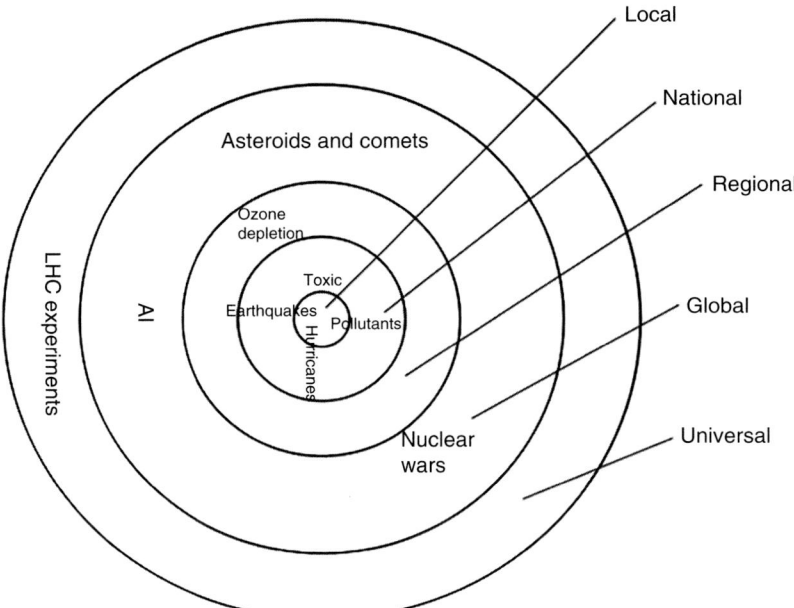

Figure 7.1 Catastrophes by the spatial scale of events.

Before we proceed to the four events, a diagram of catastrophes discriminated by the spatial scale of damages is presented in Figure 7.1 in a set of concentric circles. The innermost smallest circle indicates a local-scale catastrophe. The second circle indicates a national-scale catastrophe, the third a regional-scale catastrophe, the fourth a global-scale catastrophe, the fifth and the largest circle a universal-scale catastrophe.

Overlaid over the concentric circles are specific catastrophic events discussed throughout this book. Toxic and hazardous substances cause local- or national-scale damages, so they are placed inside the two circles. Criteria air pollutants cause local- or national-scale damages, but the damage can occur at a regional scale, i.e. a multicountry scale, in transboundary pollutants. In Figure 7.1, the category of criteria pollutants is overlaid on top of mostly local and national circles, but also on the edge of the regional circle.

Earthquakes and hurricanes are most often local-scale catastrophes, but national-scale damages can be caused in island nations such as Puerto Rico, as was the case with the strike of Hurricane Maria in the summer of 2017 (NHC 2017b). Very rarely, a regional-scale damage is caused by an earthquake or a hurricane: in a rare case that the damage is incurred in multiple countries, it is most likely not catastrophic. The category of earthquakes is overlaid on top of the three concentric circles: local, national, and regional. So is that of hurricanes.

Ozone depletion in the stratosphere is mostly a regional (multicountry)-scale catastrophe where the damage occurs in the region of the Earth under an ozone hole. The damage can be national for a country with a large land area, e.g. Australia (Molina and Rowland 1974).

A collision with a large asteroid or a comet, a nuclear war, and AI and killer robots are overlaid on the circle of a global-scale catastrophe. An LHC physics experiment is overlaid on the circle of a universal catastrophe. The scale of catastrophe from any of these categories of events could be planet-wide or universal.

A nuclear arms race which ends up with a nuclear war is predicted by a group of scientists to have the potential to disturb the global climate system, leading to nuclear winters, or to destroy the stratospheric ozone layer (Turco et al. 1983; Mills et al. 2008). The ultimate consequence could be a truly global catastrophe in which nearly all humans perish.

Although nearly all members of the UN are committed to the international nuclear treaty, i.e. the nuclear NPT, which provides a safeguards system, it can unravel by a provocation from a nonmember state with nuclear weapons or by lack of commitments by nuclear-weapon states to a full and complete disarmament (Campbell et al. 2004).

Another possible event of a global-scale catastrophe is a possible collision of a large asteroid or comet with Earth (Chapman and Morrison 1994). According to the current definition of the National Aeronautics and Space Administration (NASA), a near-Earth object (NEO) with a diameter larger than 300 m could have, upon impact, subglobal effects, and an NEO with a diameter larger than 1 km could have global effects upon collision (NASA 2014). The asteroid that caused the extinction of dinosaurs 66 million years ago is estimated to have been as large as 10 km in diameter (Kaiho and Oshima 2017).

According to the aforementioned organization, there are 200 000 NEOs that have some possibility of colliding with the Earth, and 50,000 of those NEOs have been discovered as of the end of 2017. Of the discovered NEOs, about half have a diameter smaller than 100 m, which can make only limited impacts, most likely no global-scale destruction (NASA 2017). Early detection is the key for a successful mission for diverting an asteroid on a collision course with the Earth.

The third possibility of a global-scale catastrophe is high-risk large-scale physics experiments (Dar et al. 1999; Jaffe et al. 2000). In this book, the author explained the rationale, experiments, machines, and risk of the LHC experiments by the European Nuclear Research Organization (CERN). According to the physics theory of the Big Bang, the risk, according to concerned scientists, is universal in that an unexpected sequence of events can initiate strangelets, also called black matter, which forces the absolute dissolution of the universe into a black-hole.

Although the CERN experiments have been successful and disaster-free up to this point, including the award of the Nobel Prize in physics in 2013, there remains a small probability of a very large-scale catastrophe that cannot be eliminated, on which scientists on both sides seem to agree broadly (Overbye 2013).

There are other scientific experiments that have a possibility of causing a truly catastrophic chain of events. One of the side effects of the advances in biological sciences such as the Human Genome Project and stem cell research may include unexpected effects on ecosystems and human societies through alterations of genetic codes of animals, plants, and humans (USDOE 2008; NASEM 2017).

Through gene alterations, scientists can create new species or breeds of animals and plants, e.g. genetically modified fish, genetically modified crops, and genetically modified insects (Watanabe et al. 2014). Through the stem cell research of human embryonic and nonembryonic stem cells, scientists can create duplicates of animals and human organs (Thomson et al. 1998; Nobel Prize 2012).

Ethical issues aside, these experiments and applications can have unpredictable consequences, and scientists may not grasp all elements of those consequences.

Regulation mechanisms are in place that are designed to prevent such consequences or countermeasures that are intended to control damages from such chain events. However, it may not be possible to eliminate a small probability of such an event occurring.

The fourth possibility of a global or universal catastrophe is advances in AI. The fear by some scientists is that robots are increasingly better than humans in many areas of human activities, thereby replacing human laborers. At present, robots are already perceived to be better than humans in some aspects of stock investments, medical diagnosis, driving cars, or even perhaps making coffee. Moreover, robots defeat human champions easily in Chess and Go games, barely pondering their next moves.

The most concerning aspect of AI at present, as expressed by experts, is the race among the world's powerful nations to gain superiority in AI in order to attain military dominance over competitor countries. Robots are already widely employed in both local police responses and military battles in a war zone, one example of which is military drones. These robots are, as they are called by the UN, a lethal automatic weapons system (UNODA 2017).

At the present stage of AI, robots must be ordered by humans to fight in a battle or kill enemies in a battle. A great fear by concerned scientists is that robots may start a war on their own someday without a human order, which becomes uncontrollable and cannot be stopped by humans (CNBC 2017).

That fatal day is called the moment of singularity (Kurzweil 2005; Hawking et al. 2014). The singularity is defined to be the moment when the brain capacity of an AI exceeds that of a human. Experts argue that there will be no turning back when singularity arrives, which would mean, sooner or later, human extinction.

All things described in this section concerning the four global or universal catastrophes, summarized in Table 7.3, may read a lot like a science-fiction novel or a Hollywood science-fiction movie. In fact, these stories in one adapted form or another are abundantly found in those genres of works. Readers, however, should be reminded that the author presents the four catastrophe scenarios on the presumption that each of these scenarios has a very low likelihood of materializing. These events may not materialize at all or their risks may be reduced to a manageable level when proper actions are taken some day; this is addressed more directly in Section 7.5.

Table 7.3 Global-scale or universal-scale catastrophes.

Catastrophe event	Scale of potential damage	Main channels
A nuclear war	Global or subglobal	Global cooling; ozone layer depletion
An asteroid/comet collision with its diameter greater than 10 km	Global	Impact and disturbance to Earth's systems; impact and alteration to Earth's rotation
AI	Global or humanity-scale	Singularity: robots surpass humans in brain capacity
A high-risk physics or biology experiment: LHC experiments, genetic alterations	Universal	Strangelets or black matter that create a black hole

7.5 Making Rational Decisions on Planet-wide or Universal Catastrophes

There "always" exists a "small" possibility of a global-scale or universal-scale catastrophe through one of the four events described or other routes. The fact that humanity cannot eliminate the small chance of occurrence of such a catastrophe has had enduring impacts on individuals and societies throughout human history, which can be verified by, among other things, the three fables or tales in historical records that were introduced in Chapter 1 (Mandelbrot and Hudson 2004; Posner 2006; Parson 2007; Taleb 2007; Wagoner and Weitzman 2015).

This small probability can be appropriated by world religious systems for the respective religion's prediction of a catastrophic end of life on Earth that human beings can neither escape nor control. These religions would simply regard the possibility of a catastrophic end as the divine destiny, which is obviously an egregious misunderstanding of the sciences underlying these events.

This small probability can be snowballed into a worldwide frenzy if no informed one or science is there to disentangle the psychology of fear. As in the rabbit's case introduced in Chapter 1, a rumor and unchecked spreading of the rumor does the trick of snowballing it to a worldwide fear (Cowell et al. 1895).

The fear of a small-probability catastrophic event can cause an individual to live under severe depression, as Giwoo who lived his life always screaming that the sky is falling (Wong 2001). He must have no ability of rational reasoning, but we wonder whether he would have accepted a modern scientist's explanations of the atmosphere and the fact that there is no sky to fall.

In all three salient examples in history of responses to catastrophic possibilities, we are all compelled to accept the importance of correct knowledge, i.e. sound scientific and experiential understanding of the events that concern a society. In the tale of Giwoo, atmospheric scientists may have cured his baseless fear if he had at least the ears to listen to others' words.

One may argue that the science is itself a primary cause of all these catastrophic probabilities hovering over everyday human life. Critics would say that advances in science and technology are creating and increasing the possibility of each of these catastrophic events. To extreme critics, scientists may be a public foe.

It must be heeded, however, that the science and technology does provide a means to solve the problem of asteroids and comets on a collision course with the planet. In this case, it is clear that scientific knowledge and technological capacities are a cure for the disease, not a cause of the problem.

In the case of a nuclear or chemical catastrophe, scientific advances in nuclear physics made it possible for humans to build nuclear bombs. However, it is difficult to argue that nuclear physics is the cause of a nuclear catastrophe. It is rather the humans and political systems that have control over these weapons that may cause a global-scale catastrophe.

In attempts to describe rational decisions on a tiny possibility global-scale catastrophe event, it is therefore indispensable to accept scientific knowledge and technological advances as a friend not an enemy of humanity. A flawed science, if critics are concerned about one, should be corrected by a correct science. Indeed, many of the solutions to these problems rely critically on technological possibilities (NRC 2010, 2015; Seo 2017a).

The LHC experiments provide an excellent example of the power of science in addressing humanity's biggest challenges. The scale of the LHC catastrophe is truly universal: it may create strangelets which form a stable black hole which may destroy the entire universe completely (Plaga 2009; Ellis et al. 2008). If such a possibility exists at all, even a very tiny probability of such an event would pose a gigantic policy challenge for the global community.

However, scientific observations of the naturally occurring particle collisions in the universe as well as laboratory experiments on the possibility of a strangelet creation can lead to the conclusion that the LHC experiments present "no danger" or, put differently, the possibility of the universal-scale catastrophe is "absurdly small" (APS 1997; Dar et al. 1999; Jaffe et al. 2000; Ellis et al. 2008; Peskin 2008; CERN 2017).

What the LHC–black hole controversy demonstrates is that we cannot make appropriate responses and rational decisions without the sound scientific descriptions of the event when it comes to humanity's biggest catastrophe challenges. Recent policy experiences with global warming science, AI, asteroids, toxic substances, hurricanes, and earthquakes all point to a similar conclusion (Le Treut et al. 2007; Hawking et al. 2014; Seo 2017a, 2017b).

At this point, the author will tell the rest of the tale of the frightened rabbit introduced in the very beginning of this book. To summarize the first half of the story already told, the rabbit who was always worried about the world's breakup heard a coconut falling and, startled, started to run as fast as he could, screaming "The world is breaking up." He was joined, one after another, by a second rabbit, a third rabbit, deer, foxes, and elephants, all screaming that the world is ending, who did not question the validity of the assertion by the first rabbit. The tale goes on as follows:

> At last the lion saw the animals running, and heard their cry that the earth was all breaking up. He thought there must be some mistake, so he ran to the foot of a hill in front of them and roared three times. This stopped them, for they knew the voice of the king of beasts, and they feared him.
> "Why are you running so fast?" asked the lion.
> "Oh, King Lion," they answered him, "the earth is all breaking up!"
> "Who saw it breaking up?" asked the lion.
> "I didn't," said the elephant. "Ask the fox – he told me about it."
> "I didn't," said the fox.
> "The rabbits told me about it," said the deer.
> One after another of the rabbits said: "I did not see it, but another rabbit told me about it."
> At last the lion came to the rabbit who had first said the earth was all breaking up.
> "Is it true that the earth is all breaking up?" the lion asked.
> "Yes, O lion, it is," said the rabbit. "I was asleep under a palm-tree. I woke up and thought, 'What would become of me if the earth should all break up?' At that very moment, I heard the sound of the earth breaking up, and I ran away."
> "Then," said the lion, "you and I will go back to the place where the earth began to break up, and see what is the matter."
> So the lion put the little rabbit on his back, and away they went like the wind. The other animals waited for them at the foot of the hill. The rabbit told the lion when they were near the place where he slept, and the lion saw just where the rabbit had

been sleeping. He saw, too, the coconut that had fallen to the ground nearby. Then the lion said to the rabbit, "It must have been the sound of the coconut falling to the ground that you heard. You foolish rabbit!"

And the lion ran back to the other animals, and told them all about it. If it had not been for the wise king of beasts, they might be running still.

The tale of the frightened rabbit is a story of a universal catastrophe in which the world would end, whose primary giveaway is the importance of investigation into an unverified assertion which, in this case, turned out to be false. All of the truly big catastrophes explained in Section 7.4 can only be understood by someone with advanced-level knowledge of physics and biology, for which reason scientists will play the critical role of the lion in the rabbit fable. In the case of the fear of strangelets and a black hole, scientists have done admirable work for assessing the truth of an initial cry (Dar et al. 1999; Jaffe et al. 2000; Ellis et al. 2008).

Besides the sound scientific knowledge, the second component of a rational decision on a big catastrophe challenge, which is also salient in the tale of the frightened rabbit, is psychology. Psychological aspects are known, formally or informally, to play a major role in the situations of various catastrophes besides those described above in this chapter. From the society's or global perspective, the fear created by a rumor which spreads unchecked through social media and unverified sources can be deadly. It creates a global fear based on misinformation. A rumor is an untested story which turns out eventually to be false.

There lies another important giveaway of the rabbit fable. A rumor or a fake story can spread at the speed of light in the age of social media and the internet, in which process the psychology of fear propels the story into a social frenzy. A rumor rooted on fear is often unstoppable even if scientists or investigators could at a later time disprove the rumor.

The pivotal role of psychology in the past financial market crashes has been repeatedly observed and documented. The Great Depression in 1929, the dot-com bubble burst in 2000, and the global financial crisis which started from a subprime mortgage crisis in the US in 2007 are attributed to a psychology-driven bubble in financial markets (Galbraith 1954; Shiller 2005).

This book has made extensive efforts to explain the power law or the fat-tail distribution as a potential distribution of a truly big catastrophe event (Weitzman 2009; Nordhaus 2011). Many researchers also applied the power law and scale invariance to predict financial market crashes (Mandelbrot 1997, 2001; Gabaix 2009). Another prominent theory of financial market crashes has been the theory of bubbles driven by psychological elements of market participants. Shiller defines the bubble as follows:

> A situation in which news of price increases spurs investor enthusiasm which spreads by psychological contagion from person to person, in the process amplifying stories that might justify the price increases and bringing in a larger and larger class of investors, who, despite doubts about the real value of an investment, are drawn to it partly through envy of others' successes and partly through a gambler's excitement
>
> (Shiller 2005, 2014).

A bubble is building in the financial market through stories of market successes and psychological contagion which amplify each other, but there must come a time for it to burst eventually when the bubble becomes too big to sustain, given the market fundamentals. When the bubble bursts, it yields a "10-sigma crash event," devastating individuals, families, businesses, and the government.

An important contribution of the theory of bubble to economics and policy-making is widely recognized, which is the discovery of psychology as one of the drivers of the market and market crashes. Bubble theorists discovered the underlying factor that drives one of the power law catastrophe events – as such, a potential remedy for such an event (Shiller 2005; Thaler and Sunstein 2009).

With regard to the four global-scale catastrophes, psychological experts that snowball the fear of a small-probability high-risk event may do harm more than good in addressing the big problem itself. They may argue that it is impossible for them to push for a strong governmental action on any of the four catastrophes without reliance on the cases made for fearful consequences from no governmental action. Besides the sound scientific knowledge, the second component of a rational decision on a big catastrophe challenge, which is also salient in the tale of the frightened rabbit, is psychology (see Table 7.4 for all the components).

The arguments are not well reasoned. For any future disaster event, it is indispensable to first establish the possibility of such an event, second evaluate the impacts it could have if materialized, and third estimate the value of the impacts. Without these steps taken, there is no basis for policy discussions and actions. These steps call for rational analyses, not emotional responses.

Table 7.4 Elements and functions of a rational decision on global-scale catastrophes.

Elements	Functions
Science and technological advances	Impact assessments of the LHC experiments minimize uncertainty and probability of a catastrophe
Psychology	Beware a snowballing of fear based on a rumor or an invalidated report
Spirituality and religion	Alleviate extreme mental clinging to a global-scale, mostly improbable event
Economic aspects: the least-cost approach or the social welfare optimizing approach	Choose the least-cost option from a range of possible solution mechanisms, given the potential benefit
Adaptation: behavioral and physical	Build adaptation knowledge and portfolios in parallel to technological advances: examples include a self-driving AI car, high-risk physics, and biological experiments
An ultimate stop-control technology	An ultimate stop-control switch in the case of a runaway catastrophe: examples include a robot war, global climate-induced catastrophe, and LHC experiments

In situations of global-scale or universal-scale catastrophes, psychologists' principles rooted on reason and rationality are called for rather than those based on emotions and fear. The principled psychology, if we may call it this, is needed more when there is larger uncertainty regarding the concerned global-scale catastrophic event or when the concerned event is predicted to unfold over a long period of time, say, over the scale of a century. These situations make scientific investigations and rational examinations harder, but not impossible, and more time-consuming.

Third, fear of a small chance for a truly horrific event arises in most cases, although not exclusively, from the mental attachment, or even addiction, to a specific event. A strong mental clinging to a certain probable event can lead to the extreme fear of an individual or an entire community which is unfounded in reality. A cure for the extreme panic of the individual or the social group lies in changes in the mental perspective from the one specific object of terror to a bigger picture perspective in which a whole universe of things is depicted.

Religions and spirituality traditions may play a vibrant role in this regard, given their influence on so many people's ways of life. Religious and spirituality leaders should highlight the aspects of mental propensities of individuals and societies to become obsessed by selected objects or events and point toward a broader reality that humanity is faced with. They can and are in the right position to emphasize destructive consequences of mental attachments to a particular event as well as peace of mind obtained from a broader perspective.

There is no reason for a particular religion or spirituality tradition to push fervently for a strong governmental and global action on a specific catastrophic event whose eventual realization is deemed by experts so uncertain and unlikely and whose coming into being is unrelated to any religious doctrine.

From a slightly different point of view, the vital role of the religions and spirituality traditions is perhaps unavoidable in global-scale and universal-scale catastrophes because these events are described as an end of all things as we know them. The humanity-scale catastrophes portrayed by the forecasters of these events fall into the realms of world religions, that is, the world's beginning and end. As such, there may arise an inescapable need for a world religion to craft a "right" and balanced approach for the tradition to the range of these forecasted events.

The fourth component of rational decisions concerns economic aspects. When the concerned catastrophic event is a long-term event, i.e. it takes many decades or centuries to unfold, economists and policy-makers should pay attention to the cost of time. When the catastrophe is expected to occur a hundred years from now, they should look into alternative pathways to avoid the catastrophe over time and choose the least-cost option to achieve it.

If an immediate remedy to the concerned problem is by far costlier than a graduated remedy over the course of a century, the society should rationally choose the latter. The cost in this case is the total net cost, i.e. total cost minus total benefit. In comparing the present cost to the stream of costs that is predicted to accrue in the future, we should rely on a discounting factor, the selection of which should be made with reference to the market transactions that capture revealed tradeoffs between the present and the future, and thereby a pure rate of time preference (Ainslie 1991; Arrow et al. 1996; Nordhaus 2007).

For the society that relies on reason and rationality, the least-cost approach should be chosen from a set of alternative solutions because it is the social welfare optimizing solution, given that the ultimate outcome or benefit is the same, that is, the prevention of a catastrophe of concern (von Neumann and Morgenstern 1947; Koopmans 1963, 1965).

For example, in the catastrophe problem of asteroids and comets that are on a collision course with the Earth, multiple scenarios are possible on how to prevent such NEOs from impacting the Earth. It is quite soundly known in the scientific community which of these scenarios is least costly, assuming there are multiple successful technologies. The least-cost approach according to NASA is composed of early detection and altering the velocity of the asteroid only slightly at an appropriate distance from the Earth through a gravity tractor or a kinetic impactor methodology (NRC 2010; NASA 2014).

The social welfare optimization framework has been described at length in this book, especially in Chapters 4 and 5. The descriptions in these chapters should help individuals and policy-makers who endeavor to choose the least-cost alternative among an array of possible remedy options.

The fifth element of rational decisions on global-scale catastrophes is adaptation possibilities. Rational decisions should look into the question of whether there are adaptive mechanisms in response to the small possibility of a big catastrophe. In the advances of AI and robot technologies, for example, it is expected that the control capacity of humans over AI machines and robots is also being improved in tandem (Hawking et al. 2014; Fox News 2017).

Let's consider a self-driving car as an AI. For it to be adopted by people and competitive in the market, the developers should prove that a self-driving car is safer than a human-driven car, at the core of which is the possibility of a human control over the AI when needed. That is, additional measures may include various inside-the-car safety measures, remote control of the AI driver by a third party, a possible conversion from AI driving to manual driving when needed, escape measures in emergency, and insurances, but the ultimate assurance will be given only when it is possible for a human rider to take control of the AI driver whenever necessary.

From the policy angle, it is not the best option to stop the development of AI and robots altogether based on the fear of an ultimate catastrophe through one of the mechanisms explained in this book. A more rational policy approach is to mandate additional safety measures as an adaptive mechanism in the development of self-driving AI automobiles.

Past policy responses to humanity-scale catastrophes relied on the potential for adaptive mechanisms by humans or in the physical world. The high-risk physics experiments such as the LHC experiments provide an example of the possibility of an adaptive system in the physical world which can be a determining factor in policy directions. More concretely, the LHC scientific assessment group argued that even if a microscopic black hole is produced by the LHC experiment, it would decay in the atmosphere and would be stopped by astronomical bodies, which formed the basis for the US–CERN agreement on the LHC experiments (Ellis et al. 2008; Jaffe et al. 2000; APS 1997):

> Any microscopic black holes produced at the LHC are expected to decay by Hawking radiation before they reach the detector walls. If some microscopic black holes were stable, those produced by cosmic rays would be stopped inside the Earth or other astronomical bodies. The stability of astronomical bodies constrains strongly the possible rate of accretion by any such microscopic black holes, so that they present no conceivable danger. In the case of strangelets, the

good agreement of measurements of particle production at RHIC [Relativistic Heavy Ion Collider] with simple thermodynamic models constrains severely the production of strangelets in heavy-ion collisions at the LHC, which also present no danger [Ellis et al. 2008].

The critical implication of an adaptive system has been emphasized throughout this book. In the catastrophe theories of René Thom described in Chapter 2, the author provided a modified model with a regulation parameter through which an occurrence of a catastrophe event can be controlled. In the descriptions of the power law and fat-tail catastrophe events, an adaptive system is explained in Chapter 4 to be a key factor through which the tail distribution can be moderated.

The expansive literature on adaptation behaviors and strategies is highlighted throughout this book. In Chapter 5, the author elaborates on the empirical models of adaptation behaviors as well as prominent adaptation behaviors to tropical cyclones. In Chapter 6, an extensive list of adaptation behaviors is described in the contexts of policy experiences with numerous catastrophe events.

The sixth element of a rational decision concerns a runaway catastrophe. A rational decision-making on a truly big catastrophe should be able to answer whether social welfare maximizing coupled with an adaptation paradigmatic solution offers a near-certain probability of preventing such a humanity-scale catastrophe. More specifically, a rational solution should be equipped with an ultimate control or an ultimate stop-switch for a runaway catastrophe, when needed.

A runaway catastrophe is defined as an event, once started, that cannot be stopped, whose damage magnifies exponentially to reach a global-scale catastrophe. It is often visualized in the literature as a hockey-stick hypothesis in reference to the rising end of a hockey-stick. In the catastrophe literature, a runaway global warming, a runaway AI system, and a runaway LHC accident are cited as examples of a potential runaway catastrophe (Mann et al. 1999; Posner 2004; Kurzweil 2005; Hawking et al. 2014).

In the AI self-driving automobile example, a rational solution to a catastrophe should include an ultimate stop-control switch of the AI robot by a human passenger in any circumstances. In a broader AI accident such as a robots-initiated war, a similar stop-control switch may be installed in each robot. In the case of a failure of the first stop-control switch, a secondary stop-control switch, a back-up, which is remotely controlled by a central command center may be installed at the same time.

In the case of the potentially catastrophic LHC experiments, an ultimate stop-switch may be a switch that turns off the LHC tunnel completely or a switch that destroys the LHC at once. Alternatively, a secondary "shield" that can stop the formation of a stable black hole may be established, which according to scientists seems to offer a less expensive but more effective stop-control switch (Ellis et al. 2008).

A stop-control switch has been researched quite extensively in the context of a runaway global warming catastrophe. A stop-control technology for global warming policy includes a carbon capture and storage system, climate engineering, and albedo modification through the solar reflector placed outside the Earth (Lackner et al. 2012; NRC 2015; Seo 2017a).

However, given that the event of global warming is projected to unfold gradually over many centuries, alternative energy production technologies such as solar energy and nuclear fission energy that can replace currently dominant fossil-fuel energy production technologies are regarded as a stop-control switch in global warming policy (MIT 2003, 2015; ITER 2015). In the global warming economics literature, an ultimate stop-control switch is often referred to as a backstop technology (Nordhaus 1994).

Of all the catastrophes surveyed in the book, it is the LHC experiments-caused black hole catastrophe that is perhaps the most severe runaway event: once the strangelets are created, it may not take many minutes to form a black hole which would collapse the universe. It is notable that, even in this most severe runaway catastrophe scenario, there has been no attempt to install an ultimate stop-control.

As explained above, the scientific assessments of the LHC experiments clarified that the event is a very long-tail event but also a very thin-tail event. Further, the possibility of a physical adaptation process that would be in play in natural environments in the case of an LHC accident makes it unnecessary to build a backstop system. A rational decision-making on a random catastrophe should consider the possible backstop technologies but does not require that it should always go for the ultimate option.

7.6 Conclusion

Humanity has faced a multitude of unprecedented challenges to survive on this planet for millions of years, many of which were unexpected and utterly shocking. This book has described the most serious catastrophe challenges that humanity will confront in the decades and centuries ahead, with the hope of laying a foundation for a system of theories and tools that can be relied upon in meeting with those challenges, based on a wide-ranging multidisciplinary review of sciences, mathematical modeling, philosophy, economics, empirical modeling, policy studies, and literary works.

References

Ainslie, G. (1991). Derivation of "rational" economic behavior from hyperbolic discount curves. *American Economic Review* 81: 334–340.

American Physical Society (APS) (1997). US CERN Agreement on the LHC. APS, College Park, MD. http://www.aps.org/units/dpf/governance/reports/lhc.cfm.

Arrow, K.J., Cline, W.R., Maler, K.G. et al. (1996). Intertemporal equity, discounting, and economic efficiency. In: *Climate Change 1995: Economic and Social Dimensions of Climate Change* (ed. J.P. Bruce, H. Lee and E.F. Haites). Cambridge: Cambridge University Press.

Bakkensen, L.A. and Mendelsohn, R. (2016). Risk and adaptation: evidence from global hurricane damages and fatalities. *Journal of the Association of Environmental and Resource Economists* 3: 555–587.

Blake ES, Landsea CW, Gibney EJ (2011). The deadliest, costliest, and most intense United States tropical cyclones from 1851 to 2010 (and other frequently requested hurricane facts). NOAA Technical Memorandum NWS NHC-6. NOAA, Silver Spring, MD.

Boccaletti, S., Grebogi, C., Lai, Y.-C. et al. (2000). The control of chaos: theory and applications. *Physics Reports* 329: 103–197.

Broecker, W.S. (1997). Thermohaline circulation, the Achilles' heel of our climate system: will man-made CO_2 upset the current balance? *Science* 278: 1582–1588.

Campbell, K.M., Einhorn, R.J., and Reiss, M.B. (ed.) (2004). *The Nuclear Tipping Point: Why States Reconsider their Nuclear Choices*. Washington, DC: Brookings Institution Press.

Carson, R. (1962). *Silent Spring*. Boston, MA: Houghton Mifflin.

CERN (2017). The Safety of the LHC. CERN, Geneva. https://press.cern/backgrounders/safety-lhc.

Chapman, C.R. and Morrison, D. (1994). Impacts on the earth by asteroids and comets: assessing the hazard. *Nature* 367: 33–40.

Chapman University (2017). *Survey of American Fears: Wave 4*. Orange, CA: Chapman University.

CNBC (2017). Elon Musk Says Global Race for A.I. Will Be the most Likely Cause of World War III. CNBC, New York. https://www.cnbc.com/2017/09/04/elon-musk-says-global-race-for-ai-will-be-most-likely-cause-of-ww3.html.

Collet, P. and Eckmann, J.-P. (1979). Properties of continuous maps of the interval to itself. In: *Mathematical Problems in Theoretical Physics* (ed. K. Osterwalder). New York: Springer.

Collet, P. and Eckmann, J.-P. (1980). *Iterated Maps on the Interval as Dynamical Systems*. Boston, MA: Birkhäuser.

Cowell, E.B., Chalmers, R., Rouse, W.H.D. et al. (1895). *The Jataka; or, Stories of the Buddha's Former Births*. Cambridge, UK: Cambridge University Press.

Dar, A., Rújula, A.D., and Heinz, U. (1999). Will relativistic heavy ion colliders destroy our planet? *Physics Letters B* 470: 142–148.

Edesess, M. (2014). *Catastrophe Bonds: An Important New Financial Instrument*. Paris: EDHEC-Risk Institute, EDHEC Business School.

Ellis, J., Giudice, G., Mangano, M. et al. (2008). Review of the safety of LHC collisions. *Journal of Physics G: Nuclear and Particle Physics* 35 (11).

Emanuel, K. (2005). Increasing destructiveness of tropical cyclones over the past 30 years. *Nature* 436: 686–688.

Emanuel, K. (2008). The hurricane–climate connection. *Bulletin of American Meteorological Society* 89: ES10–ES20.

Emanuel K (2013). Increased global tropical cyclone activity from global warming: results of downscaling CMIP5 climate models. Presented at the International Summit on Hurricanes and Climate Change. Kos, Greece.

Fabozzi, F.J., Modigliani, F.G., and Jones, F.J. (2009). *Foundations of Financial Markets and Institutions*, 4the. New York: Prentice Hall.

Feigenbaum, M.J. (1978). Quantitative universality for a class of non-linear transformations. *Journal of Statistical Physics* 19: 25–52.

Fox News (2017). Robots will be 100 times smarter than humans in 30 years, tech exec says. http://www.foxnews.com/tech/2017/10/27/robots-will-be-100-times-smarter-than-humans-in-30-years-tech-exec-says.html.

Fractal Foundation (2009). *Educators' Guide*. Albuquerque, NM: Fractal Foundation.

Frame M, Mandelbrot B, Neger N (2017). Fractal Geometry. Yale University, New Haven, CT. http://users.math.yale.edu/public_html/people/frame/fractals.

Gabaix, X. (2009). Power laws in economics and finance. *Annual Review of Economics* 1: 255–293.

Galbraith, J.K. (1954). *The Great Crash 1929*. Boston, MA: Houghton Mifflin.

Graham, T. Jr., (2004). *Avoiding the Tipping Point*. Washington, DC: Arms Control Association.

Hájek A (2012). Pascal's Wager. Stanford Encyclopedia of Philosophy. https://plato.stanford.edu/entries/pascal-wager.

Hansen, J., Sato, M., and Reudy, R. (2012). Perception of climate change. *Proceedings of the National Academy of Sciences of the United States of America* 109: E2415–E2423.

Hawking S, Tegmark M, Russell S, Wilczek F (2014). Transcending complacency on superintelligent machines. *Huffington Post*. https://www.huffingtonpost.com/stephen-hawking/artificial-intelligence_b_5174265.html.

Intergovernmental Panel on Climate Change (IPCC) (1990). *Climate Change: The IPCC Scientific Assessment*. Cambridge, UK: Cambridge University Press.

Intergovernmental Panel on Climate Change (IPCC) (2014). *Climate Change 2014: The Physical Science Basis, the Fifth Assessment Report of the IPCC*. Cambridge, UK: Cambridge University Press.

International Thermonuclear Experimental Reactor (ITER) (2015). The ITER Tokamak. https://www.iter.org/mach.

Jaffe, R.L., Buszaa, W., Sandweiss, J., and Wilczek, F. (2000). Review of speculative disaster scenarios at RHIC. *Review of Modern Physics* 72: 1125–1140.

Kaiho, K. and Oshima, N. (2017). Site of asteroid impact changed the history of life on earth: the low probability of mass extinction. *Scientific Reports* 7: 14855. doi: 10.1038/s41598-017-14199-x.

Kiger, M. and Russell, J. (1996). *This Dynamic Earth: The Story of Plate Tectonics*. Washington, DC: USGS.

King RO (2013) The National Flood Insurance Program: status and remaining issues for Congress. CRS Report for Congress R42850. Congressional Research Service, Washington, DC.

Knowles, S.G. and Kunreuther, H.C. (2014). Troubled waters: The National Flood Insurance Program in historical perspective. *Journal of Policy History* 26: 325–353.

Koopmans T (1963). On the concept of optimal economic growth. Cowles Foundation Discussion Papers. Yale University, New Haven, CT.

Koopmans, T.C. (1965). *On the concept of optimal economic growth*, vol. 28 (1), 1–75. Academiae Scientiarum Scripta Varia.

Kurzweil, R. (2005). *The Singularity Is Near: When Humans Transcend Biology*. New York, NY: Penguin.

Lackner, K.S., Brennana, S., Matter, J.M. et al. (2012). The urgency of the development of CO_2 capture from ambient air. *Proceedings of the National Academy of Sciences of the United States of America* 109 (33): 13156–13162.

Lanford, O.E. (1982). A computer-assisted proof of the Feigenbaum conjectures. *Bulletin of American Mathematical Society* 6: 427–434.

Le Treut, H., Somerville, R., Cubasch, U. et al. (2007). Historical overview of climate change. In: *Climate Change 2007: The Physical Science Basis* (ed. S. Solomon, D. Qin, M. Manning, et al.). Cambridge: Cambridge University Press.

Lenton, T.M., Held, H., Kriegler, E. et al. (2008). Tipping elements in the earth's climate system. *Proceedings of the National Academy of Sciences of the United States of America* 105: 1786–1793.

Leopold, A. (1949). *A Sand County Almanac: And Sketches Here and There*. Oxford, UK: Oxford University Press.

Lorenz, E.N. (1963). Deterministic nonperiodic flow. *Journal of the Atmospheric Sciences* 20: 130–141.

Lorenz, E.N. (1969a). Atmospheric predictability as revealed by naturally occurring analogues. *Journal of the Atmospheric Sciences* 26: 636–646.

Lorenz, E.N. (1969b). Three approaches to atmospheric predictability. *Bulletin of the American Meteorological Society* 50: 345–349.

Mandelbrot, B. (1983). *The Fractal Geometry of Nature*. New York: Macmillan.

Mandelbrot, B. (1997). *Fractals and Scaling in Finance*. New York: Springer.

Mandelbrot, B. (2001). Scaling in financial prices: I. Tails and dependence. *Quantitative Finance* 1: 113–123.

Mandelbrot, B. (2004). *Fractals and Chaos*. Berlin: Springer.

Mandelbrot, B. and Hudson, R.L. (2004). *The (Mis)Behaviour of Markets: A Fractal View of Risk, Ruin, and Reward*. London: Profile Books.

Mann, M.E., Bradley, R.S., and Hughes, M.K. (1999). Northern hemisphere temperatures during the past millennium: inferences, uncertainties, and limitations. *Geophysical Research Letters* 26: 759–762.

Marchant, G., Allenby, B., Arkin, R. et al. (2011). International governance of autonomous military robots. *Columbia Science and Technology Law Review* 12: 272–315.

Massachusetts Institute of Technology (MIT) (2003). *The Future of Nuclear Power: An Interdisciplinary MIT Study*. Cambridge, MA: MIT.

Massachusetts Institute of Technology (MIT) (2015). *The Future of Solar Energy: An Interdisciplinary MIT Study*. Cambridge, MA: MIT.

McAdie, C.J., Landsea, C.W., Neuman, C.J. et al. (2009). *Tropical cyclones of the North Atlantic Ocean, 1851–2006*, Historical Climatology Series, vol. 6-2. Asheville, NC: NOAA.

Mendelsohn, R., Dinar, A., and Williams, L. (2006). The distributional impact of climate change on rich and poor countries. *Environment and Development Economics* 11: 1–20.

Mills, M.J., Toon, O.B., Turco, R.P. et al. (2008). Massive global ozone loss predicted following regional nuclear conflict. *Proceedings of the National Academy of Sciences of the United States of America* 105: 5307–5312.

Molina, M.J. and Rowland, F.S. (1974). Stratospheric sink for chlorofluoromethanes: chlorine atomcatalysed destruction of ozone. *Nature* 249: 810–812.

National Academies of Sciences, Engineering, and Medicine (NASEM) (2017). *Human Genome Editing: Science, Ethics, and Governance*. Washington, DC: National Academies Press. doi: 10.17226/24623.

National Aeronautics and Space Administration (NASA) (2014). *NASA's Efforts to Identify Near-Earth Objects and Mitigate Hazards. IG-14-030*. Washington, DC: NASA Office of Inspector General.

National Aeronautics and Space Administration (NASA) (2017). Planetary Defense Coordination Office. NASA, Washington, DC. https://www.nasa.gov/planetarydefense/overview.

National Hurricane Center (NHC) (2017a). Tropical Cyclone Advisory Archive. NHC, Miami, FL. http://www.nhc.noaa.gov/archive/2017.

National Hurricane Center (NHC) (2017b). MARIA Graphics Archive: Initial Wind Field and Watch/Warning Graphic. NHC, Miami, FL. http://www.nhc.noaa.gov/archive/2017/maria_graphics.php.

National Oceanic Atmospheric Administration (NOAA) (2016). *Weather Fatalities 2016*. National Weather Service, NOAA, Silver Spring, MD.

National Research Council (2010). *Defending Planet Earth: Near-Earth-Object Surveys and Hazard Mitigation Strategies*. Washington, DC: National Academies Press.

National Research Council (NRC) (2015). *Climate Intervention: Reflecting Sunlight to Cool Earth*. Washington, DC: National Academies Press.

von Neumann, J. and Morgenstern, O. (1947). *Theory of Ggames and Eeconomic Bbehavior*, 2e. Princeton, NJ: Princeton University Press.

Nobel Prize (2012). The Nobel Prize in Physiology or Medicine 2012. Sir John B. Gurdon and Shinya Yamanaka. https://www.nobelprize.org/nobel_prizes/medicine/laureates/2012/press.html.

Nordhaus, W. (1992). An optimal transition path for controlling greenhouse gases. *Science* 258: 1315–1319.

Nordhaus, W. (1994). *Managing the Global Commons*. Cambridge, MA: The MIT Press.

Nordhaus, W.D. (2006). Paul Samuelson and global public goods. In: *Samuelsonian Economics and the Twenty-First Century* (ed. M. Szenberg, L. Ramrattan and A.A. Gottesman). Oxford, UK: Oxford Scholarship Online.

Nordhaus, W. (2007). A review of the stern review on the economics of climate change. *Journal of Economic Literature* 55: 686–702.

Nordhaus, W.D. (2008). *A Question of Balance: Weighing the Options on Global Warming Policies*. New Haven, CT: Yale University Press.

Nordhaus, W. (2010). Economic aspects of global warming in a post-Copenhagen environment. *Proceedings of the National Academy of Sciences of the United States of America* 107 (26): 11721–11726.

Nordhaus, W. (2011). The economics of tail events with an application to climate change. *Review of Environmental Economics and Policy* 5: 240–257.

Nordhaus, W. and Boyer, J.G. (1999). Requiem for Kyoto: an economic analysis of the Kyoto Protocol. *Energy Journal* 20 (Special Issue): 93–130.

Oppenheimer, M. (1998). Global warming and the stability of the West Antarctic ice sheet. *Nature* 393: 325–332.

Overbye D (2013). Chasing the Higgs. NYT 4 March.

Parson, E.A. (2007). The big one: a review of Richard Posner's catastrophe: risk and response. *Journal of Economic Literature* 45: 147–164.

Pascal, B. (1670). *Penseés* (trans. WF Trotter, 1910). London: Dent.

Peskin, M.E. (2008). The end of the world at the Large Hadron Collider? *Physics* 1 (14).

Plaga R (2009). On the potential catastrophic risk from metastable quantum-black holes produced at particle colliders. arXiv:0808.1415 [hep-ph].

Poincaré, H. (1880–1890). *Mémoire sur les Courbes Définies par les Équations Différentielles I–VI, Oeuvre I*. Paris: Gauthier-Villars.

Posner, R.A. (2004). *Catastrophe: Risk and Response*. New York: Oxford University Press.

Posner, R.A. (2006). Efficient responses to catastrophic risks. *Chicago Journal of International Law* 6: 511–525.

Samuelson, P. (1954). The pure theory of public expenditure. *Review of Economics and Statistics* 36: 387–389.

Sanghi, A., Ramachandran, S., de la Fuente, A. et al. (2011). *Natural Hazards, Unnatural Disasters: The Economics of Effective Prevention*. Washington, DC: World Bank Group.

Schmalensee, R. and Stavins, R.N. (2013). The SO_2 allowance trading system: the ironic history of a grand policy experiment. *Journal of Economic Perspectives* 27: 103–122.

Seo, S.N. (2012a). Decision making under climate risks: an analysis of sub-Saharan farmers' adaptation behaviors. *Weather, Climate and Society* 4: 285–299.

Seo, S.N. (2012b). What eludes international agreements on climate change? The economics of global public goods. *Economic Affairs* 32 (2): 74–80.

Seo, S.N. (2015a). Fatalities of neglect: adapt to more intense hurricanes? *International Journal of Climatology* 35: 3505–3514.

Seo, S.N. (2015b). Adaptation to global warming as an optimal transition process to a greenhouse world. *Economic Affairs* 35: 272–284.

Seo, S.N. (2015c). Helping low-latitude, poor countries with climate change. *Regulation* 2015–2016 (4): 6–8.

Seo, S.N. (2016a). Modeling farmer adaptations to climate change in South America: a micro-behavioral economic perspective. *Environmental and Ecological Statistics* 23: 1–21.

Seo, S.N. (2016b). The micro-behavioral framework for estimating total damage of global warming on natural resource enterprises with full adaptations. *Journal of Agricultural, Biological, and Environmental Statistics* 21: 328–347.

Seo, S.N. (2016c). *Microbehavioral Econometric Methods: Theories, Models, and Applications for the Study of Environmental and Natural Resources*. Amsterdam, The Netherlands: Academic Press (Elsevier).

Seo, S.N. (2017a). *The Behavioral Economics of Climate Change: Adaptation Behaviors, Global Public Goods, Breakthrough Technologies, and Policy-Making*. London: Academic Press.

Seo, S.N. (2017b). Beyond the Paris Agreement: climate change policy negotiations and future directions. *Regional Science Policy and Practice* 9: 121–140.

Seo, S.N. (2017c). Measuring policy benefits of the cyclone shelter program in the North Indian Ocean: protection from intense winds or high storm surges? *Climate Change Economics* 8 (4): 1–18. doi: 10.1142/S2010007817500117.

Seo, S.N. (2018). Infinity unbounded: a statistical root and critique of environmental dismalism. Working Paper, Muaebak Institute of Global Warming Studies, Seoul.

Seo, S.N. and Bakkensen, L.A. (2017). Is tropical cyclone surge, not intensity, what kills so many people in South Asia? *Weather, Climate, and Society* 9: 71–81.

Sharkey, N.E. (2012). Evitability of autonomous robot warfare. *International Journal of the Red Cross* 94: 787–799.

Shiller, R.J. (2004). *The New Financial Order: Risk in the 21st Century*. Princeton, NJ: Princeton University Press.

Shiller, R.J. (2005). *Irrational Exuberance*, 2nde. Princeton, NJ: Princeton University Press.

Shiller, R.J. (2014). Speculative asset prices. *American Economic Review* 104: 1486–1517.

Swiss Re Institute (2017). *Natural Catastrophes and Man-made Disasters in 2016: A Year of Widespread Damages*. Zurich, Switzerland: Swiss Re.

Taleb, N.N. (2005). Mandelbrot makes sense. Fat tails, asymmetric knowledge, and decision making: Nassim Nicholas Taleb's essay in honor of Benoit Mandelbrot's 80th birthday. *Wilmott Magazine* 2005: 51–59.

Taleb, N.N. (2007). *The Black Swan: The Impact of the Highly Improbable*. London: Penguin.

Thaler, R.H. and Sunstein, C.R. (2009). *Nudge: Improving Decisions about Health, Wealth and Happiness*. New Haven, CT: Yale University Press.

THE/Lindau (2017). Do great minds think alike? The THE/Lindau Nobel laureates' survey. https://www.timeshighereducation.com/features/do-great-minds-think-alike-the-the-lindau-nobel-laureates-survey#survey-answer.

Thom, R. (1975). *Structural Stability and Morphogenesis*. New York: Benjamin-Addison-Wesley.

Thomson, J.A., Itskovitz-Eldor, J., Shapiro, S.S. et al. (1998). Embryonic stem cell lines derived from human blastocysts. *Science* 282 (5391): 1145–1147.

Tipitaka (2010). Brahmajāla Sutta: The All-Embracing Net of Views (trans. Bhikkhu Bodhi). https://www.accesstoinsight.org/tipitaka/dn/dn.01.0.bodh.html.

Tol, R. (2009). The economic effects of climate change. *Journal of Economic Perspectives* 23: 29–51.

Turco, R.P., Toon, O.B., Ackerman, T.P. et al. (1983). Nuclear winter: global consequences of multiple nuclear explosions. *Science* 222: 1283–1292.

United Nations Framework Convention on Climate Change (UNFCCC) (1992). The United Nations Framework Convention on Climate Change. UNFCCC, New York.

United Nations Framework Convention on Climate Change (UNFCCC) (2010). *Cancun Agreements*. New York: UNFCCC.

United Nations Framework Convention on Climate Change (UNFCCC) (2015) The Paris Agreement. Conference of the Parties (COP) 21. . UNFCCC, New York.

United Nations Office for Disarmament Affairs (UNODA) (2017). Background on Lethal Autonomous Weapons Systems in the CCW. https://www.unog.ch/80256EE600585943/(httppages)/8fa3c2562a60ff81c1257ce600393df6?opendocument.

United States Department of Energy (US DOE) (2008). *Genomics and its Impact on Science and Society: The Human Genome Project and beyond. Human Genome Program.* Washington, DC: US DOE.

United States Environmental Protection Agency (2017). *EPA Takes Another Step to Advance President Trump's America First Strategy, Proposes Repeal of "Clean Power Plan"*. Washington, DC: US EPA.

United States Geological Survey (USGS) (2017). Measuring the Size of an Earthquake. USGS, Washington, DC. https://earthquake.usgs.gov/learn/topics/measure.php.

Vogel, S.A. and Roberts, J.A. (2011). Why the Toxic Substances Control Act needs an overhaul, and how to strengthen oversight of chemicals in the interim. *Health Affairs* 30: 898–905.

Wagoner, G. and Weitzman, M. (2015). *Climate Shock: The Economic Consequences of a Hotter Planet*. Princeton, NJ: Princeton University Press.

Watanabe, J., Hattori, M., Berriman, M. et al. (2014). Genome sequence of the tsetse fly (*Glossina morsitans*): vector of African trypanosomiasis. *Science* 344 (6182): 380–386.

Weitzman, M.L. (2009). On modeling and interpreting the economics of catastrophic climate change. *Review of Economics and Statistics* 91: 1–19.

White House (2017). *Statement by President Trump on the Paris Climate Accord*. Washington, DC: The White House.

Wong, E. (2001). *Lieh-Tzu: A Taoist Guide to Practical Living*. Boston, MA: Shambhala.

World Meteorological Organization (WMO) (2014). *Scientific Assessment of Ozone Depletion 2014*. Global Ozone Research and Monitoring Project—Report No. 55. Geneva, Switzerland: World Meteorological Organization.

Zeeman, E.C. (1977). *Catastrophe Theory—Selected Papers 1972–1977*. Reading, MA: Addison-Wesley.

Index

a
Abatement cost   123
ACE   149
Adaptation   42, 133
Anthropogenic interference   205
Antibody   42
Archeology   80
Area-wide index   109
Artificial intelligence   209
Asbestos   200
Asteroid   178
Atlantic Ocean   150
Attachment point   115

b
Bayes theorem   131
Benefit–cost analysis   123
Benoit Mandelbrot   46
Bernoulli trial   105
Bernoulli utility   52
Best-shot technology   180
Bifurcation theory   57
Biggert–Waters Flood Insurance Reform and Modernization Act   185
Biological Weapons Convention   189
Blackhole   208
Black swan   128
Bubble   236
Burden of proof   200
Butterfly effect   44
Buy-up policies   109

c
Call options   111
Cap and trade   196
Carbon dioxide   204
Carbon monoxide   191
Carbon tax   123
Carcinogens   200
Catastrophe theory   39
Catastrophic coverage   104
CAT bonds   114
CAT bond spread   117
Category   5, 149
Cauchy distribution   126
CFCs (chlorofluorocarbons)   201
Chaos theory   43
Chemical Weapons Convention   189
Chernobyl disaster   187
Clean Air Act   191
Clean Air Interstate Rule (CAIR)   197
Climate change   9
CMIP (Climate Model Inter-comparison Project)   211
Coastline   47
Collapsiology   79
Comet   178
Comprehensive Environmental Response, Compensation, and Liability Act   199
Comprehensive Nuclear Test Ban Treaty (CTBT)   191
Conference of Parties (COP)   204
Control rate   123
Conway–Maxwell model   161
Corrupt government   229
Cotton price   47
Count data   161
Criteria pollutant   191
Crop insurance   107

Crop Insurance Reform Act   107
CRRA (constant relative risk aversion)   130
Cusp catastrophe   39
Cut-off points   179
Cyclone Bhola   165
Cyclone cellar   168
Cyclone Nargis   21
Cyclone shelter   165

d
Damage function   122
Dark matter   207
Dinosaur extinction   179
Dismal theorem   129
Dispersion   161
Doomsday   76
Dust Bowl   5

e
Early warning   166
Earthquake   156
Easter Island   78
Elasticity   52
El Nino Southern Oscillation (ENSO)   148
Envy   236
European Chemicals Agency   201
Evacuation order   168
Exhaustion point   115

f
Fatalities   160
Fat-tail   102
Fear   228
Federal Emergency Management Agency (FEMA)   181
Feigenbaum constant   56
Fold catastrophe   40
Fortuitous originationists   85
Fractal   46
Fractional dimension   49
Frank R. Lautenberg Chemical Safety for the, 21st Century Act   201
Frequency   148
Frightened rabbit   2
Fukushima disaster   187
Fundamentalism   230
Futures   111

g
Gambler   236
Gaussian distribution   125
Gene alterations   232
Genesis Potential Index   148
Gi-woo   1
Global public good   120
Global warming   9
Global Warming Potential   202
Grandfathered policies   185
Gravity tractor   179
Great earthquake   159
Great flood   1
Great Irish migration   5
Great Kanto earthquake   3
Green Climate Fund   205
Growth function   53

h
Haiti earthquake   3
Heavy-tail   102
HFCs (hydrofluorocarbons)   202
Higgs Boson   207
Hockey-stick hypothesis   9
Humanity-ending catastrophe   1
Hurricane   147
Hurricane Andrew   114
Hurricane Katrina   185
Hydrocarbons   202

i
Ice-core   77
Ignorance   230
Indemnity trigger   116
Indeterminacy rule   87
Indian Ocean earthquake   3
Infectious diseases   230
Infinity   125
Insurance   104
Intended Nationally Determined Contributions (INDC)   205
Intensity scale   148
Interest rate   123
Intergovernmental Panel on Climate Change   204

Index | 251

International Business Machines (IBM)   46
International Nuclear Events Scale (INES)   187

j

Joint Typhoon Warning Center   183
Julia Set   49

k

Kashmir earthquake   3
Kigali Amendment   202
Killer robots   209
Kinetic impactor   179
Kobe earthquake   3
Koch snowflake   48
Kyoto Protocol   203

l

Landfall   148
Large Hadron Collider   207
Law of large numbers   105
Lead   195
Least burdensome approach   199
Lethal Autonomous Weapons System   209
Lithosphere   3
Logistic map   56
Long-tail   128
Lorenz attractor   45
Lorenz equation   45

m

Mandelbrot set   49
Market crash   60
Maximum Wind Speeds   149
Mayan civilization   77
Medium-tail   101
Melanoma   201
Minimum Central Pressure   162
Moment scale   157
Montreal Protocol   201
Multiperil   107
Mutagens   200
Mycenaean civilization   77

n

National Ambient Air Quality Standards (NAAQS)   192
National Earthquake Hazards Reduction Program   181
National Flood Insurance Program   184
Natural gas   112
Near-Earth Object   178
Negative binomial   161
Nitrogen dioxide   194
No arbitrage   111
No data, no market rule   201
Non-Proliferation Treaty (NPT)   188
Normal yield   109
North Indian Ocean   184
Northwest Pacific Ocean   184
NPT *see* Non-Proliferation Treaty (NPT)
Nuclear accident   187
Nuclear disarmament   190
Nuclear umbrella   191

o

Ocean basins   183
Option pricing model   112
Options   111
Order   55
Ozone Depleting Substances (ODS)   202
Ozone layer   201

p

Pareto distribution   51
Pareto optimality   119
Paris Agreement   205
Participation rate   190
Particulate matter   195
Pascal's wager   82
Peak perils   118
Pensees   82
Period doubling   57
Photon   87
Planetary defense   178
Poisson distribution   161
Polder   168
Polychlorinated biphenyls (PCBs)   200
Population equation   58
Posterior distribution   131
Posterior-predictive   131
Potential function   40
Power dissipation index   150
Power law   51

Precursor   194
Pre-FIRM   185
Premium   105
Prior PDF   131
Psychology   234
Public goods   120
Put option   111

q
Quantum physics   87

r
Radiated Energy scale   158
Randomness   68
Rational   238
REACH (Registration, Evaluation, Authorisation, and Restrictions of Chemicals)   201
Regional Specialized Meteorology Center   183
Regulation mechanism   42
Reinsurance   114
René Thom   43
Richter scale   156

s
Saffir–Simpson scale   149
Sand county   73
Satellite monitoring   172
Scale invariance   51, 131
Seeding   155
Seismology   156
Self-affinity   49
Selfishness   230
Self similarity   51
Severe tropical cyclone   4
Silent spring   69
Singularity   209
Smog   192
Societal collapses   77
Southern Hemisphere   153
Stalagmites   77
State variable   40
Stem cell   232
Stochastic discounting factor   129
Storm surge   152
Strangelets   207
Subsidy rate   109

Sulfur dioxide allowance trading   197
Superfund   199

t
τ-sigma event   99
Tectonic plates   182
Temperature ceiling   204
Teratogens   200
Time preference   122
Tipping point   41
Toxicology   98
Toxic substances   198
Triangular distribution   101
Trigger point   115
Tropical cyclone   147
Tropical cyclone trajectory projection techniques   168
Trump administration   206
Typhoon   147

u
Ultraviolet radiation   188
Uncertainty   124
Undefined moments   125
United Nations Framework Convention on Climate Change   204
Universal catastrophe   7
Unreasonable risk of injury   199

v
Value   67
Value of Statistical Life   129
Volatile Organic Compounds (VOCs)   192

w
Weapons of mass destruction   189
Weather   44
Wilderness   73
Wildlife   73

y
Yield-based   109
Yucatan Peninsula   77

z
Zero-inflated   162
Zipf's law   51